METHODS IN MOLECULA

Series Editor
John M. Walker
School of Life and Medical Sciences
University of Hertfordshire
Hatfield, Hertfordshire, AL10 9AB, UK

For further volumes:
http://www.springer.com/series/7651

Phagocytosis and Phagosomes

Methods and Protocols

Edited by

Roberto Botelho

Department of Chemistry and Biology, Ryerson University, Toronto, ON, Canada

Humana Press

Editor
Roberto Botelho
Department of Chemistry and Biology
Ryerson University
Toronto, ON, Canada

ISSN 1064-3745   ISSN 1940-6029   (electronic)
Methods in Molecular Biology
ISBN 978-1-4939-8241-7   ISBN 978-1-4939-6581-6   (eBook)
DOI 10.1007/978-1-4939-6581-6

Printed on acid-free paper

This Humana Press imprint is published by Springer Nature
The registered company is Springer Science+Business Media LLC
The registered company address is: 233 Spring Street, New York, NY 10013, U.S.A.

Preface

I am pleased to present the current volume in Methods in Molecular Biology entitled *Phagocytosis and Phagosomes* that offers 23 chapters detailing experimental approaches used to investigate *Phagocytosis and Phagosomes*. This volume also encases a chapter by Gray and Botelho that reviews past and recent literature on these exquisite biological processes and which may be particularly useful for new investigators venturing into these areas.

Phagocytosis is the cellular engulfment of particulate matter and it is of broad biological importance, being employed by unicellular organisms and by numerous cell types within multicellular organisms. Unicellular organisms, exemplified by hundreds of Protista species, employ phagocytosis to ingest food particles. In multicellular organisms, phagocytosis plays an essential function in tissue remodeling and homeostasis by removing millions upon millions of apoptotic bodies, senescent cells, and cell fragments every single day. In addition, phagocytosis is a paramount immune weapon utilized by leukocytes to ingest and eliminate a plethora of pathogens and other foreign particulates. Internalization is only half the story. Once engulfed, the particle is now sequestered within a new organelle—the phagosome. The newly formed phagosome is an innocuous organelle but is equipped with molecular cues that guide its maturation into a phagolysosome, an acidic and hydrolytic organelle that ultimately degrades the particle. This transformation requires sequential fusion with endosomes and lysosomes. Not surprisingly, pathogens have evolved myriad mechanisms to control engulfment and phagosome maturation to hijack and replicate within hosts cells—these pathogens remain a large source of human morbidity globally.

Phagocytosis and phagosome maturation are complex, diverse, and highly dynamic processes. Thus, a large array of complementary methods and tools have been developed and are required to study engulfment, maturation, and pathogen manipulation. In this volume, the reader will find detailed instructions and tips in the form of "Notes" on using many of these tools. For example, several chapters offer methodology to quantify uptake and maturation specific to certain phagocytes, particles, or pathogens, while other chapters offer methods that can be applied generically across the field. Methods are presented to study phagosome maturation biophysically by manipulating individual phagosomes, or biochemically by cell fractionation, Western blotting, or proteomics. Due to the dynamic and rapid nature of phagocytosis and maturation, these processes are especially amenable to analysis by microscopy and flow cytometry. Not surprisingly, a large number of chapters detail the use of imaging methods to analyze these processes in fixed and live cells, as well as employing high-throughput microscopy and image analysis.

Overall, we believe this volume will be an important resource for both experts in the field and those investigators delving into *phagocytosis and phagosomes* for the first time. As a last word, it is important to realize that *phagocytosis* is really a collection of distinct processes, rather than a single phenomenon. Indeed, the mechanics, outcomes, and accompanying events triggered by *phagocytosis and phagosomes* differ depending on the phagocyte, phagocytic receptors, and target particles engaged. This presents a significant challenge to our understanding of *phagocytosis and phagosomes* and is often a source of confusion in the literature. Therefore, one should resist pronouncing sweeping statements about these processes.

Toronto, ON, Canada   *Roberto Botelho*

Contents

Preface. . . . .   *v*

Contributors. . . . .   *ix*

1   Phagocytosis: Hungry, Hungry Cells   1
    Matthew Gray and Roberto J. Botelho

2   Analysis of Human and Mouse Neutrophil Phagocytosis by Flow Cytometry   17
    Noah Fine, Oriyah Barzilay, and Michael Glogauer

3   Quantitative Efferocytosis Assays   25
    Amanda L. Evans, Jack W.D. Blackburn, Charles Yin, and Bryan Heit

4   Quantifying Phagocytosis by Immunofluorescence and Microscopy.   43
    Christopher H. Choy and Roberto J. Botelho

5   Single Cell Analysis of Phagocytosis, Phagosome Maturation,
    Phagolysosomal Leakage, and Cell Death Following Exposure
    of Macrophages to Silica Particles   55
    Gaurav N. Joshi, Renée M. Gilberti, and David A. Knecht

6   Quantitative Live-Cell Fluorescence Microscopy During Phagocytosis.   79
    Stella M. Lu, Sergio Grinstein, and Gregory D. Fairn

7   Intracellular Manipulation of Phagosomal Transport and Maturation
    Using Magnetic Tweezers.   93
    Shashank Shekhar, Vinod Subramaniam, and Johannes S. Kanger

8   Quantitative Immunofluorescence to Study Phagosome Maturation   113
    Roya M. Dayam and Roberto J. Botelho

9   Using Fluorescence Resonance Energy Transfer-Based Biosensors
    to Probe Rho GTPase Activation During Phagocytosis   125
    Veronika Miskolci, Louis Hodgson, and Dianne Cox

10 Analysis of LC3-Associated Phagocytosis and Antigen Presentation.   145
    Laure-Anne Ligeon, Susana Romao, and Christian Münz

11 Quantitative Spatiotemporal Analysis of Phagosome Maturation
    in Live Cells   169
    Laura Schnettger and Maximiliano G. Gutierrez

12 Measuring Phagosomal pH by Fluorescence Microscopy.   185
    Johnathan Canton and Sergio Grinstein

13 Image-Based Analysis of Phagocytosis: Measuring Engulfment
    and Internalization   201
    *Nicholas D. Condon, Adam A. Wall, Jeremy C. Yeo, Nicholas A. Hamilton,
    and Jennifer L. Stow*

14 Fluorometric Approaches to Measuring Reductive and Oxidative
    Events in Phagosomes   215
    Dale R. Balce and Robin M. Yates

15 Simultaneous Analysis of Multiple Lumenal Parameters of Individual
    Phagosomes Using High-Content Imaging.   227
    Samuel Cheung, Catherine Greene, and Robin M. Yates

16 Isolation and Western Blotting of Latex-Bead Phagosomes
    to Track Phagosome Maturation.   241
    Anetta Härtlova, Julien Peltier, Orsolya Bilkei-Gorzo, and Matthias Trost

17 Assessing the Phagosome Proteome by Quantitative Mass Spectrometry   249
    Julien Peltier, Anetta Härtlova, and Matthias Trost

18 Dissecting Phagocytic Removal of Apoptotic Cells in *Caenorhabditis elegans*.   265
    Shiya Cheng, Kai Liu, Chonglin Yang, and Xiaochen Wang

19 Measurement of *Salmonella enterica* Internalization and Vacuole
    Lysis in Epithelial Cells.   285
    Jessica A. Klein, TuShun R. Powers, and Leigh A. Knodler

20 Bacterial Binding, Phagocytosis, and Killing: Measurements Using
    Colony Forming Units   297
    Kyle E. Novakowski, Dessi Loukov, Vikash Chawla, and Dawn M.E. Bowdish

21 Filamentous Bacteria as Targets to Study Phagocytosis   311
    Akriti Prashar, Sana I. S. Khan, and Mauricio R. Terebiznik

22 Growing and Handling of *Mycobacterium tuberculosis* for Macrophage
    Infection Assays   325
    Evgeniya V. Nazarova and David G. Russell

23 *Mycobacterium tuberculosis*: Readouts of Bacterial Fitness
    and the Environment Within the Phagosome   333
    Shumin Tan, Robin M. Yates, and David G. Russell

24 Using Flow Cytometry to Analyze *Cryptococcus* Infection of Macrophages   349
    Robert J. Evans, Kerstin Voelz, Simon A. Johnston, and Robin C. May

Index.   359

Contributors

DALE R. BALCE • *Department of Comparative Biology and Experimental Medicine, Faculty of Veterinary Medicine, University of Calgary, Calgary, AB, Canada*

ORIYAH BARZILAY • *Department of Dentistry, Matrix Dynamics Group, University of Toronto, Toronto, ON, Canada*

ORSOLYA BILKEI-GORZO • *MRC Protein Phosphorylation and Ubiquitylation Unit, College of Life Sciences, University of Dundee, Dundee, UK*

JACK W.D. BLACKBURN • *Department of Microbiology and Immunology and the Centre for Human Immunology, The University of Western Ontario, London, ON, Canada*

ROBERTO J. BOTELHO • *Molecular Science Graduate Program, Ryerson University, Toronto, ON, Canada; Department of Chemistry and Biology, Ryerson University, Toronto, ON, Canada*

DAWN M.E. BOWDISH • *Department of Pathology and Molecular Medicine, McMaster University, Hamilton, ON, Canada; McMaster Immunology Research Centre, McMaster University, Hamilton, ON, Canada; M.G. DeGroote Institute for Infectious Disease Research, Hamilton, ON, Canada*

JOHNATHAN CANTON • *Program in Cell Biology, Peter Gilgan Centre for Research and Learning, Hospital for Sick Children, Toronto, ON, Canada*

VIKASH CHAWLA • *Department of Pathology and Molecular Medicine, McMaster University, Hamilton, ON, Canada; McMaster Immunology Research Centre, McMaster University, Hamilton, ON, Canada; M.G. DeGroote Institute for Infectious Disease Research, Hamilton, ON, Canada*

SHIYA CHENG • *National Institute of Biological Sciences, Zhongguancun Life Science Park, Beijing, China*

SAMUEL CHEUNG • *Department of Comparative Biology and Experimental Medicine, Faculty of Veterinary Medicine, University of Calgary, Calgary, AB, Canada*

CHRISTOPHER H. CHOY • *Molecular Science Graduate Program, Ryerson University, Toronto, ON, Canada; Department of Chemistry and Biology, Ryerson University, Toronto, ON, Canada*

NICHOLAS D. CONDON • *Institute for Molecular Bioscience, The University of Queensland, Brisbane, QLD, Australia*

DIANNE COX • *Departments of Anatomy and Structural Biology, Albert Einstein College of Medicine, Bronx, NY, USA; Gruss-Lipper Biophotonics Center, Albert Einstein College of Medicine, Bronx, NY, USA; Developmental and Molecular Biology, Albert Einstein College of Medicine, Bronx, NY, USA*

ROYA M. DAYAM • *Department of Chemistry and Biology, Ryerson University, Toronto, ON, Canada; The Graduate Program in Molecular Science, Ryerson University, Toronto, ON, Canada*

AMANDA L. EVANS • *Department of Microbiology and Immunology and the Centre for Human Immunology, The University of Western Ontario, London, ON, Canada*

ROBERT J. EVANS • *Institute of Microbiology and Infection, School of Biosciences, College of Life and Environmental Sciences, University of Birmingham, Birmingham, UK;*

Bateson Centre, Firth Court, University of Sheffield, Western Bank, Sheffield, UK; Department of Infection, Immunity and Cardiovascular Disease, Medical School, University of Sheffield, Sheffield, UK

GREGORY D. FAIRN • Department of Biochemistry, University of Toronto, Toronto, ON, Canada; Program in Cell Biology, Peter Gilgan Centre for Research and Learning, Hospital for Sick Children, Toronto, ON, Canada; Keenan Research Centre of the Li Ka Shing Knowledge Institute, St. Michael's Hospital, Toronto, ON, Canada

NOAH FINE • Department of Dentistry, Matrix Dynamics Group, University of Toronto, Toronto, ON, Canada

RENÉE M. GILBERTI • Department of Molecular and Cellular Biology, University of Connecticut, Storrs, CT, USA; Center for Academic Programs, University of Connecticut, Storrs, CT, USA

MICHAEL GLOGAUER • Department of Dentistry, Matrix Dynamics Group, University of Toronto, Toronto, ON, Canada

MATTHEW GRAY • Molecular Science Graduate Program, Ryerson University, Toronto, ON, Canada

CATHERINE GREENE • Department of Biochemistry and Molecular Biology, Faculty of Medicine, University of Calgary, Calgary, AB, Canada

SERGIO GRINSTEIN • Program in Cell Biology, Peter Gilgan Centre for Research and Learning, Hospital for Sick Children, Toronto, ON, Canada; Keenan Research Centre of the Li Ka Shing Knowledge Institute, St. Michael's Hospital, Toronto, ON, Canada

MAXIMILIANO G. GUTIERREZ • Host–Pathogen Interactions in Tuberculosis Laboratory, The Francis Crick Institute, London, UK

NICHOLAS A. HAMILTON • Institute for Molecular Bioscience, The University of Queensland, Brisbane, QLD, Australia; Research Computing Centre, The University of Queensland, Brisbane, QLD, Australia

ANETTA HÄRTLOVA • MRC Protein Phosphorylation and Ubiquitylation Unit, College of Life Sciences, University of Dundee, Dundee, UK

BRYAN HEIT • Department of Microbiology and Immunology and the Centre for Human Immunology, The University of Western Ontario, London, ON, Canada

LOUIS HODGSON • Departments of Anatomy and Structural Biology, Albert Einstein College of Medicine, Bronx, NY, USA; Gruss-Lipper Biophotonics Center, Albert Einstein College of Medicine, Bronx, NY, USA

SIMON A. JOHNSTON • Bateson Centre, Firth Court, University of Sheffield, Western Bank, Sheffield, South Yorkshire, UK; Department of Infection, Immunity and Cardiovascular Disease, Medical School, University of Sheffield, Sheffield, South Yorkshire, UK

GAURAV N. JOSHI • Department of Molecular and Cellular Biology, University of Connecticut, CT, USA; Department of Pharmaceutical Sciences, University of Connecticut, Storrs, CT, USA

JOHANNES S. KANGER • Nanobiophysics Group, University of Twente, Enschede, The Netherlands

SANA I.S. KHAN • Department of Biological Sciences, University of Toronto Scarborough, Toronto, ON, Canada

JESSICA A. KLEIN • *Paul G. Allen School for Global Animal Health, College of Veterinary Medicine, Washington State University, Pullman, WA, USA*

DAVID A. KNECHT • *Department of Molecular and Cellular Biology, University of Connecticut, Storrs, CT, USA*

LEIGH A. KNODLER • *Paul G. Allen School for Global Animal Health, College of Veterinary Medicine, Washington State University, Pullman, WA, USA*

LAURE-ANNE LIGEON • *Viral Immunobiology, Institute of Experimental Immunology, University of Zürich, Zürich, Switzerland*

KAI LIU • *State Key Laboratory of Molecular Developmental Biology, Institute of Genetics and Developmental Biology, Chinese Academy of Sciences, Beijing, China*

DESSI LOUKOV • *Department of Pathology and Molecular Medicine, McMaster University, Hamilton, ON, Canada; McMaster Immunology Research Centre, McMaster University, Hamilton, ON, Canada; M.G. DeGroote Institute for Infectious Disease Research, Hamilton, ON, Canada*

STELLA M. LU • *Department of Biochemistry, University of Toronto, Toronto, ON, Canada; Program in Cell Biology, Peter Gilgan Centre for Research and Learning, Hospital for Sick Children, Toronto, ON, Canada; Keenan Research Centre of the Li Ka Shing Knowledge Institute, St. Michael's Hospital, Toronto, ON, Canada*

ROBIN C. MAY • *Institute of Microbiology and Infection, School of Biosciences, College of Life and Environmental Sciences, University of Birmingham, Birmingham, UK*

VERONIKA MISKOLCI • *Departments of Anatomy and Structural Biology, Albert Einstein College of Medicine, Bronx, NY, USA*

CHRISTIAN MÜNZ • *Viral Immunobiology, Institute of Experimental Immunology, University of Zürich, Zürich, Switzerland*

EVGENIYA V. NAZAROVA • *Microbiology and Immunology, College of Veterinary Medicine, Cornell University, Ithaca, NY, USA*

KYLE E. NOVAKOWSKI • *Department of Pathology and Molecular Medicine, McMaster University, Hamilton, ON, Canada; McMaster Immunology Research Centre, McMaster University, Hamilton, ON, Canada; M.G. DeGroote Institute for Infectious Disease Research, Hamilton, ON, Canada*

JULIEN PELTIER • *MRC Protein Phosphorylation and Ubiquitylation Unit, College of Life Sciences, University of Dundee, Dundee, UK*

TUSHUN R. POWERS • *Paul G. Allen School for Global Animal Health, College of Veterinary Medicine, Washington State University, Pullman, WA, USA*

AKRITI PRASHAR • *Department of Biological Sciences, University of Toronto Scarborough, Toronto, ON, Canada*

SUSANA ROMAO • *Viral Immunobiology, Institute of Experimental Immunology, University of Zürich, Zürich, Switzerland*

DAVID G. RUSSELL • *Microbiology and Immunology, College of Veterinary Medicine, Cornell University, Ithaca, NY, USA*

LAURA SCHNETTGER • *Host–Pathogen Interactions in Tuberculosis Laboratory, The Francis Crick Institute, London, UK*

SHASHANK SHEKHAR • *Cytoskeleton Dynamics Group, CNRS, Gif-sur-Yvette, France*

JENNIFER L. STOW • *Institute for Molecular Bioscience, The University of Queensland, Brisbane, QLD, Australia*

VINOD SUBRAMANIAM • *Nanobiophysics Group, University of Twente, Enschede, The Netherlands; Vrije Universiteit Amsterdam, Amsterdam, The Netherlands*

SHUMIN TAN • *Microbiology and Immunology, College of Veterinary Medicine, Cornell University, Ithaca, NY, USA; Molecular Biology and Microbiology, School of Medicine, Tufts University, Boston, MA, USA*

MAURICIO R. TEREBIZNIK • *Department of Biological Sciences, University of Toronto Scarborough, Toronto, ON, Canada*

MATTHIAS TROST • *MRC Protein Phosphorylation and Ubiquitylation Unit, College of Life Sciences, University of Dundee, Dundee, UK*

KERSTIN VOELZ • *Institute of Microbiology and Infection, School of Biosciences, College of Life and Environmental Sciences, University of Birmingham, Birmingham, UK*

ADAM A. WALL • *Institute for Molecular Bioscience, The University of Queensland, Brisbane, QLD, Australia*

XIAOCHEN WANG • *National Institute of Biological Sciences, Zhongguancun Life Science Park, Beijing, China*

CHONGLIN YANG • *State Key Laboratory of Molecular Developmental Biology, Institute of Genetics and Developmental Biology, Chinese Academy of Sciences, Beijing, China*

ROBIN M. YATES • *Department of Comparative Biology and Experimental Medicine, Faculty of Veterinary Medicine, University of Calgary, Calgary, AB, Canada; Department of Biochemistry and Molecular Biology, Faculty of Medicine, University of Calgary, Calgary, AB, Canada*

JEREMY C. YEO • *Institute for Molecular Bioscience, The University of Queensland, Brisbane, QLD, Australia*

CHARLES YIN • *Department of Microbiology and Immunology and the Centre for Human Immunology, The University of Western Ontario, London, ON, Canada*

Chapter 1

Phagocytosis: Hungry, Hungry Cells

Matthew Gray and Roberto J. Botelho

Abstract

Phagocytosis is the cellular internalization and sequestration of particulate matter into a `phagosome, which then matures into a phagolysosome. The phagolysosome then offers a specialized acidic and hydrolytic milieu that ultimately degrades the engulfed particle. In multicellular organisms, phagocytosis and phagosome maturation play two key physiological roles. First, phagocytic cells have an important function in tissue remodeling and homeostasis by eliminating apoptotic bodies, senescent cells and cell fragments. Second, phagocytosis is a critical weapon of the immune system, whereby cells like macrophages and neutrophils hunt and engulf a variety of pathogens and foreign particles. Not surprisingly, pathogens have evolved mechanisms to either block or alter phagocytosis and phagosome maturation, ultimately usurping the cellular machinery for their own survival. Here, we review past and recent discoveries that highlight how phagocytes recognize target particles, key signals that emanate after phagocyte-particle engagement, and how these signals help modulate actin-dependent remodeling of the plasma membrane that culminates in the release of the phagosome. We then explore processes related to early and late stages of phagosome maturation, which requires fusion with endosomes and lysosomes. We end this review by acknowledging that little is known about phagosome fission and even less is known about how phagosomes are resolved after particle digestion.

Key words Phagocytosis, Phagosome maturation, Myeloid cells, Pathogen, Apoptotic bodies, Lysosomes, Review, Endosomes, Receptors, Immunity

1 Phagocytic Cells and Physiological Roles of Phagocytosis

In order to protect their bodies as a whole, organisms must be able to remove unwanted particulate matter. These particles can be of exogenous origin including a large variety of opportunistic and pathogenic bacteria, fungi, and protists, or polluting debris such as asbestos and silicate particles, or endogenous to the organism such as apoptotic and senescent cells [1, 2]. In order to deal with these threats and challenges, many cells have developed the ability to engulf the particles and break them down, thereby removing the threat. This process is phagocytosis, a specialized form of endocytosis that employs receptor-mediated actin remodeling of the plasma membrane to engulf particles greater than 0.5 μm in diameter [2].

Roberto Botelho (ed.), *Phagocytosis and Phagosomes: Methods and Protocols*, Methods in Molecular Biology, vol. 1519, DOI 10.1007/978-1-4939-6581-6_1, © Springer Science+Business Media New York 2017

Many cell types in multicellular organisms, including in simple organisms like *C. elegans*, are capable of performing phagocytosis, primarily to clear apoptotic cells: these include epithelial cells of the airway, the gastrointestinal tract, in the mammary tissue, hepatocytes in liver, skeletal muscle satellite cells, Sertoli cells in the testes, retinal epithelial cells, and neuronal progenitor cells, among others [1]. In addition to these cells, there are cells that are specialized for phagocytosis and are thus termed professional phagocytes. These tend to be immune-associated cells, though they can also play a role in tissue homeostasis and include tissue-resident macrophages such as microglia and Kupffer cells, as well as migratory and blood borne phagocytes such as macrophages, dendritic cells, and neutrophils [1, 3, 4]. Resident macrophages are derived originally from yolk-sac-derived erythro-myeloid progenitors, before being eventually replaced by hematopoietic stem cell derived macrophages [5]. The migratory phagocytes are also hematopoietic stem cell derived cells [6].

2   Phagocytic Receptors and Ligands Mediate Target Selection

Phagocytosis is initiated when the plasma membrane of a phagocyte contacts the surface of a target particle. This contact allows phagocytic receptors to engage cognate ligands on the particle surface, leading to signal transduction that will ultimately cause particle engulfment. This process typically requires receptor–ligand clustering, which helps to select for particulate targets rather than soluble ones [7]. Interestingly, while nonprofessional phagocytes tend to engulf directly juxtaposed targets [8, 9], professional phagocytes such as macrophages and dendritic cells can probe the local environment with long actin-rich plasma membrane extensions to actively capture targets [10]. These projections are dynamic, moving about the environment through rounds of actin polymerization and depolymerization within their core, as well as the contractile actions of myosin interacting with the actin core [10].

Phagocytes are equipped with a variety of receptors to recognize specific ligands associated with a plethora of phagocytic targets. Moreover, these receptors then engage signaling modalities to elicit different particle processing and response, such as whether an inflammatory response and change in the gene expression profile should be concurrent with phagocytosis [11, 12]. For example, apoptotic cells can be recognized by the presence of several surface markers that are unique to apoptotic cells and lacking on viable cells, termed apoptotic-cell associated molecular patterns (ACAMPs) [13]. ACAMPs include normally intracellular proteins such as calreticulin and annexin I, modified versions of cell surface proteins such as ICAM-3, and altered membrane composition, each of which are recognized by specific pattern recognition receptors (PRRs) on the phagocyte's surface [14–17]. The best studied ACAMP is phosphatidylserine, a phospholipid that is present in the inner leaflet of the cell membrane in viable cells [18]. Upon initiation of apoptosis, the mechanisms that maintain phosphatidylserine on the inner leaflet are inhibited, allowing phosphatidylserine to flip to the outer leaflet. Phosphatidylserine is then recognized by a variety of PRRs on the phagocyte, such as BAI1, TIM-4, and MerTK, each of which signal for phagocytosis initiation [1]. As well, ACAMP recognition results in inhibition of inflammation [1, 13].

Like apoptotic cells, pathogens have a set of unique molecules that allow phagocytes to recognize them as being foreign to the body. These molecules are termed pathogen associated molecular patterns (PAMPs), and include bacterial wall and membrane components, bacterial flagellin, viral double-stranded RNA (dsRNA), and yeast wall components [11]. Like their ACAMP counterparts, PAMPs are recognized by PRRs tailored to each ligand type, allowing for recognition of the pathogen. The earliest pathogen associated PRR identified was the macrophage mannose receptor (MMR), a C-type lectin that recognize mannose residues on glycoproteins found on various pathogens [19, 20]. Further studies identified several other PRRs, including the Toll-like receptor family (TLR), whose members are able to recognize a wide array of PAMPs, including the prototypical PAMP lipopolysaccharide (LPS) found in gram-negative bacteria [21]. PAMP recognition leads to receptor clustering of additional molecules of the same receptor type, as well as other receptor types, which are recruited to receptor rafts selectively based on the initial PAMP recognized [22].

Recognition of the molecular patterns, both on apoptotic cells and pathogens, by phagocytes can also occur through indirect means. This occurs through opsonization of the target particle, whereby an intermediate protein, produced by the host, acts as a receptor ligand after coating the pathogens surface [11, 23]. This allows for a smaller number of receptors to recognize a broad spectrum of pathogens, as specificity is generated by the opsonins. The opsonin proteins include members of the immunoglobulin (Ig) proteins and complement system, which are produced by B cells and the liver, respectively [24]. Ig molecules are generated by B-cells in response to specific antigens on pathogens; this allows for rapid detection and opsonization of the pathogen [24]. The Ig molecules are then recognized by Fc receptors, which then cluster upon binding to multivalent Ig, leading to their phosphorylation by various associated tyrosine kinases [25]. The complement system recognizes pathogens vialectin motifs of the C1 complex [24]. This leads to activation of a proteolytic cascade that culminates in the generation of C3b, which is covalently attached to proteins on the pathogens surface. The bound C3b is then recognized by one of several complement receptors, leading to receptor clustering and phagocytosis [24].

3   Signaling and Internalization Mechanics

Upon receptor clustering, internal signals are initiated that lead to particle internalization through phagocytosis (Fig. 1). The ultimate goal of these early phagocytic signals is to allow particle internalization to proceed and invariably leads to remodeling of the actin cytoskeleton [12]. The specific signals leading to changes in actin

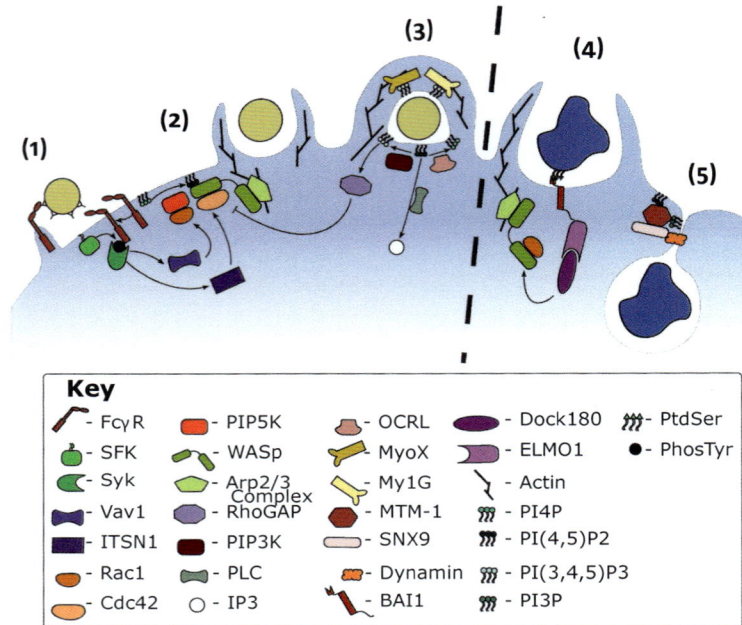

Key

- FcγR	- PIP5K	- OCRL	- Dock180	- PtdSer
- SFK	- WASp	- MyoX	- ELMO1	- PhosTyr
- Syk	- Arp2/3 Complex	- My1G	- Actin	
- Vav1	- RhoGAP	- MTM-1	- PI4P	
- ITSN1	- PIP3K	- SNX9	- PI(4,5)P2	
- Rac1	- PLC	- Dynamin	- PI(3,4,5)P3	
- Cdc42	- IP3	- BAI1	- PI3P	

Fig. 1 Initiation of phagocytosis and the formation of the phagosome. Phagocytosis is a multistep process that involves a variety of protein–protein interactions and protein–lipid interactions. (**1**) IgG-mediated phagocytosis is initiated by IgG-opsonized particles binding FcγRs during environmental probing by phagocytes. This leads to phosphorylation of the FcγR ITAM by SFKs, leading to docking and activation of Syk. (**2**) Syk activation stimulates Rac and Cdc42 GTPases through GEF proteins. GTP-bound GTPases in cooperation with PIP2 then stabilize WASp family proteins to activate the Arp2/3 complex. This leads to the generation of branched actin filaments that drive pseudopods around the particle. (**3**) The turnover of PIP2 to PIP3 leads to phagosome closure. PIP3 recruits RhoGAPs that turn-off Rac and Cdc42, allowing actin remodeling at the base of the phagosome. PIP3 also recruits myosins that provide the contractile forces needed to seal the phagosome. (**4**) Apoptotic cell clearance begins with recognition of ACAMPs by their corresponding receptors, exemplified here by PtdSer and BAI1, respectively. Activation of the receptor leads to activation of the ELMO1-Dock180 complex, allowing Dock180 to active Rac through its GEF activity. This leads to activation of WASp proteins and the Arp2/3 complex, allowing for phagocytosis to proceed as described for FcγR mediated phagocytosis. (**5**) PIP2 recruits the phosphatase MTM-1 to control PI3P levels before phagosome scission. MTM-1 also interacts with SNX9, which recruits dynamin, which scissions the phagosome from the plasma membrane. This process has been shown in *C. elegans* during apoptotic cell phagocytosis [65], and may be representative of sealing for other phagocytic processes

dynamics is dependent on the receptor initiating the signal. In the case of Fc receptors, such as FcγR bound to IgG, the first signaling event to occur after receptor–ligand binding is phosphorylation of tyrosine residues in the immunoreceptor tyrosine-based activation motifs (ITAMs) on the cytosolic tail of the receptor by Src family kinases (SFKs) [26]. The SFKs are activated through dephosphorylation upon receptor activation, allowing them to adopt an active conformation to phosphorylate the ITAMs at two tyrosine residues. These phosphorylated residues act as docking sites for additional SFKs, which further phosphorylate the ITAM to generate docking sites for spleen tyrosine kinase (Syk), another tyrosine kinase [26]. The accumulation of phosphorylation sites on the ITAM is offset by dephosphorylation by CD45, a membrane integrated phosphatase. During phagocytosis, CD45 becomes largely excluded from the phagocytic cup through the formation of a diffusional barrier generated by activated integrins [27]. This allows for the accumulation of ITAM phosphorylation sites for SFK and Syk activation.

Syk becomes active upon ITAM docking, leading to eventual activation of members of the Rho family of small GTPases, specifically Rac1/2 (Rac) and Cdc42 [28]. The activation of Rac and Cdc42 occurs by GTP-loading, which requires an active guanine exchange factor (GEF). Seemingly, SFK and/or Syk activate Vav1 and intersectin-1, which are exchange factors for Rac1 and Cdc42, respectively. The activation of Rac and Cdc42 leads to concerted assembly and turnover of filamentous actin structures that drive pseudopod formation to engulf the particle. Both Rac and Cdc42 control actin filament dynamics by stimulating Wiskott–Aldrich syndrome protein (WASp) family members through indirect and direct interactions, respectively [29]. WASp proteins then stimulate the actin related protein 2/3 complex (Arp2/3) to nucleate de novo branched actin branching network that helps mold the plasma membrane during engulfment [30]. Activation of WASp-family members at the phagocytic cup requires the adaptor protein Nck 1, which in turn is recruited through interactions with the phosphorylated ITAM [31, 32].

Phagocytosis initiated through PRRs ultimately leads to the same changes in the actin cytoskeleton that are initiated through FcγR phagocytosis, though the pathways that bring them about may differ. One of the better studied phagocytosis inducing PRRs is Dectin-1, a C-type lectin that recognizes β-glucans found in fungal cell walls [33]. Like FcγRs, Dectin-1 contains an ITAM, though it differs from the FcγR ITAM in that it requires a phosphorylation of a single tyrosine by SFKs to induce phagocytosis rather than two [34]. This single tyrosine phosphorylation results in a requirement for receptor clustering to allow Syk to bind two phosphorylated tyrosines, as is needed for its activation. The clustering of Dectin-1 results in the formation of a "phagocytic synapse" that ensures phagocytosis is only induced by particle binding and not through Dectin-1 binding of soluble β-glucans [35]. After Syk activation by

Dectin-1, many of the same signals that are utilized by FcγRs to induce phagocytosis, such as Vav1, Rac, and Cdc42 activation, occur [36, 37].

Other PRRs induce actin cytoskeleton rearrangement through means that are independent of ITAMs, SFKs, and Syk. Several ACAMP PRRs utilize a pathway that activates Rac through a different Rac GEF called the dedicator of cytokinesis 180 (Dock180) [38]. Many of these PRRs recognize phosphatidylserine, including BAI1, MerTK and the integrins $\alpha_v b_3$ and $\alpha_v \beta_5$, where the last three recognize phosphatidylserine via bridging proteins [39–42]. Interestingly, Dock180 must interact with Engulfment and Cell Motility Protein 1 (ELMO1) to stimulate its GEF activity towards Rac [43]. The ELMO1/Dock180 complex is activated by interactions with the PRR cytoplasmic tails, either through direct interactions, such as with BAI1, or through interactions mediated by the adaptor protein CrkII [39–42].

Initiation of phagocytosis after particle docking and recognition does not only rely on changes in protein activity, but also alterations in membrane composition. One of the best known membrane components that are involved in phagocytosis are the phosphoinositides (PIPs). These phospholipids have an inositol head group that can be phosphorylated at three positions, positions 3, 4, and 5 of the inositol ring, allowing for generation of seven phosphorylated species that help to give different membranes unique signatures [44–46]. During initiation of FcγR-induced phagocytosis, phosphatidylinositol-4,5-bisphosphate (PIP2) undergoes dynamics changes at the phagocytic cup [47]. First, PI(4)P5-kinases (PIP5Ks) are recruited to the phagocytic cup leading to an increase in PIP2 [47–50]. As phagocytosis progresses, PIP2 is locally depleted from the phagocytic cup through cleavage by phospholipase C (PLC), dephosphorylation into PI(4)P by OCRL1 or Inpp5B, or converted to phosphatidylinositol-3,4,5-trisphosphate (PIP3) through the actions of phosphatidylinositol 3-kinases (PI3Ks) [47, 51–54]. This dynamic behavior of PIP2 during phagocytosis helps to coordinate actin polymerization and turnover [55]. For example, PIP2 is important in WASP activation, as it helps maintain WASP in its active conformation alongside Cdc42 [56]. By comparison, PIP3 undergoes a burst of synthesis at the phagocytic cup [54]. As with PIP2, PIP3 likely coordinates multiple processes during phagocytosis. For example, PIP3 recruits Rho GTPase-activating proteins (GAPs) to inactivate Rac and Cdc42 to help remodel the actin cytoskeleton during uptake [57, 58]. In addition, PIP3 regulates localized secretion and recruitment of myosin X to help drive pseudopod extensions around the target particle [53, 59].

The final step of internalization, the closure of the phagocytic arms around the particle, is less well understood. Recent work suggests a role for Myosin 1G, which appears to be sandwiched between the plasma membrane and the actin network via interaction with

PIP3 and the actin [60]. Myosin 1G then likely exerts force by pushing against the actin network to pull the leading pseudopod edges towards each other, closing the phagocytic cup. In addition, loss of PIP2 and PIP3 may be important for phagosome closure, likely to facilitate membrane and actin remodeling [61–64]. Additional clues from *C. elegans* suggest that PIP2 and PI3P synthesis and turnover are coordinated to help time phagosome sealing [65].

4   Phagosome Maturation Mechanisms

As the newly formed phagosome separates from the plasma membrane, it begins to undergo a series of changes that promotes degradation of the internalized particle. This process is termed phagosome maturation (Fig. 2), and involves the stepwise fusion of the phagosome to other cellular compartments. These fusion events alter the phagosomes' internal environment, changing it from the neutral pH present in the intercellular space to an acidic one, which aids in the digestion of internalized particles. Digestion itself is carried out by wide array of hydrolytic enzymes that work optimally at an acidic environment (pH < 5) and which are delivered to the phagosome from lysosomes during the final phase of phagosome maturation, described later [2, 66].

Phagosome maturation is initiated with conversion of the newly formed phagosome into an early phagosome, which shares the same hallmarks as early endosomes. One of the hallmarks of early phagosome maturation is acquisition of the small GTPase Rab5. However, Rab5 recruitment is preceded by the phagosomal acquisition of Rab20, which in turn recruits the Rab5 GEF, Rabex-5 [67, 68]. Rabex-5 then recruits and activates Rab5, which then engages a positive-feedback loop by stimulating Radaptin-5, which then enhances Rabex-5 activity [69]. Alternatively, Rab5 is activated on apoptotic body-containing phagosomes through the GEF Gapex-5 that is delivered to early phagosomes via microtubules [70].

GTP-loaded Rab5 recruits a variety of other effectors that drive phagosome maturation process forward. One of the initial effectors recruited by Rab5 to the surface of the early phagosome is Early Endosome Antigen-1 (EEA1) protein, which interacts with Rab5 through a Zn^{2+}-finger domain and an N-terminal extension, as well as through a C-terminal domain similar to the Rab5-interaction region of Radaptin-5 [71, 72]. Rab5 also recruits, indirectly, the Class III PI3K, hVps34, through its regulatory subunit p150 [73]. hVps34 synthesizes PI3P to further recruit EEA1 via PI3P interaction with its FYVE domain [72]. EEA1 acts as a tether between Rab5-associated membranes and helps mediate fusion by coordinating with the endosomal SNARE proteins, syntaxin 6 and syntaxin-13 [74–76]. Overall EEA1 is necessary for phagosome maturation and may be usurped by pathogens such as *Mycobacterium* to block phagosome maturation [77–79].

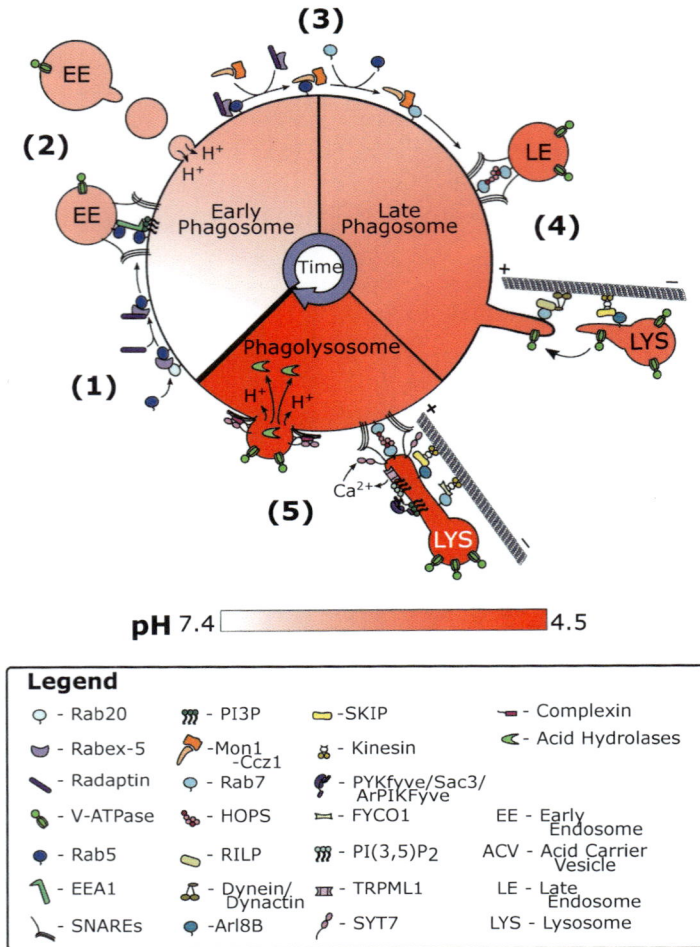

Fig. 2 Phagosome maturation and phagolysosome biogenesis. The phagosome is remodeled during phago-some maturation, gaining and losing proteins and lipid markers through fusion with endosomes and lyso-somes. (**1**) The early phagosome is typified by the acquisition of the small GTPase Rab5, which is activated by the GEF Rabex-5 that is recruited to the phagosome by Rab20. (**2**) The early phagosome is acidified slightly through the acquisition of the V-ATPase proton pump and delivery of protons during fusion with early endo-somes in a Rab5-EEA1 dependent manner. (**3**) The conversion from early to late phagosome is characterized by the transition from Rab5 to Rab7 on the phagosome surface. This process occurs through the recruitment of the Mon1-Ccz1 by Rab5; Mon1-Ccz1 then recruits and acts as a GEF for Rab7, activating it. (**4**) The late phagosome is further acidified through the acquisition of additional copies of V-ATPase. This occurs either through fusion with late endosomes in a Rab7-HOPS complex dependent manner, or through direct acquisition from tubulated lysosomes, which require microtubule motors. (**5**) Fusion with lysosomes builds the phagolyso-some, a highly acidic and proteolytic organelle that culminates with particle degradation. Phagosome–lyso-some fusion requires lysosomal Ca^{2+} released by TRPML1

The next major step in phagosome maturation is the transition from an early phagosome form to a phagolysosome. This transition is denoted by the conversion of the maturing phagosome from a Rab5-marked to a Rab7-positive organelle. Rab conversion on endosomes is initiated by the displacement of Rabex-5 and

Radaptin-5 by a complex of Mon1 and Ccz1 [80]. Mon1 is a Rab5 effector, being recruited to phagosomes by the GTP-bound form of Rab5 [81]. By recruiting the Mon1-Ccz1 complex to the phagosome, Rab5 turns off its own positive feedback loop, allowing a shift towards GTP hydrolysis by Rab5, presumably aided by a Rab5-GAP [82]. Mon1-Ccz1 then recruits and activates Rab7 by serving as a Rab7 GEF [81, 83, 84].

GTP-loaded Rab7 primes the phagosome for fusion with late endosomes and lysosomes, likely through various mechanisms [85]. First, activated Rab7 directs the inward movement of the maturing phagosome through interaction with the Rab7-interacting lysosomal protein (RILP) and oxysterol-binding protein-related protein 1L (ORP1L) [86]. The recruitment of RILP and ORP1L tethers the Rab7-positive phagosome to dynein-dynactin, transporting phagosomes inward via microtubules and towards lysosomes [87, 88]. Rab7-RILP also coordinates tubular membrane extensions that facilitate phagosome–lysosome fusion and acidification [87, 89]. In addition, GTP-loaded Rab7 likely promotes phagosome–lysosome fusion via the *Hom*otypic fusion and *p*rotein *s*orting (HOPS) complex [85, 90, 91]. HOPS is a six-subunit complex composed of a four-subunit core (Vps11, Vps13, Vps16, and Vps33) and two Rab7-interacting subunits, Vps39 and Vps41 [85]. Thus, HOPS can tether two Rab7-positive compartments to initiate fusion. Once bound to GTP-bound Rab7, HOPS chaperones and proofreads SNAREs, directly interacting with the quaternary SNARE complex consisting of a single R-SNARE and 3 Q-SNARES [92, 93]. Overall, HOPS seems to play a role in phagosome maturation in *C. elegans* during the clearance of apoptotic bodies [90], though its role in the maturation of phagosomes containing other particles has yet to be examined directly.

However, it is unlikely that HOPS and trans-SNARE complexes are sufficient to complete phagosome–lysosome fusion alone. Inhibition of Transient Receptor Potential Cation Channel of the Mucolipin subfamily, member 1 (TRPML1; or Mucolipin-1) allowed lysosomes to dock with phagosomes but they failed to fuse together [94]. TRPML1 is a lysosome-localized Ca^{2+} channel that facilitates diffusion of lysosomal Ca^{2+} into the cytosol [95]. Since chelation of Ca^{2+} led to similarly futile docked phagosome-lysosomes and that Ca^{2+} ionophores could rescue fusion, this suggests that lysosomal Ca^{2+} is required to trigger fusion [94]. The exact role of Ca^{2+} in phagosome–lysosome fusion is unclear, but it may mimic its role in activation of Synaptotagmin VII during lysosome exocytosis, including with nascent phagosomes [96, 97]. TRPML1 is also gated by the phosphatidylinositol-3,5-bisphosphate [$PI(3,5)P_2$] [98]. The absence of $PI(3,5)P_2$ dramatically enlarges and disrupts lysosome function [99, 100]. $PI(3,5)P_2$ is synthesized by the type III PIP kinase PIKfyve [101]. PIKfyve localizes to the lysosome as part of a complex that contains its opposing enzyme, the phosphoinositide 5-phosphatase Sac3/Fig4, along with the scaffold protein ArPIKfyve/Vac14 [102, 103]. Inhibition of PIKfyve blocked phagosome and endosome maturation in macrophages, likely by various additional mechanisms in addition to disrupting TRPML1 activity [104].

In addition to Rab7 and $PI(3,5)P_2$-mediated pathways, the ADP-ribosylation factor-like protein 8B (Arl8B)GTPase is fast becoming a central player in lysosome function and dynamics [105–107]. Arl8 links lysosomes to plus-end directed kinesins through SifA and kinesin-interacting protein (SKIP) when GTP-bound, and allows lysosomes to move outward from the microtubule organizing center [108]. Interestingly, Arl8b appears to play an important role in phagolysosome biogenesis in macrophages and *C. elegans* during microbial and apoptotic body removal, respectively [109, 110]. The role in phagosome maturation, may not simply be due to control of kinesin-dependent movement—Arl8b appears to be necessary for membrane association of the HOPS complex, suggesting a complex interplay between Arl8b and Rab7 [110, 111].

A corollary of phagolysosome biogenesis is the dramatic restructuring of the phagosomal lumen. The phagolysosomal lumen undergoes a deep acidification, going from pH of 7.4 to <4.5 by the concerted action of V-ATPase and ion transporters. This drop in pH is critical to optimal activity of more than 60 different acid-optimized hydrolytic enzymes such as proteases, peptidases, phosphatases, nucleases, glycosidases, sulfatases, and lipases [66]. Each of the individual hydrolases degrades a subset of internalized macromolecules, with the resulting components being available to be utilized by the cell. Ultimately, the particle is digested. Thus, it is not surprising that many human pathogens ranging from *Salmonella*, *Mycobacterium*, *Legionella*, *Neisseria*, and fungal species have evolved mechanisms to arrest or modify phagosome maturation to avoid being killed within the phagocyte. For the purpose of this review, we suggest the following excellent review articles on how pathogens alter phagosome maturation [112, 113].

5   Phagosome Fission and Phagolysosome Resolution

The least well-characterized events in the phagosome life cycle are membrane fission from the phagosome and the eventual resolution of the phagolysosome. Membrane fission likely occurs throughout phagosome maturation, starting with the recycling of early endosome proteins like transferrin receptor [114], and export of antigen-bound presenting receptors to promote the immune response [115]. After digestion, peptide antigens are loaded onto the major histocompatibility complexes II (MHCII) within the phagolysosome [115]. The loaded MHCs are then trafficked from the phagolysosome to the plasma membrane to be presented to immune cells. MHCII trafficking to the plasma membrane has been shown to be carried out through tubular structures, which are derived from lysosome-like organelles and phagosomes [116–119]. Although not tested on phagosome maturation, lysosome tubulation required Arl8b and the mammalian target of rapamycin (mTOR), a kinase that responds to cellular energy levels, nutrient levels, and growth signals to modulate cellular growth and division [119, 120]. Tubulation also occurs from phagosomes in a manner that required intact actin and microtubule networks and aided in content transfer out of the phagosome [118].

Compared to phagosome fission, even less is known about phagosome resolution, or the complete breakdown of the phagosome, which must occur after particle digestion. During a late stage of phagosome maturation, small vesicles appear to leave the phagolysosome in an mTOR dependent process [121]. In particular, mTOR appears to be recruited to phagolysosomes by amino acid released during particle breakdown [121], which might mimic mTOR activation lysosomal amino acid efflux [122, 123]. mTOR then appears to promote fission of vesicles form the phagosome, followed by vesicular trafficking to the remaining lysosomal network [121]. This process appears to have parallels to lysosome reformation from autophagolysosomes [124]. The molecular players downstream of mTOR that control the fission of either phagolysosomes or autophagolysosomes are unknown.

6   Concluding Remarks

The uptake of particles through phagocytosis, be they pathogenic or an apoptotic neighbor, starts a complex journey for the nascent phagosome. After properly selecting the target particle through receptor interactions, a phagocytic cell must coordinate a vast array of proteins and membrane lipids to internalize the particle. Once sequestered within a phagosome, the particle is subjugated to increasing hostile conditions, through the process of phagosome maturation, ultimately resulting in its disassembly. This whole process of phagocytosis and phagosome maturation has been extensively studied, though as outlined in this piece, there is still much work to be done. Several gaps still exist in our understanding of phagocytosis, specifically in terms of how uptake is completed, as well as how the phagolysosome is ultimately resolved.

Acknowledgements

M.G. is funded through support from Ryerson University, a Canada Research Chair and Early Researcher Awards to R.J.B. Research in the laboratory of R.J.B. is supported by grants from the Canadian Institutes of Health Research and the Natural Sciences and Engineering Research Council of Canada.

References

1. Arandjelovic S, Ravichandran KS (2015) Phagocytosis of apoptotic cells in homeostasis. Nat Immunol 16:907–917

2. Flannagan RS, Jaumouillé V, Grinstein S (2012) The cell biology of phagocytosis. Annu Rev Pathol 7:61–98

3. Dale DC, Boxer L, Liles WC (2008) The phagocytes: neutrophils and monocytes. Blood 112:935–945

4. Davies LC, Jenkins SJ, Allen JE, Taylor PR (2013) Tissue-resident macrophages. Nat Immunol 14:986–995

5. Gomez Perdiguero E, Klapproth K, Schulz C et al (2014) Tissue-resident macrophages originate from yolk-sac-derived erythro-myeloid progenitors. Nature 518:547–551

6. Akashi K, Traver D, Miyamoto T, Weissman IL (2000) A clonogenic common myeloid progenitor that gives rise to all myeloid lineages. Nature 404:193–197

7. Jaumouillé V, Farkash Y, Jaqaman K et al (2014) Actin cytoskeleton reorganization by Syk regulates Fcγ receptor responsiveness by increasing its lateral mobility and clustering. Dev Cell 29:534–546

8. Monks J, Rosner D, Jon Geske F et al (2005) Epithelial cells as phagocytes: apoptotic epithelial cells are engulfed by mammary alveolar epithelial cells and repress inflammatory mediator release. Cell Death Differ 12:107–114

9. Ravichandran KS, Lorenz U (2007) Engulfment of apoptotic cells: signals for a good meal. Nat Rev Immunol 7:964–974

10. Flannagan RS, Harrison RE, Yip CM et al (2010) Dynamic macrophage "probing" is required for the efficient capture of phagocytic targets. J Cell Biol 191:1205–1218

11. Stuart LM, Ezekowitz RAB (2005) Phagocytosis: elegant complexity. Immunity 22:539–550

12. Freeman SA, Grinstein S (2014) Phagocytosis: receptors, signal integration, and the cytoskeleton. Immunol Rev 262:193–215

13. Gregory CD (2000) CD14-dependent clearance of apoptotic cells: relevance to the immune system. Curr Opin Immunol 12:27–34

14. Arur S, Uche UE, Rezaul K et al (2003) Annexin I is an endogenous ligand that mediates apoptotic cell engulfment. Dev Cell 4:587–598

15. Moffatt OD, Devitt A A, Bell ED et al (1999) Macrophage recognition of ICAM-3 on apoptotic leukocytes. J Immunol 162:6800–6810

16. Torr EE, Gardner DH, Thomas L et al (2012) Apoptotic cell-derived ICAM-3 promotes both macrophage chemoattraction to and tethering of apoptotic cells. Cell Death Differ 19:671–679

17. Vandivier RW, Ogden CA, Fadok VA et al (2002) Role of surfactant proteins A, D, and C1q in the clearance of apoptotic cells in vivo and in vitro: calreticulin and CD91 as a common collectin receptor complex. J Immunol 169:3978–3986

18. Fadok VA, Voelker DR, Campbell PA et al (1992) Exposure of phosphatidylserine on the surface of apoptotic lymphocytes triggers specific recognition and removal by macrophages. J Immunol 148:2207–2216

19. Ezekowitz RA, Sastry K, Bailly P, Warner A (1990) Molecular characterization of the human macrophage mannose receptor: demonstration of multiple carbohydrate recognition-like domains and phagocytosis of yeasts in Cos-1 cells. J Exp Med 172:1785–1794

20. Ezekowitz RA, Williams DJ, Koziel H et al (1991) Uptake of Pneumocystis carinii mediated by the macrophage mannose receptor. Nature 351:155–158

21. Jiménez-Dalmaroni MJ, Gerswhin ME, Adamopoulos IE (2016) The critical role of toll-like receptors—from microbial recognition to autoimmunity: a comprehensive review. Autoimmun Rev 15:1–8

22. Pfeiffer A, Bttcher A, Ors E et al (2001) Lipopolysaccharide and ceramide docking to CD14 provokes ligand-specific receptor clustering in rafts. Eur J Immunol 31:3153–3164

23. Groves E, Dart AE, Covarelli V, Caron E (2008) Molecular mechanisms of phagocytic uptake in mammalian cells. Cell Mol Life Sci 65:1957–1976

24. Murphy K, Travers P, Walport M (2008) Janeway's immunobiology. Garland Science, New York

25. Suzuki T, Kono H, Hirose N et al (2000) Differential involvement of Src family kinases in Fc gamma receptor-mediated phagocytosis. J Immunol 165:473–482

26. Underhill DM, Goodridge HS (2007) The many faces of ITAMs. Trends Immunol 28:66–73

27. Freeman SA, Goyette J, Furuya W et al (2016) Integrins form an expanding diffusional barrier that coordinates phagocytosis. Cell 164:128–140

28. Park H, Cox D (2011) Syk regulates multiple signaling pathways leading to CX3CL1 chemotaxis in macrophages. J Biol Chem 286:14762–14769

29. Tomasevic N, Jia Z, Russell A et al (2007) Differential regulation of WASP and N-WASP by Cdc42, Rac1, Nck, and PI(4,5)P2. Biochemistry 46:3494–3502

30. Pollard T (2002) Structure and function of the Arp2/3 complex. Curr Opin Struct Biol 12:768–774

31. Dart AE, Donnelly SK, Holden DW et al (2012) Nck and Cdc42 co-operate to recruit N-WASP to promote Fc R-mediated phagocytosis. J Cell Sci 125:2825–2830

32. Humphries AC, Donnelly SK, Way M (2014) Cdc42 and the Rho GEF intersectin-1 collaborate with Nck to promote N-WASP-dependent actin polymerisation. J Cell Sci 127:673–685

33. Brown GD, Gordon S (2001) Immune recognition. A new receptor for beta-glucans. Nature 413:36–37

34. Rogers NC, Slack EC, Edwards AD et al (2005) Syk-dependent cytokine induction by dectin-1 reveals a novel pattern recognition pathway for C type lectins. Immunity 22:507–517

35. Goodridge HS, Reyes CN, Becker CA et al (2011) Activation of the innate immune receptor Dectin-1 upon formation of a "phagocytic synapse". Nature 472:471–475

36. Herre J, Marshall ASJ, Caron E et al (2004) Dectin-1 uses novel mechanisms for yeast phagocytosis in macrophages. Blood 104:4038–4045

37. Shah VB, Ozment-Skelton TR, Williams DL, Keshvara L (2009) Vav1 and PI3K are required for phagocytosis of beta-glucan and subsequent superoxide generation by microglia. Mol Immunol 46:1845–1853

38. Côté J-F, Vuori K (2002) Identification of an evolutionarily conserved superfamily of DOCK180-related proteins with guanine nucleotide exchange activity. J Cell Sci 115:4901–4913

39. Akakura S, Singh S, Spataro M et al (2004) The opsonin MFG-E8 is a ligand for the αvβ5 integrin and triggers DOCK180-dependent Rac1 activation for the phagocytosis of apoptotic cells. Exp Cell Res 292:403–416

40. Albert ML, Kim JI, Birge RB (2000) alphavbeta5 integrin recruits the CrkII-Dock180-rac1 complex for phagocytosis of apoptotic cells. Nat Cell Biol 2:899–905

41. Park D, Tosello-Trampont A-C, Elliott MR et al (2007) BAI1 is an engulfment receptor for apoptotic cells upstream of the ELMO/Dock180/Rac module. Nature 450:430–434

42. Wu Y, Singh S, Georgescu M-M, Birge RB (2005) A role for Mer tyrosine kinase in alphavbeta5 integrin-mediated phagocytosis of apoptotic cells. J Cell Sci 118:539–553

43. Brugnera E, Haney L, Grimsley C et al (2002) Unconventional Rac-GEF activity is mediated through the Dock180-ELMO complex. Nat Cell Biol 4:574–582

44. Behnia R, Munro S (2005) Organelle identity and the signposts for membrane traffic. Nature 438:597–604

45. Di Paolo G, De Camilli P (2006) Phosphoinositides in cell regulation and membrane dynamics. Nature 443:651–657

46. Botelho RJ (2009) Changing phosphoinositides "on the fly": how trafficking vesicles avoid an identity crisis. Bioessays 31:1127–1136

47. Botelho RJ, Teruel M, Dierckman R et al (2000) Localized biphasic changes in phosphatidylinositol-4,5-bisphosphate at sites of phagocytosis. J Cell Biol 151:1353–1367

48. Mao YS, Yamaga M, Zhu X et al (2009) Essential and unique roles of PIP5K-gamma and -alpha in Fcgamma receptor-mediated phagocytosis. J Cell Biol 184:281–296

49. Szymańska E, Korzeniowski M, Raynal P et al (2009) Contribution of PIP-5 kinase Iα to raft-based FcγRIIA signaling. Exp Cell Res 315:981–995

50. Coppolino MG, Dierckman R, Loijens J et al (2002) Inhibition of phosphatidylinositol-4-phosphate 5-kinase Iα impairs localized actin remodeling and suppresses phagocytosis. J Biol Chem 277:43849–43857

51. Scott CC (2005) Phosphatidylinositol-4,5-bisphosphate hydrolysis directs actin remodeling during phagocytosis. J Cell Biol 169:139–149

52. Bohdanowicz M, Balkin DM, De Camilli P, Grinstein S (2012) Recruitment of OCRL and Inpp5B to phagosomes by Rab5 and APPL1 depletes phosphoinositides and attenuates Akt signaling. Mol Biol Cell 23:176–187

53. Cox D, Tseng CC, Bjekic G, Greenberg S (1999) A requirement for phosphatidylinositol 3-kinase in pseudopod extension. J Biol Chem 274:1240–1247

54. Marshall JG, Booth JW, Stambolic V et al (2001) Restricted accumulation of phosphatidylinositol 3-kinase products in a plasmalemmal subdomain during Fc gamma receptor-mediated phagocytosis. J Cell Biol 153:1369–1380

55. Levin R, Grinstein S, Schlam D (2015) Phosphoinositides in phagocytosis and macropinocytosis. Biochim Biophys Acta 1851:805–823

56. Higgs HN, Pollard TD (2000) Activation by Cdc42 and PIP(2) of Wiskott-Aldrich syndrome protein (WASp) stimulates actin nucleation by Arp2/3 complex. J Cell Biol 150:1311–1320

57. Beemiller P, Zhang Y, Mohan S et al (2010) A Cdc42 activation cycle coordinated by PI 3-kinase during fc receptor-mediated phagocytosis. Mol Biol Cell 21:470–480

58. Schlam D, Bagshaw RD, Freeman SA et al (2015) Phosphoinositide 3-kinase enables phagocytosis of large particles by terminating actin assembly through Rac/Cdc42 GTPase-activating proteins. Nat Commun 6:8623

59. Cox D, Berg JS, Cammer M et al (2002) Myosin X is a downstream effector of PI(3)K during phagocytosis. Nat Cell Biol 4:469–477

60. Dart AE, Tollis S, Bright MD et al (2012) The motor protein myosin 1G functions in Fc R-mediated phagocytosis. J Cell Sci 125:6020–6029

61. Horan KA, Watanabe K-I, Kong AM et al (2007) Regulation of FcgammaR-stimulated phagocytosis by the 72-kDa inositol polyphosphate 5-phosphatase: SHIP1, but not the 72-kDa 5-phosphatase, regulates complement receptor 3 mediated phagocytosis by differential recruitment of these 5-phosphatases to the phagocytic cup. Blood 110:4480–4491

62. Ai J, Maturu A, Johnson W et al (2006) The inositol phosphatase SHIP-2 down-regulates FcγR-mediated phagocytosis in murine macrophages independently of SHIP-1. Blood 107:813–820

63. Bohdanowicz M, Cosío G, Backer JM, Grinstein S (2010) Class I and class III phosphoinositide 3-kinases are required for actin polymerization that propels phagosomes. J Cell Biol 191:999–1012

64. Cox D, Dale BM, Kashiwada M et al (2001) A regulatory role for Src homology 2 domain-containing inositol 5'-phosphatase (SHIP) in phagocytosis mediated by Fc gamma receptors and complement receptor 3 (alpha(M)beta(2); CD11b/CD18). J Exp Med 193:61–71

65. Cheng S, Wang K, Zou W et al (2015) PtdIns(4,5)P(2) and PtdIns3P coordinate to regulate phagosomal sealing for apoptotic cell clearance. J Cell Biol 210:485–502

66. Appelqvist H, Wäster P, Kågedal K, Öllinger K (2013) The lysosome: from waste bag to potential therapeutic target. J Mol Cell Biol 5:214–226

67. Pei G, Repnik U, Griffiths G, Gutierrez MG (2014) Identification of an immune-regulated phagosomal Rab cascade in macrophages. J Cell Sci 127:2071–2082

68. Egami Y, Araki N (2012) Rab20 regulates phagosome maturation in RAW264 macrophages during Fc gamma receptor-mediated phagocytosis. PLoS One 7, e35663

69. Zhang Z, Zhang T, Wang S et al (2014) Molecular mechanism for Rabex-5 GEF activation by Rabaptin-5. Elife 3

70. Kitano M, Nakaya M, Nakamura T et al (2008) Imaging of Rab5 activity identifies essential regulators for phagosome maturation. Nature 453:241–245

71. Mishra A, Eathiraj S, Corvera S, Lambright DG (2010) Structural basis for Rab GTPase recognition and endosome tethering by the C2H2 zinc finger of Early Endosomal Autoantigen 1 (EEA1). Proc Natl Acad Sci U S A 107:10866–10871

72. Simonsen A, Lippé R, Christoforidis S et al (1998) EEA1 links PI(3)K function to Rab5 regulation of endosome fusion. Nature 394:494–498

73. Murray JT, Panaretou C, Stenmark H et al (2002) Role of Rab5 in the recruitment of hVps34/p150 to the early endosome. Traffic 3:416–427

74. Simonsen A, Gaullier JM, D'Arrigo A, Stenmark H (1999) The Rab5 effector EEA1 interacts directly with syntaxin-6. J Biol Chem 274:28857–28860

75. McBride HM, Rybin V, Murphy C et al (1999) Oligomeric complexes link Rab5 effectors with NSF and drive membrane fusion via interactions between EEA1 and syntaxin 13. Cell 98:377–386

76. Christoforidis S, McBride HM, Burgoyne RD, Zerial M (1999) The Rab5 effector EEA1 is a core component of endosome docking. Nature 397:621–625

77. Fratti RA, Backer JM, Gruenberg J et al (2001) Role of phosphatidylinositol 3-kinase and Rab5 effectors in phagosomal biogenesis and mycobacterial phagosome arrest. J Cell Biol 154:631–644

78. Hackam DJ, Rotstein OD, Zhang WJ et al (1997) Regulation of phagosomal acidification. J Biol Chem 272:29810–29820

79. Sturgill-Koszycki S, Schlesinger P, Chakraborty P et al (1994) Lack of acidification in Mycobacterium phagosomes produced by exclusion of the vesicular proton-ATPase. Science 263:678–681

80. Poteryaev D, Datta S, Ackema K et al (2010) Identification of the switch in early-to-late endosome transition. Cell 141:497–508

81. Kinchen JM, Ravichandran KS (2010) Identification of two evolutionarily conserved genes regulating processing of engulfed apoptotic cells. Nature 464:778–782

82. Haas AK, Fuchs E, Kopajtich R, Barr FA (2005) A GTPase-activating protein controls Rab5 function in endocytic trafficking. Nat Cell Biol 7:887–893

83. Nordmann M, Cabrera M, Perz A et al (2010) The Mon1-Ccz1 complex is the GEF of the late endosomal Rab7 homolog Ypt7. Curr Biol 20:1654–1659

84. Cabrera M, Nordmann M, Perz A et al (2014) The Mon1-Ccz1 GEF activates the Rab7 GTPase Ypt7 via a longin-fold-Rab interface and association with PI3P-positive membranes. J Cell Sci 127:1043–1051

85. Balderhaar HJK, Ungermann C (2013) CORVET and HOPS tethering complexes—coordinators of endosome and lysosome fusion. J Cell Sci 126:1307–1316

86. Johansson M, Rocha N, Zwart W et al (2007) Activation of endosomal dynein motors by stepwise assembly of Rab7-RILP-p150Glued, ORP1L, and the receptor βIII spectrin. J Cell Biol 176:459–471

87. Harrison RE, Bucci C, Vieira OV et al (2003) Phagosomes fuse with late endosomes and/or lysosomes by extension of membrane protrusions along microtubules: role of Rab7 and RILP. Mol Cell Biol 23:6494–6506

88. Wyroba E, Surmacz L, Osinska M, Wiejak J (2007) Phagosome maturation in unicellular eukaryote Paramecium: the presence of RILP, Rab7 and LAMP-2 homologues. Eur J Histochem 51:163–172

89. Sun-Wada G-H, Tabata H, Kawamura N et al (2009) Direct recruitment of H+-ATPase from lysosomes for phagosomal acidification. J Cell Sci 122:2504–2513

90. Kinchen JM, Doukoumetzidis K, Almendinger J et al (2008) A pathway for phagosome maturation during engulfment of apoptotic cells. Nat Cell Biol 10:556–566

91. Akbar MA, Tracy C, Kahr WHA, Krämer H (2011) The full-of-bacteria gene is required for phagosome maturation during immune defense in Drosophila. J Cell Biol 192:383–390

92. Krämer L, Ungermann C (2011) HOPS drives vacuole fusion by binding the vacuolar SNARE complex and the Vam7 PX domain via two distinct sites. Mol Biol Cell 22:2601–2611

93. Lobingier BT, Merz AJ (2012) Sec1/Munc18 protein Vps33 binds to SNARE domains and the quaternary SNARE complex. Mol Biol Cell 23:4611–4622

94. Dayam RM, Saric A, Shilliday RE, Botelho RJ (2015) The phosphoinositide-gated lysosomal Ca²⁺ channel, TRPML1, is required for phagosome maturation. Traffic 16:1010–26

95. Wang W, Zhang X, Gao Q, Xu H (2014) Trpml1: an ion channel in the lysosome. Handb Exp Pharmacol 222:631–645

96. Czibener C, Sherer NM, Becker SM et al (2006) Ca2+ and synaptotagmin VII-dependent delivery of lysosomal membrane to nascent phagosomes. J Cell Biol 174:997–1007

97. Südhof TC (2013) Neurotransmitter release: the last millisecond in the life of a synaptic vesicle. Neuron 80:675–690

98. Dong X, Shen D, Wang X et al (2010) PI(3,5)P(2) controls membrane trafficking by direct activation of mucolipin Ca(2+) release channels in the endolysosome. Nat Commun 1:38

99. Ho CY, Alghamdi TA, Botelho RJ (2012) Phosphatidylinositol-3,5-bisphosphate: no longer the poor PIP 2. Traffic 13:1–8

100. McCartney AJ, Zhang Y, Weisman LS (2014) Phosphatidylinositol 3,5-bisphosphate: low abundance, high significance. Bioessays 36:52–64

101. Ikonomov OC, Sbrissa D, Shisheva A (2001) Mammalian cell morphology and endocytic membrane homeostasis require enzymatically active phosphoinositide 5-kinase PIKfyve. J Biol Chem 276:26141–26147

102. Sbrissa D, Ikonomov OC, Fenner H, Shisheva A (2008) ArPIKfyve homomeric and heteromeric interactions scaffold PIKfyve and Sac3 in a complex to promote PIKfyve activity and functionality. J Mol Biol 384:766–779

103. Jin N, Chow CY, Liu L et al (2008) VAC14 nucleates a protein complex essential for the acute interconversion of PI3P and PI(3,5)P(2) in yeast and mouse. EMBO J 27:3221–3234

104. Kim GHE, Dayam RM, Prashar A et al (2014) PIKfyve inhibition interferes with phagosome and endosome maturation in macrophages. Traffic 15:1143–1163

105. Hofmann I, Munro S (2006) An N-terminally acetylated Arf-like GTPase is localised to lysosomes and affects their motility. J Cell Sci 119:1494–1503

106. Bagshaw RD, Callahan JW, Mahuran DJ (2006) The Arf-family protein, Arl8b, is involved in the spatial distribution of lysosomes. Biochem Biophys Res Commun 344:1186–1191

107. Khatter D, Sindhwani A, Sharma M (2015) Arf-like GTPase Arl8: moving from the periphery to the center of lysosomal biology. Cell Logist 5, e1086501

108. Rosa-Ferreira C, Munro S (2011) Arl8 and SKIP act together to link lysosomes to kinesin-1. Dev Cell 21:1171–1178

109. Sasaki A, Nakae I, Nagasawa M et al (2013) Arl8/ARL-8 functions in apoptotic cell removal by mediating phagolysosome formation in Caenorhabditis elegans. Mol Biol Cell 24:1584–1592

110. Garg S, Sharma M, Ung C et al (2011) Lysosomal trafficking, antigen presentation, and microbial killing are controlled by the Arf-like GTPase Arl8b. Immunity 35:182–193

111. Khatter D, Raina VB, Dwivedi D et al (2015) The small GTPase Arl8b regulates assembly of the mammalian HOPS complex on lysosomes. J Cell Sci 128:1746–1761

112. do Vale A, Cabanes D, Sousa S (2016) Bacterial toxins as pathogen weapons against phagocytes. Front Microbiol 7:42

113. Smith LM, May RC (2013) Mechanisms of microbial escape from phagocyte killing. Biochem Soc Trans 41:475–490

114. Botelho RJ, Hackam DJ, Schreiber AD, Grinstein S (2000) Role of COPI in phagosome maturation. J Biol Chem 275:15717–15727

115. Mantegazza AR, Magalhaes JG, Amigorena S, Marks MS (2013) Presentation of phagocytosed antigens by MHC Class I and II. Traffic 14:135–152

116. Boes M, Bertho N, Cerny J et al (2003) T cells induce extended class II MHC compartments in dendritic cells in a toll-like receptor-dependent manner. J Immunol 171:4081–4088

117. Vyas JM, Kim Y-M, Artavanis-Tsakonas K et al (2007) Tubulation of class II MHC compartments is microtubule dependent and involves multiple endolysosomal membrane proteins in primary dendritic cells. J Immunol 178:7199–7210

118. Mantegazza AR, Zajac AL, Twelvetrees A et al (2014) TLR-dependent phagosome tubulation in dendritic cells promotes phagosome cross-talk to optimize MHC-II antigen presentation. Proc Natl Acad Sci U S A 111:15508–15513

119. Saric A, Hipolito VEB, Kay JG et al (2016) mTOR controls lysosome tubulation and antigen presentation in macrophages and dendritic cells. Mol Biol Cell 27:321–333

120. Zoncu R, Efeyan A, Sabatini DM (2011) mTOR: from growth signal integration to cancer, diabetes and ageing. Nat Rev Mol Cell Biol 12:21–35

121. Krajcovic M, Krishna S, Akkari L et al (2013) mTOR regulates phagosome and entotic vacuole fission. Mol Biol Cell 24:3736–3745

122. Sancak Y, Bar-Peled L, Zoncu R et al (2010) Ragulator-Rag complex targets mTORC1 to the lysosomal surface and is necessary for its activation by amino acids. Cell 141:290–303

123. Zoncu R, Bar-Peled L, Efeyan A et al (2011) mTORC1 senses lysosomal amino acids through an inside-out mechanism that requires the vacuolar H(+)-ATPase. Science 334:678–683

124. Yu L, McPhee CK, Zheng L et al (2010) Termination of autophagy and reformation of lysosomes regulated by mTOR. Nature 465:942–946

Chapter 2

Analysis of Human and Mouse Neutrophil Phagocytosis by Flow Cytometry

Noah Fine, Oriyah Barzilay, and Michael Glogauer

Abstract

Neutrophils are primary phagocytes that recognize their targets through surface chemistry, either through Pattern Recognition Receptor (PPR) interaction with Pathogen-Associated Molecular Patterns (PAMPs) or through immunoglobulin (Ig) or complement mediated recognition. Opsonization can be important for target recognition, and phagocytosis by neutrophils in whole blood can be greatly enhanced due to the presence of blood serum components and platelets. Powerful and sensitive flow cytometry based methods are presented to measure phagocytosis by human blood neutrophils and mouse peritoneal neutrophils.

Key words Neutrophil, Phagocytosis, pHrodo, Peritonitis, Flow cytometry

1   Introduction

Ingestion of inflammatory targets by neutrophils is an essential aspect of the innate immune response [1]. Sensitive measurement of this phenomenon is essential for a thorough understanding of the molecular and physiological mechanisms that control this process. pHrodo™ BioParticles® conjugates from Invitrogen allow for sensitive measurement of phagocytosis by flow cytometry [2]. pHrodo™ becomes highly fluorescent when exposed to acidic pH after internalization, and the process can be measured by conventional detection modalities including fluorescence imaging microscopy, fluorescent microplate reader and flow cytometry. Here we describe protocols for flow cytometric analysis of phagocytosis using pHrodo™ Green *E. coli* BioParticles® conjugates. This method is easily adapted for use of Zymosan or *S. aureus* bioparticle conjugates, and pHrodo™ Red conjugates, which are also available.

Neutrophils are among the fastest moving cells in the body, and their responses to inflammatory stimuli are nearly instantaneous. These characteristics, which are appropriate for the effectiveness of neutrophils in combating microbes, are detrimental with respect to the in vitro manipulation of these cells for

Roberto Botelho (ed.), *Phagocytosis and Phagosomes: Methods and Protocols*, Methods in Molecular Biology, vol. 1519, DOI 10.1007/978-1-4939-6581-6_2, © Springer Science+Business Media New York 2017

experimental purposes. Because neutrophils can be easily activated by experimental manipulation [3–6] it is desirable to keep these interactions to a minimum if possible. We have developed a protocol whereby neutrophils in whole human blood, or peritoneally elicited mouse neutrophils are fluorescently labeled with pHrodo™ BioParticles® and fixed in order to analyze phagocytosis by flow cytometry with a minimum amount of manipulation. For the in vitro blood neutrophil assay, the endpoint of the phagocytosis reaction is achieved by fixing the samples with paraformaldehyde (PFA), and samples are subsequently labeled for flow cytometric analysis. In the mouse, pHrodo™ BioParticles® are used to induce peritonitis, and neutrophils are recovered and immediately fixed with PFA. A combination of CD16 and SSC-A is sufficient to gate on neutrophils in whole human blood (unpublished results), while Ly6G^{+vc}/F4/80^{-vc} gating is used for mouse neutrophils. In addition to being simpler and faster than conventional approaches, we have found that our approach minimizes the activation of neutrophils that would otherwise occur during the purification of neutrophils from blood. Furthermore, fixation of neutrophils does not interfere with subsequent antibody labeling and multicolor flow cytometry analysis. Our protocol focuses on sample acquisition, preparation, labeling, flow cytometric acquisition and analysis.

2 Materials

1. pHrodo™ Green *E. coli* BioParticles® conjugates from Invitrogen.

2. Phosphate-buffered saline (PBS): 137 mM NaCl, 2.7 mM KCl, 10 mM Na_2HPO_4, 1.8 mM KH_2PO_4, pH 7.4.

3. Red blood cell lysis buffer such as BD Pharm lyse solution.

4. Hanks buffered salt solution (HBSS): 138 mM NaCl, 5.3 mM KCl, 0.34 mM Na_2HPO_4, 0.44 mM KH_2PO_4, 4.2 mM $NaHCO_3$, 5.6 mM glucose, pH 7.4.

5. FACS buffer: HBSS with 2 mM EDTA, 1% BSA.

6. Vacutainers containing sodium citrate.

7. 16% PFA ampoule.

8. 1 mg/ml rat serum.

9. 1 mg/ml mouse IgG.

10. Mouse anti-human CD16-PE (Clone: 3G8).

11. Mouse anti-human CD66a-APC (Clone: CD66a-B1.1).

12. Anti-mouse Ly6G-PerCP-Cy5.5 (Clone: 1A8).

13. Anti-mouse F4/80-BV421 (Clone: BM8).

14. Flow cytometry compensation beads (example OneComp eBeads from eBiosciences).

15. Sonicator.

16. Coulter counter or hemocytometer.

17. Wild type C57 black six mice (8–16 weeks old).

---

3   Methods

Unless noted otherwise, all samples are maintained on ice and centrifugations are at 4 °C.

3.1   Phagocytosis by Human Blood Neutrophils

1. Resuspend pHrodo™ fluorescent bioparticles (2 mg lyophilized per vial) in 2 ml of sterile PBS.

2. Sonicate on ice for 5 min in a glass tube (*see* **Note 1**).

3.1.1   Preparation of pHrodo Particles

3.1.2   Acquisition of Blood Samples

1. Blood should be drawn by a trained phlebotomist (*see* **Note 2**).

2. Blood samples are drawn from the median cubital vein in the crook of the elbow into a vacutainer containing 0.1 volumes of sodium citrate anticoagulant.

3. Mix the blood by gentle inversion and maintain at room temperature (*see* **Note 3**).

3.1.3   Blood Neutrophil Stimulation

1. Aliquot 50 μl of blood into flow cytometry tubes.

2. Add pHrodo™ (0.5, 2, 10 μl) or vehicle control to the blood and vortex the tubes gently to mix (Fig. 1).

3. Cover the tubes with paraffin wax and incubate them at 37 °C for the appropriate amount of time (*see* **Note 4**).

4. Add 1/10th volume of PFA (1.6% final) to each tube to stop the phagocytosis reaction (*see* **Note 5**).

5. Gently vortex each tube and incubate them on ice for 15 min.

6. Add 1 ml of PBS to dilute the fixative.

7. Centrifuge the tubes for 5 min at $1000 \times g$ at 4 °C.

8. Decant the supernatant.

9. Lyse the red blood cells (RBCs) by resuspending each pellet in 1 ml of red blood cell lysis buffer.

10. Incubate the tubes on ice for 15 min.

11. Centrifuge the tubes for 5 min at $1000 \times g$ at 4 °C.

12. Decant the supernatant.

13. Repeat the RBC lysis steps (**steps 9–12**) until pellets are white (*see* **Note 6**).

14. Resuspend the pellets in 1 ml of PBS.

15. Use a Coulter counter or hemocytometer to count the cells (*see* **Note 7**).

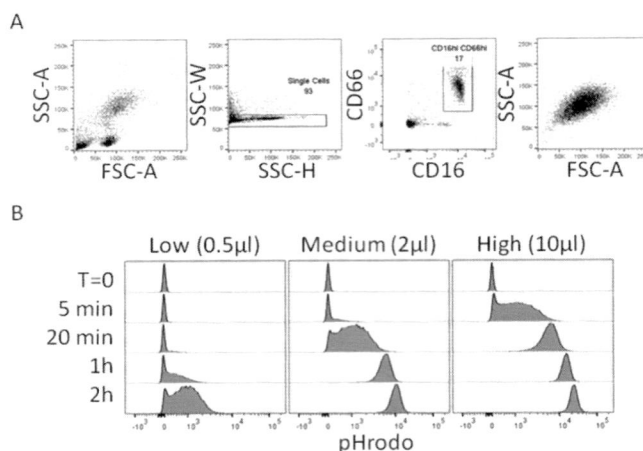

Fig. 1 Phagocytosis by human blood neutrophils. (**a**) Cells passing through the cytometer in pairs (doublets) were excluded using side scatter height (SSC-H) × side scatter width (SSC-W) and neutrophils were gated in whole blood based on high expression of CD16 and CD66a. (**b**) Three dosages of pHrodo were added to 50 μl of human blood and incubated at 37 °C for the indicated times. At least 2×10^4 gated neutrophil events were acquired

3.1.4 Fluorescent Labeling

1. Resuspend the cells in FACS buffer so that the final volume, including antibody, will be 50 μl.

2. To block Fc-receptors, add 1 μl rat serum and 2 μl mouse IgG to each tube, vortex gently, and incubate on ice for 20 min.

3. Prepare the mastermix of fluorescently conjugated antibodies (*see* **Notes 8** and **9**).

4. Do not use FITC-conjugated antibodies, since this is the same fluorescent channel as pHrodo Green bioparticles.

5. Add the antibody mastermix, vortex gently and incubate for 30 min on ice protected from light.

6. Wash each pellet with 1 ml of FACS buffer.

7. Vortex briefly and pellet the cells at $1000 \times g$ for 5 min at 4 °C.

8. Decant the supernatant.

9. Repeat the wash steps (**steps 6–8**) two additional times.

10. Resuspend the pellets in 250 μl of FACS buffer and vortex briefly.

11. Prepare single stained compensation beads controls according to manufacturers' instructions.

12. Resuspend labeled compensation beads in 400 μl of FACS buffer and vortex briefly.

13. Cover the tubes with paraffin wax and store in the fridge protected from the light.

3.1.5 Sample Acquisition

1. Perform instrument setup according to the protocols of your flow cytometry facility.

2. Vortex the first sample and load onto the cytometer.

3. Adjust the forward scatter (FSC) and side scatter (SSC) settings so that the population of interest (neutrophils) is displayed in the center of the scatterplot.

4. Ensure fluorescent signals are on scale.

5. Perform automated compensation using single stained beads controls and pHrodo-labeled cells.

6. Run each sample and acquire at least 2×10^4 gated events.

7. Export FCS files.

8. Data can be analyzed using FlowJo or other commercially available software.

3.2 Mouse Peritoneal Recruitment and Phagocytosis

3.2.1 Preparation of pHrodo Particles and Intraperitoneal Injection

1. Resuspend pHrodo™ fluorescent bioparticles (2 mg lyophilized per vial) in 2 ml of sterile PBS.

2. Sonicate on ice for 5 min in a glass tube.

3. Inject 100 μl into the peritoneum using a 0.5 cc syringe and a 25–27 gauge needle (*see* **Note 10**).

4. Inject between the midline and the right knee.

3.2.2 Peritoneal Lavage

1. Recover the cellular infiltrate from the peritoneal cavity (*see* **Note 11**) as follows.

2. Euthanize the mouse, and sterilize with ethanol.

3. Cut the skin to expose the peritoneal cavity, pinning down the skin to the sides.

4. Carefully inject 3 ml of cold 1× PBS into the peritoneum using a 30 gauge needle.

5. Gently massage the peritoneum for 40 s.

6. Carefully draw out up to 2.5 ml of peritoneal lavage using the same needle and syringe and place it in a 15 ml conical centrifuge tube on ice.

3.2.3 Sample Fixation

1. Add PFA to the lavage solution at a final concentration of 1.6%.

2. Vortex each tube briefly, and allow the cells to fix on ice for 15 min.

3. Top up the conical centrifuge tubes with cold 1×PBS to dilute the PFA, and centrifuge for 5 min at $1000 \times g$.

4. Discard the supernatant.

5. If red blood cells are present resuspend the pellet in 1 ml of 1× BD Pharm lyse solution.

6. Incubate for 5 min on ice.

7. Repeat the centrifugation step.

8. Resuspend the pellets in 1 ml of PBS.

9. Use a Coulter counter or hemocytometer to count the cells.

3.2.4 Labeling

1. Resuspend 0.5×10^6 cells in FACS buffer so that the final volume, including antibody, will be 50 µl, and aliquot to flow cytometry tubes.

2. To block the Fc-receptors, add 1 µl of rat serum and 2 µl of mouse IgG and incubate for 20 min on ice.

3. Prepare a mastermix of fluorescently conjugated antibodies (*see* **Notes 9** and **12**).

4. Do not use FITC-conjugated antibodies, since this is the same fluorescent channel as pHrodo Green bioparticles.

5. Add the antibody mastermix, vortex gently and incubate for 30 min on ice protected from light.

6. To wash, add 1 ml of FACS buffer.

7. Vortex briefly and pellet the cells at $1000 \times g$ for 5 min at 4 °C.

8. Decant the supernatant.

9. Repeat the wash steps (**steps 6–8**) two more times.

10. Resuspend cells in 250 µl of FACS buffer and vortex briefly.

11. Prepare single stained compensation beads controls according to manufacturers' instructions.

12. Resuspend labeled compensation beads in 400 µl of FACS buffer and vortex briefly.

13. Cover the tubes with paraffin wax and store in the fridge protected from the light.

14. Perform flow cytometric acquisition and analysis as outlined in Subheading 3.1.5.

4 Notes

1. Excess unused pHrodo™ can be frozen at −20 °C in aliquots, however repeat freeze thawing was not tested.

2. Human studies require institutional Research Ethics Board (REB) approval to ensure safety and ethical considerations. Prior to inclusion, signed informed consent should be obtained from all participants.

3. Fresh blood is aliquoted immediately at room temperature and is not maintained on ice prior to stimulation at 37 °C.

4. Some degree of neutrophil stimulation occurs at 37 °C and therefore we recommend a 20 min stimulation, or less, depending on the dosage of pHrodo™ bioparticles (Fig. 1b). This is sufficient

to observe phagocytosis of *E. coli* bioparticles, while minimizing metabolic activation of neutrophils at 37 °C.

5. One limitation of our approach is that flow cytometric assessment and exclusion of dead cells is not possible, since all of the neutrophils are dead after fixation with PFA. However, by cytological staining of human blood leukocytes and mouse peritoneal neutrophils we found that only a small percentage (<5 %) of cells had a morphology typical of necrotic cells.

6. Lysis of RBCs takes more lysis steps when the samples are fixed compared to unfixed cells. Usually two to three lysis steps will be sufficient for whole blood. It is not necessary to completely eliminate the RBCs, since any remaining RBCs can be gated out during flow cytometry. However presence of many RBCs will throw off cell counts using a coulter counter.

7. Fifty microliters of human blood should contain more than enough leukocytes for the labeling procedure, and sufficient neutrophils to acquire at least 20,000 events by flow cytometry. For optimal labeling in a 50 µl volume ensure that $0.5–1 \times 10^6$ leukocytes are present in each tube.

8. In our experience, labeling of leukocytes with CD16 and CD66a is sufficient to gate on neutrophils in whole blood leukocyte populations (Fig. 1a). Other neutrophil CD markers can also be included for multicolor flow cytometry applications and for gating on alternate phagocytic populations (e.g., monocytes).

9. Antibody concentrations should be titrated appropriately. Prepare enough mastermix for 1.1× the number of tubes to be

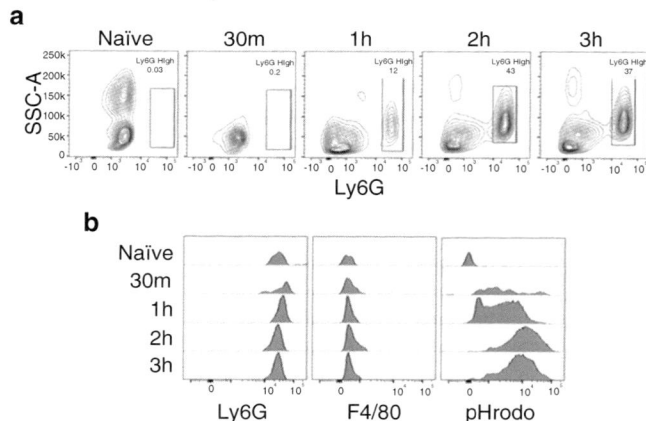

Fig. 2 Peritoneal injection of pHrodo. (**a**) C57BL6 mice were injected with 100 µl of pHrodo and peritoneal cells were recovered after 0.5, 1, 2, and 3 h. Neutrophils were gated based on high expression of Ly6G. F4/80 positive cells and doublets were excluded. Percentages of Ly6G high-gated populations are indicated. (**b**) Histograms showing expression of pHrodo by peritoneal neutrophils. At least 2×10^4 Ly6G high-gated events were acquired at the 1 h, 2 h, and 3 h time points

labeled. Be sure to prepare extra tubes for unlabelled as well as fluorescence minus one (FMO) controls. For compensation, a control tube with pHrodo™ positive cells should be prepared. Commercially available beads can be used for compensation of each fluorescent antibody channel.

10. Animal studies need to obtain institutional research ethics approval to ensure proper and ethical treatment of animals.

11. Significant numbers of neutrophils begin to appear in the peritoneal cavity within 1 h of pHrodo™ injection, and most of these cells are pHrodo™ positive. Peak pHrodo™ fluorescence occurs at the 2 h time point (Fig. 2a).

12. Labeling of leukocytes with Ly6G and F4/80 is sufficient to gate on neutrophils in mouse peritoneal exudates (Fig. 2). Other neutrophil CD markers can also be included for multicolor flow cytometry applications and for gating on alternate phagocytic populations (e.g., macrophages).

References

1. Borregaard N (2010) Neutrophils, from marrow to microbes. Immunity 33(5):657–670

2. Neaga A, Lefor J, Lich KE et al (2013) Development and validation of a flow cytometric method to evaluate phagocytosis of pHrodo BioParticles(R) by granulocytes in multiple species. J Immunol Methods 390(1-2):9–17

3. Kuijpers TW, Tool AT, van der Schoot CE et al (1991) Membrane surface antigen expression on neutrophils: a reappraisal of the use of surface markers for neutrophil activation. Blood 78(4):1105–1111

4. Berends C, Dijkhuizen B, de Monchy JG et al (1994) Induction of low density and up-regulation of CD11b expression of neutrophils and eosinophils by dextran sedimentation and centrifugation. J Immunol Methods 167(1-2):183–193

5. Sandilands GP, Ahmed Z, Perry N et al (2005) Cross-linking of neutrophil CD11b results in rapid cell surface expression of molecules required for antigen presentation and T-cell activation. Immunology 114(3):354–368

6. Mosca T, Forte WC (2015) Comparative efficiency and impact on the activity of blood neutrophils isolated by Percoll, Ficoll and spontaneous sedimentation methods. Immunol Invest 23:1–9

Chapter 3

Quantitative Efferocytosis Assays

Amanda L. Evans, Jack W.D. Blackburn, Charles Yin, and Bryan Heit

Abstract

Efferocytosis, the phagocytic removal of apoptotic cells, is a dynamic process requiring recruitment of numerous regulatory proteins to forming efferosomes in a tightly regulated manner. Herein we describe microscopy-based methods for the enumeration of efferocytic events and characterization of the spatio-temporal dynamics of signaling molecule recruitment to efferosomes, using genetically encoded probes and immunofluorescent labeling. While these methods are illustrated using macrophages, they are applicable to any efferocytic cell type.

Key words Efferocytosis, Phagocytosis, Method, Microscopy, Immunostaining, Macrophage

1 Introduction

Apoptosis (programmed cell-death) is a ubiquitous process in multicellular organisms required for the normal turnover of superfluous cells generated during development, and for removal of damaged or senescent cells [1]. In contrast to other forms of cell death such as necrosis and necroptosis, apoptosis is non-inflammatory and prevents release of intracellular materials through packaging cellular contents into discrete membrane-bound vesicles know as apoptotic bodies [2, 3]. Efferocytosis, the phagocytosis of apoptotic bodies, mediates the timely removal of these apoptotic bodies, preventing onset of secondary necrosis and release of inflammatory and toxic cellular contents into the extracellular milieu [3, 4]. As an integral aspect of tissue homeostasis, normal development and in resolution of inflammation, defects in efferocytosis can result in developmental defects and chronic inflammatory and autoimmune disease states [5–10].

Efferocytosis occurs through a unique phagocytic mechanism reliant on the successful completion of three distinct steps. First, "find-me" signals such as lysophosphatidylcholine are released from apoptotic bodies, inducing recruitment of phagocytes to the apoptotic cell [11]. Next, phagocytes bind apoptotic corpses

Roberto Botelho (ed.), *Phagocytosis and Phagosomes: Methods and Protocols*, Methods in Molecular Biology, vol. 1519, DOI 10.1007/978-1-4939-6581-6_3, © Springer Science+Business Media New York 2017

through recognition of "eat-me" signals (e.g., phosphatidylserine (PS), oxidized low-density lipoprotein, and ICAM3) by apoptotic cell binding opsonins (MFG-E8 and GAS6) and phagocytic receptors (such as CD36, MERTK, and $\alpha_5\beta_3$ integrin) [12–15]. Finally, the phagocyte engulfs the apoptotic corpse into an intracellular membrane-bound compartment termed the efferosome, where it is degraded [16]. While primarily performed by macrophages, nonprofessional phagocytes in numerous tissue types are capable of efferocytosis, as illustrated by apoptotic cell removal by kidney epithelia [17]. Despite its highly conserved and ubiquitous nature, many of the molecular mechanisms, ligands, and receptors involved in efferocytosis remain unelucidated.

Investigation into the molecular mechanisms regulating efferocytosis requires methods that allow for clear and concise quantification of apoptotic cell uptake and recruitment of effector molecules to the forming efferosome. To be quantitative such assays must clearly delineate bound versus internalized apoptotic targets, be applicable to both multi-receptor and receptor-specific models, and be compatible with conventional microscopy and molecular biology techniques. Herein we present protocols for in vitro efferocytosis assays, which can be applied to studies of receptors and ligands involved in apoptotic cell recognition, as well as signaling and vesicular trafficking events regulating degradation of efferocytosed apoptotic cells. We detail two distinct approaches—the first using highly defined apoptotic cell mimics, and the second using bona fide apoptotic cells. For studies requiring receptor-specific or spatially defined targets, and for assessing intracellular effector recruitment, the use of synthetic apoptotic mimics and transfected cell lines provides a robust model due to their ability to be respectively loaded or transfected with a range of molecules. In contrast, the use of apoptotic cells and primary human macrophages mirrors physiological conditions, and provides the ability to investigate the role and impact of efferocytic molecules in a biologically relevant system. When applied to microscopy based methods capable of using both primary human macrophages and macrophage cell lines, these assays provide a powerful set of tools for elucidation of efferocytic signaling processes.

2 Materials

2.1 Primary Human Macrophage Preparation

1. #1.5 thickness 18 mm circular glass coverslips.
2. Human blood (*see* **Note 1**).
3. 12-well tissue culture plate.
4. Roswell Park Memorial Institute 1640 medium with 2 mMl-glutamine, 25 mM HEPES and sodium bicarbonate (RPMI 1640), serum-free.

5. RPMI 1640 + 10 % fetal bovine serum (FBS).

6. Phosphate-buffered saline (PBS: 137 mM NaCl, 2.7 mM KCl, 10 mM Na_2HPO_4, 1.8 mM KH_2PO_4, pH 7.4).

7. Benchtop centrifuge.

8. Water bath or heat block.

9. Humidified 5 % CO_2 incubator.

10. 50 mL centrifuge tubes.

11. 15 mL centrifuge tubes.

12. Cytokine and growth factor stock solutions in PBS: 10 µg/ mL M-CSF, 10 µg/mL GM-CSF,10 µg/mL LPS, 100 µg/ mL IFN-γ, and 10 ng/mL IL-4 (see **Note 2**).

13. Dexamethasone, 100 µM in dimethyl sulfoxide (see **Note 2**).

14. Peripheral-blood mononuclear cell (PBMC) isolation medium, e.g., Lympholyte-poly (Cedarlane Labs) (see **Note 3**).

15. 100× antibiotic–antimycotic solution: 10,000 U/mL penicillin, 10 mg/mL streptomycin, and 25 µg/mL amphotericin B.

2.2 J774.2 Cell Culture and Transfection

1. J744.2 macrophage cell line (TIB-67, American Type Culture Collection).

2. PBS.

3. Dulbecco's Modified Eagle Medium (DMEM), serum-free.

4. DMEM + 10 % FBS.

5. Polystyrene tissue culture treated flasks, T25.

6. 5 mL pipettes.

7. Cell scrapers.

8. Hemocytometer.

9. pLAMP1-mCherry plasmid (Addgene plasmid #45147).

10. Lipofectamine 3000 Transfection Reagent Kit (Thermo Fisher Scientific) (see **Note 4**).

2.3 Synthetic Efferocytic Targets

1. 2.50–4.99 µm silica beads (Bangs Laboratories, Inc.).

2. 25 mg/mL phosphatidylserine (PS) in chloroform (Avanti Polar Lipids).

3. 25 mg/mL phosphatidylcholine (PC) in chloroform (Avanti Polar Lipids).

4. 10 mg/mL biotinylated phosphatidylethanolamine in chloroform (Avanti Polar Lipids).

5. 1 mg/mL rhodamine phosphatidylethanolamine in chloroform (Avanti Polar Lipids).

6. 25 µL and 50 µL glass Hamilton syringes.

7. 2 mL glass vials.

8. Chloroform.

9. Beakers.

10. Fume hood.

11. PBS.

12. 1.5 mL microcentrifuge tubes.

13. RPMI 1640.

14. Minicentrifuge.

2.4 Preparation of Apoptotic Jurkat T Cells

1. Jurkat T cells (TIB-152, American Type Culture Collection).

2. RPMI 1640.

3. RPMI 1640 + 10 % FBS.

4. Staurosporine.

5. Optional: Annexin V FITC (eBioscience, *see* **Note 5**).

6. Optional: Cell Proliferation Dye eFluor 670 (eBioscience, *see* **Note 6**).

2.5 Preparation of Apoptotic Neutrophils

1. Residual neutrophil layer from primary macrophage preparation, in PBS.

2. RPMI 1640.

3. RPMI 1640 + 10 % FBS.

4. Potassium chloride solution, 0.6 M KCl.

5. 50 mL centrifuge tubes.

6. 1.5 mL microcentrifuge tubes.

7. Optional: Annexin V FITC (*see* **Note 5**).

8. Optional: Cell Proliferation Dye eFluor 670 (*see* **Note 6**).

2.6 Efferocytosis Assays

1. RPMI 1640.

2. RPMI 1640 + 10 % FBS.

3. 12-well plate.

4. #1.5 thickness 18 mm coverslips.

5. Synthetic efferocytosis targets.

6. Humidified 5 % CO_2 incubator.

7. Benchtop centrifuge.

8. 4 % paraformaldehyde in PBS.

9. Streptavidin-Alexa 647 (Thermo Fisher Scientific).

10. PBS.

11. Optional: DNA dye, Hoechst 33342, DAPI, or other.

12. Optional: Antifade mounting reagent such as Permafluor Aqueous Mounting Medium (Thermo Fisher Scientific) and glass slides (*see* **Note 7**).

13. Optional: 18 mm Leiden Chamber (Quorum Instruments) or equivalent live cell imaging chamber (*see* **Note 7**).

2.7 Immunostaining

1. 4% PFA in PBS.

2. 5% bovine serum albumin (BSA) in PBS.

3. PBS.

4. 0.1% Triton X-100 in PBS.

5. Primary antibody (or antibodies) (e.g., mouse anti-human LAMP1: Developmental Studies Hybridoma Bank).

6. Fluorescently labeled secondary antibody (or antibodies) (e.g., Cy3-labeled goat anti-mouse: Cedarlane Labs).

2.8 Microscopy

1. Widefield or confocal microscope equipped with appropriate white light optics (phase contrast or differential interference contrast), fluorescent channels, EM-CCD or CMOS camera, a minimum 60× objective lens, and image-capture software.

2. For live-cell imaging the microscope also should be equipped with a heated/CO_2 perfused imaging chamber.

3 Methods

3.1 Cell Culture and Cell Transfection

Efferocytosis can be assessed in any cultured primary or immortalized cell type which is capable of efferocytosis. However, professional phagocytes such as macrophages are often preferred due to their potent phagocytic capacity, well-established cell culture protocols and the availability of easily transfected immortalized cells lines. In this chapter we provide protocols for culturing efferocytosis-competent primary human macrophages and transgene-expressing J774.2 cell lines.

3.1.1 Primary Human Macrophage Preparation

1. Collect human blood into heparin. 10 mL of blood will produce sufficient macrophages for 12 wells of a 12-well tissue culture plate (*see* **Note 1**).

2. For every 5 mL of human blood, warm 5 mL of Lympholyte-poly in a 15 mL centrifuge tube to room temperature in the dark (*see* **Note 3**).

3. Slowly layer 5 mL of human blood over Lympholyte-poly (*see* **Note 3**).

4. Centrifuge at $300 \times g$ for 35 min using medium acceleration and no break.

5. Remove upper PBMC-rich band using plastic pipettor and transfer to a 50 mL centrifuge tube.

6. Bring volume to 50 mL with PBS.

7. Centrifuge at $300 \times g$ for 8 min.

8. During **step 7** aseptically place 18 mm coverslips into a 12-well plate.

9. Aspirate supernatant and resuspend PBMC pellet in 300 µL serum-free RPMI 1640 per desired well of macrophages.

10. Pipette 300 µL of cell suspension onto the coverslips and incubate for 1 h at $37\ ^{\circ}C + 5\%\ CO_2$.

11. Wash coverslips three times with 1 mL sterile PBS to remove non-adherent cells.

12. Add 1 mL of RPMI 1640 + 10% FBS and antibiotic/antimytoic solution, along with cytokines for macrophage sub-type differentiation (*see* **Notes 2** and **8**).

    (a) M0, M2, and M2c differentiated macrophages: 10 ng/mL M-CSF.

    (b) M1 differentiated macrophages: 20 ng/mL GM-CSF.

13. Incubate for 5 days at $37\ ^{\circ}C + 5\%\ CO_2$.

14. Wash cells three times with 1 mL of sterile PBS.

15. Add 1 mL of RPMI 1640 + 10% FBS with antibiotic/antimycotic to each well, along with the appropriate cytokines to complete differentiation:

    (a) M0: 10 ng/mL M-CSF

    (b) M1: 20 ng/mL GM-CSF + 100 ng/mL IFN-γ and 250 µg/mL LPS.

    (c) M2 10 ng/mL M-CSF + 10 ng/mL IL-4

    (d) M2c: 10 ng/mL M-CSF + 100 nM dexamethasone

16. Incubate for an additional 2 days at $37\ ^{\circ}C + 5\%\ CO_2$.

17. Cells should be used within the next 3 days.

3.1.2 Culture and Transfection of the J774.2 Macrophage Cell Line

1. Culture J774.2 cells in DMEM + 10% FBS in T25 flasks at $37\ ^{\circ}C + 5\%\ CO_2$.

2. Passage cells once the culture reaches 80–90% confluency. Cells are passaged by rinsing once with PBS and scraping cells into 5 mL of fresh DMEM + 10% FBS.

3. Dilute cells 1:5 and replate 5 mL of diluted cells into a fresh flask.

4. Place coverslips into a 12-well plate.

5. Collect J774.2 cells at 70–80% confluency in a 15 mL conical centrifuge tube using a cell scraper.

6. Count cells using a hemocytometer and dilute to 5×10^4 cells/mL. Aliquot 1 mL of cell suspension into each well.

7. Incubate plate for 18 h at $37\ ^{\circ}C + 5\%\ CO_2$ to allow cells to recover and adhere to coverslips.

8. Prepare the DNA for transfection following the manufacturer's protocol (*see* **Note 4**). If using Lipofectamine 3000 prepare two 1.5 mL centrifuge tubes containing:

   (a) 1 μg of desired DNA construct(s) (e.g., pLAMP1-mCherry to demark late endosomes and lysosomes) and 2 μL of Lipofectamine P3000 Reagent in serum-free DMEM to a total volume of 40 μL.

   (b) 2 μL of Lipofectamine 3000 reagent in 38 μL of serum-free DMEM.

   (c) Add DNA-containing solution into Lipofectamine 3000-mixture. Incubate at room temperature for 5 min.

9. Add DNA:transfection solution dropwise to wells.

10. Incubate plate for 8 h at 37 °C + 5% CO_2, then replace media with fresh DMEM + 10% FBS.

11. Incubate plate overnight at 37 °C + 5% CO_2 to allow for cell recovery and transgene expression (*see* **Note 9**).

3.2 Generation of Synthetic and Natural Efferocytic Targets

Several efferocytic ligands have been characterized (*see* ref. 18), with PS representing the best understood ligand found on apoptotic cells. Normally restricted to the inner leaflet of the plasmalemma, during apoptosis PS is scrambled across both leaflets, with exofacial PS acting as an "eat-me" signal [19, 20]. In efferocytosis assays synthetic targets containing a lipid mixture mimicking the exofacial exposure of PS can be generated, as can bona fide apoptotic cells which contain both exofacial PS as well as other "eat-me" signals. Synthetic targets are generated using the classical method of coating small beads of defined size (1–8 μm) in known phagocytic ligands or opsonins [21–23]. For efferocytosis, silica beads are coated in lipid mixtures which mimic either healthy cells (100% phosphatidylcholine [PC]) or apoptotic bodies (20% PS/80% PC). The addition of small quantities of biotinylated-phosphatidylethanolamine (biotin-PE) and/or rhodamine-PE (Rhod-PE) adds the capacity for inside-out labeling and target-intrinsic fluorescence, respectively. Furthermore, due to their consistent size, the use of silica beads eases the analysis of signal molecule recruitment to efferosomes, and—due to the inability of cells to degrade them—allows for reliable imaging of internalized beads. In contrast to these synthetic mimics, the use of apoptotic cells more completely recapitulates the diversity of ligands and receptors involved in in vivo efferocytosis, providing the ability to observe and quantify efferocytosis in a more biologically relevant system. However, these natural targets are less amenable for quantification of efferocytic indices compared to synthetic targets due to their irregular size and tendency to fragment following engulfment. Below, the generation of both synthetic mimics and apoptotic cells compatible with efferocytosis assays are described.

3.2.1 Synthetic Apoptotic Cell Targets

All steps should be performed in a fume hood and samples should be protected from light (*see* **Note 9**).

1. Vortex silica beads for 30 s to ensure even suspension.
2. Pipette 10 μL of silica beads into a glass vial.
3. Wash Hamilton syringe three times with chloroform.
4. Using a Hamilton syringe, add 145 μL (control beads) or 114 μL (apoptotic cell mimics) of phosphatidylcholine to glass vial. Flick vial repeatedly until mixture is free of bead clumps.
5. Wash Hamilton syringe three times with chloroform.
6. Using a Hamilton syringe, add 84 μL of phosphatidylserine to the apoptotic cell mimic beads. Mix by flicking the side of the vial vigorously. *See* **Note 10** for generating fluorescent targets compatible with inside-out staining.
7. Wash Hamilton syringe three times with chloroform.
8. Evaporate chloroform by gently rotating vial under a steady flow of nitrogen gas.
9. Resuspend beads in 1 mL of PBS and transfer to a 1.5 mL microcentrifuge tube.
10. Centrifuge at $4500 \times g$ for 1 min and discard the supernatant.
11. Wash beads by resuspending in 1 mL PBS and centrifuging at $4500 \times g$ for 1 min. Discard supernatant.
12. Repeat **step 11**.
13. Resuspend in 100 μL PBS. *See* **Note 11** if opsonized targets are required.
   For fluorescent labeling and opsonin binding *see* **Notes 9–11**.

3.2.2 Preparation of Apoptotic Jurkat T Cells

1. Culture Jurkats in RPMI 1640 + 10 % FBS at 37 °C + 5 % CO_2. Jurkats are non-adherent and can be maintained by diluting 1:5 in fresh media every 5 days.
2. Transfer 10 mL of cells to a 50 mL centrifuge tube, centrifuge at $300 \times g$ for 5 min, and resuspend in 5 mL of serum-free RPMI 1640 medium.
3. To induce apoptosis, treat a 1 mL aliquot of cells with 1 μM staurosporine for 4 h at 37 °C + 5 % CO_2.
4. Wash cells three times with PBS to remove residual staurosporine. Apoptotic Jurkat cells should be used on the same day they are prepared. Examples of apoptotic and non-apoptotic Jurkat cells stained with FITC-conjugated Annexin V are shown in Fig. 1a, **Note 5**. Apoptotic cells can also be fluorescently stained to improve their detection in subsequent efferocytosis assays (*see* **Notes 6** and **9**).
   For fluorescent labeling and staining of apoptotic cells *see* **Notes 5, 6** and **9**.

1. Perform **steps 1–5** of the primary human macrophage culture protocol (Subheading 3.1.1). After removing the PBMC-rich upper band, recover the lower neutrophil band and transfer to a 50 mL centrifuge tube. Bring volume to 50 mL with ice-cold PBS.

2. Centrifuge at $300 \times g$ for 5 min. Discard supernatant.

3. If erythrocyte contamination is present—indicated by a red pellet—lyse erythrocytes by resuspending pellet 12 mL of ice-cold ddH_2O. Incubate for 15–20 s, then add 4 mL of ice-cold 0.6 M KCl to stop lysis reaction. Bring volume to 50 mL with ice-cold PBS and centrifuge at $300 \times g$ for 5 min. Discard supernatant.

4. Wash cells once with ice-cold PBS and centrifuge at $300 \times g$ for 5 min.

5. Resuspend cells in 1 mL of serum-free RPMI 1640 and transfer to a 1.5 mL microcentrifuge tube.

6. Induce apoptosis by incubating neutrophils for 1 h at 43 °C, followed by incubation for 3 h at 37 °C to allow for exposure of efferocytic ligands on the exofacial leaflet of the plasmalemma.

7. Wash cells once with ice-cold PBS and centrifuge at $300 \times g$ for 5 min. Apoptotic neutrophils should be used on the same day they are prepared. Examples of apoptotic and non-apoptotic neutrophils stained with FITC-conjugated Annexin V are shown in Fig. 1b, **Note 5**. Apoptotic cells can also be fluorescently stained to improve their detection in subsequent efferocytosis assays (*see* **Notes 6** and **9**).
   For fluorescent labeling and staining of apoptotic cells *see* **Notes 5, 6,** and **9**.

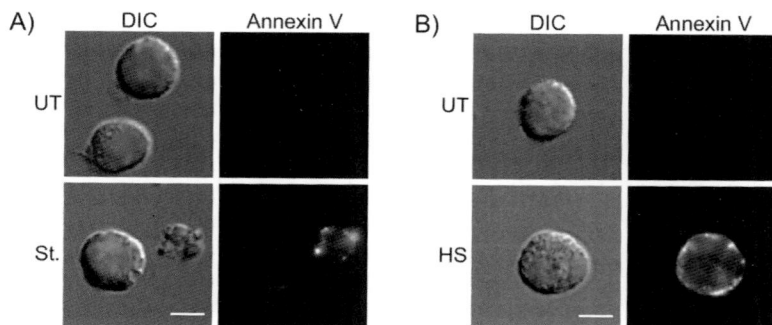

Fig. 1 Annexin V staining of apoptotic and non-apoptotic Jurkat cells and neutrophils. (**a**) Annexin V staining of Jurkat T cells following 4 h treatment with 1 μM staurosporine. (**b**) Annexin V staining of human neutrophils following 1 h heat shock at 43 °C followed by a 3 h recovery period at 37 °C. *UT* untreated, *St.* Staurosporine, *HS* Heat-Shock. Images are representative of 30 cells per condition, captured in two independent experiments. Apoptotic cells are stained with FITC-conjugated Annexin V, scale bars are 5 μm

3.3 Efferocytosis Assays with Synthetic and Natural Targets

The efferocytic targets produced in Subheading 3.2 can be applied to a variety of microscopy-based assays, ranging from simple quantification of the rates of target engulfment, through to immunostaining or live-cell microscopy of transgene-expressing cells to assess the recruitment of regulatory molecules to the maturing efferosome.

3.3.1 Synthetic Target -Based Efferocytosis Assay

1. Seed primary macrophages (Subheading 3.1.1) or cell lines (Subheading 3.1.2) on glass coverslips placed in a 12-well plate.

2. Add 1 mL of 37 °C RPMI 1640 and 3 µL of synthetic target solution (Subheading 3.2.1) to each well.

3. Centrifuge at $300 \times g$ for 1 min to force beads into contact with the cells (*see* **Note 12**).

4. Incubate for 10 min at 37 °C + 5 % CO_2 to allow macrophages to bind the beads.

5. Wash cells three times with 1 mL of PBS.

6. Add 1 mL of 37 °C RPMI 1640 to each well.

7. For fixed-cell imaging, incubate at 37 °C + 5 % CO_2 for the desired time to allow for bead engulfment (typically 20–60 min) and continue this protocol. For live cell imaging, mount coverslip in a Leiden Chamber, or equivalent cell-imaging chamber (*see* **Note 7**), and proceed to **step 3** of protocol 3.3.5.

8. Wash cells three times with 1 mL of PBS.

9. If beads were prepared with biotin-PE (*see* **Note 10**), non-internalized beads can be stained by adding a 1:1000 dilution of fluorescent streptavidin in PBS for 4 min. Co-staining with Hoechst 33342 (1:10,000) or other dyes can be performed at this time.

10. Wash cells three times with 1 mL of PBS.

11. Fix cells using 4% PFA in PBS for 15 min at room temperature.

12. If required, additional immunostaining can be performed (Subheading 3.3.3) to demark proteins or cellular structures of interest. Samples can be mounted on a glass slide with an anti-fade mounting reagent, or imaged immediately in a Leiden chamber (*see* **Note 7**).

3.3.2 Efferocytosis Assay Using Apoptotic Cells

1. Seed primary macrophages (Subheading 3.1.1) or cell lines (Subheading 3.1.2) on glass coverslips placed in a 12-well plate.

2. Count the numbers of apoptotic cells prepared in Subheading 3.2.2 or Subheading 3.2.3 with a hemocytometer. Add 1×10^6 apoptotic cells in 1 mL of 37 °C media to each well of macrophages in a 12-well plate.

3. Centrifuge at $300 \times g$ for 1 min to force apoptotic cells into contact with the macrophages (*see* **Note 12**).

4. For fixed-cell analysis, incubate plates at 37 °C + 5 % CO_2 for the desired period of time (typically 20–60 min) to allow for binding and uptake of apoptotic targets, and continue this protocol. For live cell imaging, mount coverslip in a Leiden Chamber and proceed to **step 3** of protocol 3.3.5.

5. Wash each coverslip with 1 mL of PBS.

6. Fix cells using 4 % PFA in PBS for 15 min.

7. If required, additional immunostaining can be performed (Subheading 3.3.3) to demark proteins or cellular structures of interest. Samples can be mounted on a glass slide with an anti-fade mounting reagent or imaged immediately in a Leiden chamber.

3.3.3 Immunostaining

1. Begin with PFA-fixed cells samples (Subheading 3.3.1 or Subheading 3.3.2).

2. Wash cells three times with PBS.

3. Permeabilize cells with 0.1 % Triton X-100 in PBS for 10 min (*see* **Note 13**).

4. Block cells with 5 % BSA in PBS for 1 h (*see* **Note 13**).

5. Label with primary antibody at an appropriate dilution (typically 1:250 to 1:5000) in PBS + 5 % BSA for 1 h.

6. Wash cells three times with 1 mL of PBS.

7. Label with an appropriate dilution of secondary antibody (typically 1:500 to 1:10,000) in PBS + BSA for 1 h.

8. Wash cells three times with 1 mL of PBS.

9. Samples can then be mounted on a glass slide with an antifade mounting reagent or imaged immediately in a Leiden chamber.

3.3.4 Fixed Sample Imaging and Quantification of Efferocytosis

1. To ensure detection of all apoptotic targets and cells in the sample it is recommended that the microscope's white light optics be aligned (Kohler alignment), as per manufacturer's instructions, prior to each imaging session.

2. Configure microscope acquisition software to capture a white light (phase contrast or differential interference contrast) image, as well as images of all florescent labels.

3. Place samples, prepared in Subheading 3.3.1 or Subheading 3.3.2 on microscope stage

4. Using a 60× or 100× objective, collect a sufficient number of images to get a statistically representative sample (30–50 cells per condition).

5. For enumerating efferocytosis it is recommended that images be captured using beads labeled with biotin-PE and Rhod-PE and post-stained with fluorescent streptavidin, to ensure a clear delineation between internalized versus bound beads. Internalized beads will be rhodamine-positive, while non-internalized beads

will be stained with both rhodamine and streptavidin (Fig. 2a). Alternatively, white-light images can be used to identify all beads in the image and streptavidin staining used to detect non-internalized beads. Using images captured under these conditions, two measures of efferocytic capacity can be enumerated:

(a) Efferocytic index (a measure of total efferocytic capacity): average number of internalized beads per cell (Fig. 2b).

(b) Efferocytic efficiency (a measure of the efficiency of receptor-mediated internalization following binding): # internalized beads/(internalized beads + bound beads).

6. For characterizing recruitment of signaling molecules or other cellular effectors, image samples that have been immunostained (Subheading 3.3.3) or which are ectopically expressing a fluorescent transgene (Subheading 3.1.2). Localization of the fluorescently labeled molecules relative to the efferosome can be determined from the resulting images (Fig. 3a).

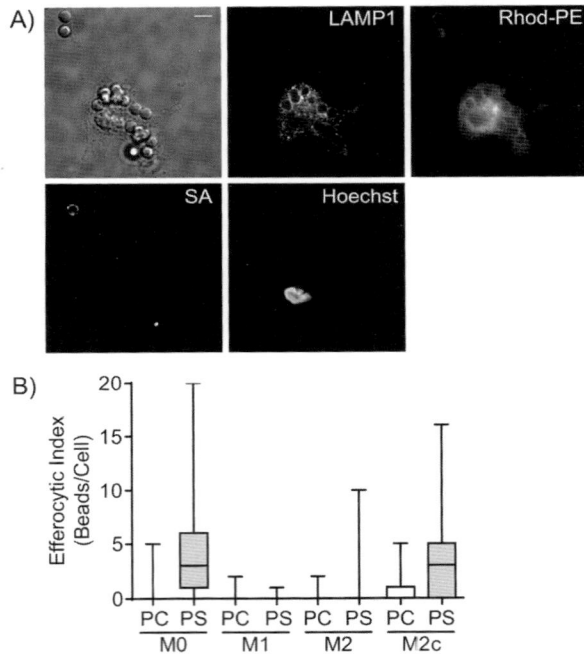

Fig. 2 Efferocytosis of synthetic targets by primary human macrophages. (**a**) Efferocytosis of targets containing 79.8 % PC, 20 % PS, 0.1 % Biotin-PE, and 0.1 % Rhod-PE by unpolarized (M0) primary human macrophages. Non-internalized beads stained with Alexa 647 labeled streptavidin (SA), with cells immunostained for LAMP1 and nuclei stained with Hoechst 33342. Images are representative of 20 images captured in two independent experiments. Scale bar is 5 μm. (**b**) Efferocytic index of M0, M1, M2, and M2c-polarized macrophages. Representative data from one experiment. Data is plotted as interquartile range ± minimum/maximum, minimum of 30 cells/condition. *PC* beads containing 99.9 % PC + 0.1 % biotin-PE, *PS* beads containing 79.9 % PC + 20 % PS + 0.1 % biotin-PE

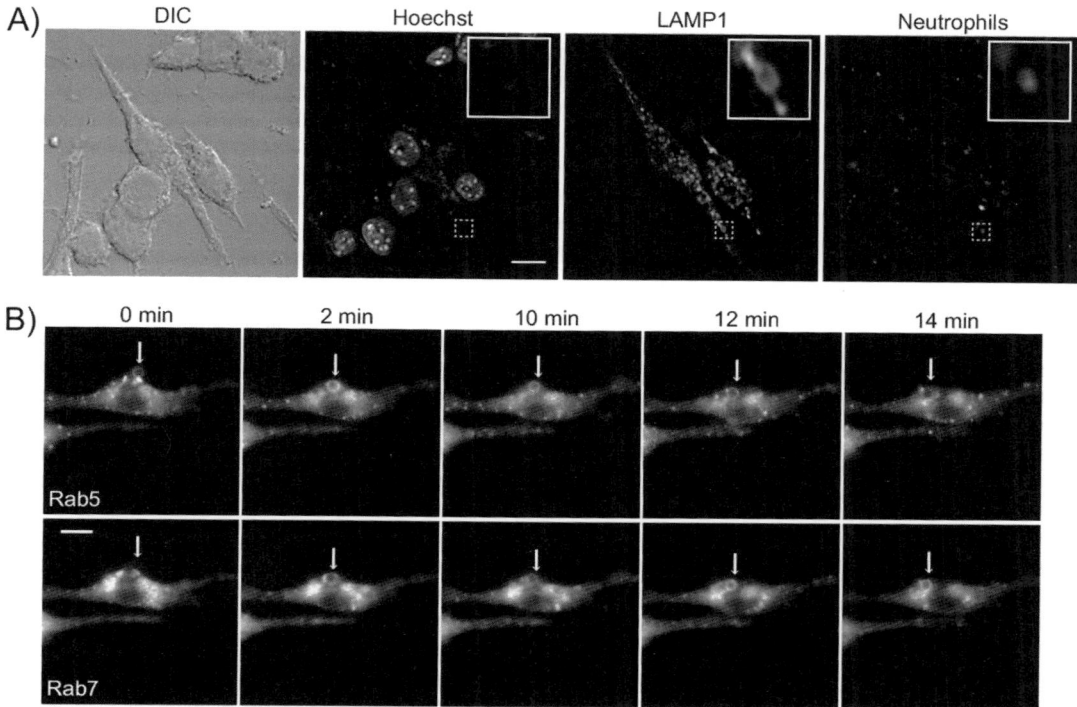

Fig. 3 Immunofluorescent staining of macrophages following efferocytosis of apoptotic neutrophils and apoptotic cell mimics. (**a**) Efferocytosis of Cell Proliferation Dye eFluor 670 labeled apoptotic neutrophils by J774.2 macrophages expressing the lysosomal marker LAMP1-mCherry and stained with Hoechst 33342. *Inset* is 5.2 μm × 5.2 μm (**b**) Time-lapse of the uptake of an 80 % PC + 20 PS apoptotic cell mimic by a J774.2 macrophage expressing Rab5-GFP and Rab7-RFP. *Arrow* tracks an efferosome that began forming at 0 min. Images are representative of ten images captured in three independent experiments. Scale bars are 10 μm

3.3.5 Live Cell Imaging

1. To ensure detection of all apoptotic targets and cells in the sample it is recommended that the microscope's white light optics be aligned (Kohler alignment), as per manufacturer's instructions, prior to each imaging session.

2. Configure microscope acquisition software to capture a time-series containing a white light (phase contrast or DIC) image, as well as images of all florescent labels at each time point. Frame rates vary with the process under investigation, but as a general guideline use 1 frame/10 s for rapid events such as phagocytic cup formation and 1 frames/2 min for slower processes such as efferosome maturation (Fig. 3b). If available, point-visiting can be used with slower acquisitions to collect images from several cells during a single acquisition (*see* **Note 14**).

3. Place Leiden chamber prepared in **step 7** of Subheading 3.3.1 or **step 4** of Subheading 3.3.2 in a live cell imaging chamber pre-warmed to 37 °C and perfused with 5 % CO_2 and then mount the imaging chamber on the microscope stage. If your imaging chamber lacks CO_2 perfusion capabilities *see* **Note 15**.

4. Using a 60× or 100× objective, collect a sufficient number of images to get a statistically representative sample (*see* **Note 16**).

4 Notes

1. Human blood is collected from healthy volunteers via venipuncture into a heparinized syringe or tube. To maximize macrophage yield the PBMC isolation must be initiated within 20 min of drawing blood. Human blood should be treated as a biosafety level 2 substance. Ethics approval and donor consent is required prior to blood collection.

2. Cytokines and growth factors should be stored as 10 µL aliquots at –20 °C. Individual aliquots are thawed immediately before use. Partially used aliquots can be stored at 4 °C for up to 1 week, but should not be refrozen.

3. Any PBMC isolation medium that produces viable mononuclear cells can be used. However, reagents which produce a separate neutrophil layer are preferred.

4. J774.2 cells, and other macrophage cell lines, can be difficult to transfect. Transfection reagents designed for hard-to-transfect cell lines, or macrophage-specific transfection reagents, need to be used. In our experience Lipofectamine 3000 works well for most murine and human macrophage cell lines.

5. To assess apoptosis in Jurkats or neutrophils resuspend prepared apoptotic cells in 100 µL of serum-free RPMI 1640 medium and stain with 5 µL of FITC-conjugated Annexin V for 10 min at room temperature. Examples of Annexin V stained apoptotic and non-apoptotic Jurkat and neutrophils cells are shown in Fig. 1.

6. To fluorescently label apoptotic cells prior to efferocytosis, resuspend prepared apoptotic cells in PBS + 2 µL/mL of Cell Proliferation Dye eFluor 670 for 5 min on ice and then wash three times with RPMI 1640.

7. If prepared on coverslips, fixed samples can be imaged by either mounting the coverslip on a glass slide or by placing the coverslip into a Leiden chamber filled with PBS. Alternatively, all procedures described in this chapter can be performed using macrophages grown in live-cell imaging chambers rather than on coverslips.

8. M0 macrophages are non-polarized and unactivated macrophages similar to tissue-resident macrophages. M1 macrophages are pro-inflammatory macrophages such as those which differentiate at the sites of bacterial infections, secrete a range of pro-inflammatory mediators, and are

highly bactericidal. M2 and M2c macrophages are anti-inflammatory macrophages which secrete a range of anti-inflammatory compounds and growth factors [24].

9. Fluorescently labeled samples should be protected from light during all stages of preparation to reduce photobleaching.

10. Biotin-PE (4 μL/0.1% mol) and/or Rhod-PE (60 μL/0.1% mol) can be added to the bead mixtures prior to drying under nitrogen. This will produce beads which can be labeled with fluorescent streptavidin or intrinsically fluorescent beads, respectively. Permeabilization using methanol or detergents can lead to some redistribution of Rhod-PE to cellular membranes.

11. Beads can be opsonized prior to efferocytosis assays. For each well, add 3 μL of beads to a solution of human serum, or purified opsonin. Incubate at room temperature for 30 min, under constant rotation.

12. If your centrifuge lacks a plate adaptor, add efferocytic targets to macrophage-containing wells and incubate on a flat surface for 20 min at room temperature to allow the targets to settle and bind the macrophages. Warm the sample to 37 °C to initiate efferocytosis.

13. Triton X-100 is a modestly strong detergent that can disrupt cellular membranes. To better preserve cellular membranes, permeabilize with 0.5 mg/mL saponin in PBS for 30–60 min, while 0.1% SDS or NP-40 in PBS can be used if more robust permeabilization is required. If background immunostaining is observed an alternate blocking reagents should be used. Options include 1% skim milk powder, or 5% serum from a species which will not cross-react with the primary or secondary antibodies. Blocking reagents should be diluted in PBS. Ice-cold methanol or methanol/acetone should be avoided in most situations due to their highly disruptive effect on membranes and epitopes.

14. Cellular motion, photobleaching and phototoxicity limit the use of z-stacks during live-cell imaging. If performing z-stacks over multiple time points, minimize excitation damage and cellular motion artifacts through use of bright fluorophores, short exposure times and rapid z-slicing. Generally, this requires a microscope equipped with a high-sensitivity back-thinned EM-CCD camera and a high-speed piezoelectric stage.

15. If the live cell imaging chamber lacks CO_2 perfusion use RPMI 1640 or DMEM buffered with HEPES, in place of sodium bicarbonate, in order to maintain physiological pH in air.

16. Due to the heterogeneous nature of cellular responses it is important to collect microscopy data in a manner consisted with good statistical practices. Sufficient replicates (individually

imaged cells) and repeats (independent experiments) should be collected to ensure statistically representative samples [25]. Images should be analyzed by applying robust quantification of efferocytic events (Subheading 3.3.4) and/or colocalization [26], followed by statistical analysis of the resulting data with the appropriate statistical test [27]. As with other data types, it is important to select a statistical test compatible with the number of experimental groups (single-group: t-Test or Mann–Whitney U test; multi-group: ANOVA or Kruskal–Wallis test) and the distribution of the data (parametric: t-test or ANOVA; nonparametric: Mann–Whitney U test or Kruskal–Wallis test). The application of a robust statistical approach to image analysis ensures that these data support objective conclusions based on reproducible and statistically validated results.

References

1. Elmore S (2007) Apoptosis: a review of programmed cell death. Toxicol Pathol 35:495–516. doi:10.1080/01926230701320337

2. Wallach D, Kang T-B, Rajput A et al (2010) Anti-inflammatory functions of the "apoptotic" caspases. Ann N Y Acad Sci 1209:17–22. doi:10.1111/j.1749-6632.2010.05742.x

3. Wickman G, Julian L, Olson MF (2012) How apoptotic cells aid in the removal of their own cold dead bodies. Cell Death Differ 19:735–742. doi:10.1038/cdd.2012.25

4. Eligini S, Crisci M, Bono E et al (2012) Human monocyte-derived macrophages spontaneously differentiated in vitro show distinct phenotypes. J Cell Physiol. doi:10.1002/jcp.24301

5. Heo K-S, Cushman HJ, Akaike M et al (2014) ERK5 activation in macrophages promotes efferocytosis and inhibits atherosclerosis. Circulation 130:180–191. doi:10.1161/CIRCULATIONAHA.113.005991

6. Thorp E, Cui D, Schrijvers DM et al (2008) Mertk receptor mutation reduces efferocytosis efficiency and promotes apoptotic cell accumulation and plaque necrosis in atherosclerotic lesions of apoe-/- mice. Arterioscler Thromb Vasc Biol 28:1421–1428. doi:10.1161/ATVBAHA.108.167197

7. Thorp EB (2010) Mechanisms of failed apoptotic cell clearance by phagocyte subsets in cardiovascular disease. Apoptosis 15:1124–1136. doi:10.1007/s10495-010-0516-6

8. Wigren M, Nilsson J, Kaplan MJ (2015) Pathogenic immunity in systemic lupus erythematosus and atherosclerosis: common mechanisms and possible targets for intervention. J Intern Med 278:494–506. doi:10.1111/joim.12357

9. Kimani SG, Geng K, Kasikara C et al (2014) Contribution of defective PS recognition and efferocytosis to chronic inflammation and autoimmunity. Front Immunol 5:566. doi:10.3389/fimmu.2014.00566

10. Recarte-Pelz P, Tàssies D, Espinosa G et al (2013) Vitamin K-dependent proteins GAS6 and Protein S and TAM receptors in patients of systemic lupus erythematosus: correlation with common genetic variants and disease activity. Arthritis Res Ther 15:R41. doi:10.1186/ar4199

11. Peter C, Waibel M, Radu CG et al (2008) Migration to apoptotic "find-me" signals is mediated via the phagocyte receptor G2A. J Biol Chem 283:5296–5305. doi:10.1074/jbc.M706586200

12. Qingxian L, Qiutang L, Qingjun L (2010) Regulation of phagocytosis by TAM receptors and their ligands. Front Biol (Beijing) 5:227–237

13. Savill J, Hogg N, Ren Y, Haslett C (1992) Thrombospondin cooperates with CD36 and the vitronectin receptor in macrophage recognition of neutrophils undergoing apoptosis. J Clin Invest 90:1513–1522. doi:10.1172/JCI116019

14. Wu Y, Singh S, Georgescu M-M, Birge RB (2005) A role for Mer tyrosine kinase in alphavbeta5 integrin-mediated phagocytosis of apoptotic cells. J Cell Sci 118:539–553. doi:10.1242/jcs.01632

15. Nandrot EF, Anand M, Almeida D et al (2007) Essential role for MFG-E8 as ligand for alphav-beta5 integrin in diurnal retinal phagocytosis. Proc Natl Acad Sci U S A 104:12005–12010. doi:10.1073/pnas.0704756104

16. Kinchen JM, Doukoumetzidis K, Almendinger J et al (2008) A pathway for phagosome maturation during engulfment of apoptotic cells. Nat Cell Biol 10:556–566. doi:10.1038/ncb1718

17. Ichimura T, Asseldonk EJPV, Humphreys BD et al (2008) Kidney injury molecule-1 is a phosphatidylserine receptor that confers a phagocytic phenotype on epithelial cells. J Clin Invest 118:1657–1668. doi:10.1172/JCI34487

18. Ravichandran KS (2010) Find-me and eat-me signals in apoptotic cell clearance: progress and conundrums. J Exp Med 207:1807–1817. doi:10.1084/jem.20101157

19. Suzuki J, Umeda M, Sims PJ, Nagata S (2010) Calcium-dependent phospholipid scrambling by TMEM16F. Nature 468:834–838. doi:10.1038/nature09583

20. Fadeel B (2004) Plasma membrane alterations during apoptosis: role in corpse clearance. Antioxid Redox Signal 6:269–275. doi:10.1089/152308604322899332

21. Leffell MS, Spitznagel JK (1975) Fate of human lactoferrin and myeloperoxidase in phagocytizing human neutrophils: effects of immunoglobulin G subclasses and immune complexes coated on latex beads. Infect Immun 12:813–820

22. Yeung T, Heit B, Dubuisson J-F et al (2009) Contribution of phosphatidylserine to membrane surface charge and protein targeting during phagosome maturation. J Cell Biol 185:917–928. doi:10.1083/jcb.200903020

23. Flannagan RS, Harrison RE, Yip CM et al (2010) Dynamic macrophage "probing" is required for the efficient capture of phagocytic targets. J Cell Biol 191:1205–1218. doi:10.1083/jcb.201007056

24. Zhou D, Huang C, Lin Z et al (2014) Macrophage polarization and function with emphasis on the evolving roles of coordinated regulation of cellular signaling pathways. Cell Signal 26:192–197. doi:10.1016/j.cellsig.2013.11.004

25. Vaux DL, Fidler F, Cumming G (2012) Replicates and repeats-what is the difference and is it significant? A brief discussion of statistics and experimental design. EMBO Rep 13:291–296. doi:10.1038/embor.2012.36

26. Bolte S, Cordelières FP (2006) A guided tour into subcellular colocalization analysis in light microscopy. J Microsc 224:213–232. doi:10.1111/j.1365-2818.2006.01706.x

27. Nayak BK, Hazra A (2011) How to choose the right statistical test? Indian J Ophthalmol 59:85–86. doi:10.4103/0301-4738.77005

Chapter 4

Quantifying Phagocytosis by Immunofluorescence and Microscopy

Christopher H. Choy and Roberto J. Botelho

Abstract

Phagocytosis is the actin-driven internalization of solid particles, utilized by phagocytic immune cells to sequester potentially infectious microorganisms. Aided by the innate and adaptive immune system, the activation of various phagocytic receptors triggers a cascade of downstream signaling mediators that drive actin and plasma membrane remodeling. Modulation of these molecular players can lead to distinct changes in the capacity and rates of phagocytosis. Here, we present a fluorescence microscopy based technique to quantify phagocytosis using a macrophage-like cell line. We exemplify the technique through the phagocytosis of antibody-opsonized polystyrene beads. This method can be extended to other phagocytes and phagocytic particles.

Key words Phagocytosis, Phagosome, Phagocytic index, Microscopy, Immunofluorescence, Quantitation

1 Introduction

Phagocytosis is an essential process utilized by various cells to engulf particulate matter. In particular, immune cells employ phagocytosis to clear potentially infectious microorganisms by sequestering these organisms into phagosomes, which then fuse with lysosomes to digest the microorganism [1, 2]. Due to their phagocytic capacity and proficiency, macrophages, neutrophils, and dendritic cells are referred to as professional phagocytes. These cells can form dozens of phagosomes by recognizing numerous ligands through the expression of an array of phagocytic receptors [3]. Receptor–ligand engagement initializes a series of tightly regulated downstream signals encoded by lipid and protein modifications and recruitment in both space and time that help coordinate particle engulfment [2]. Thus, the capacity and rate of phagocytosis can be affected by modulating the activity of any of these signaling intermediates such as the tyrosine kinases Syk and Src-family, Rho-family GTPases that control actin remodeling, and lipid-metabolizing enzymes such as PLC and phosphatidylinositol 3-kinase (PI3K) [4–8].

Roberto Botelho (ed.), *Phagocytosis and Phagosomes: Methods and Protocols*, Methods in Molecular Biology, vol. 1519, DOI 10.1007/978-1-4939-6581-6_4, © Springer Science+Business Media New York 2017

In addition, phagocytes can be activated by the presence of factors such as cytokines (interferon-γ), host-derived peptides (LL-37), and bacterial peptides (N-formylmethionine-leucyl-phenylalanine) [9–13]. The adaptive immune system cooperates by producing immunoglobulin G (IgG) antibodies that then bind to the particle's surface to tag it for phagocytosis through the Fcγ receptors, highly efficient phagocytic receptors [14]. To understand the regulation of phagocytosis and the impact that the cellular environment has on phagocytosis, we must be able to quantify particle engulfment. There are several methods available to quantify phagocytosis including those described in other chapters in this volume, each with their own advantages and disadvantages. Here, we describe the use of immunofluorescence to quantify phagocytosis. This method differentiates internalized particles from bound but external particles. The methods below employ the RAW 264.7 macrophage-like cell line and three particle models: polystyrene beads, *Escherichia coli,* and red blood cells—but the method can easily be adapted to other phagocytes and particle types if antibodies to the particles are available.

2 Materials

2.1 Cell Culture

1. RAW 264.7 macrophages (*see* **Note 1**).

2. Serum-free medium: Dulbecco's Modified Eagle Medium with 4 g/L glucose, 4 mM l-glutamine, and 1 mM pyruvate; store at 4 °C.

3. Full growth medium: Serum-free medium supplemented with 10% heat-inactivated fetal bovine serum; store at 4 °C.

4. T-25 flasks.

5. CO_2 cell culture incubator.

6. Cell scrapers.

7. Glass coverslips (No. 1 thickness).

8. 6-well tissue culture plate (*see* **Note 2**).

9. Plasmid DNA.

10. FuGene HD transfection reagent.

11. siRNA: nontargeting siRNA as control and siRNA against gene of interest.

12. Neon Transfection System: System includes Neon transfection device, Neon pipette, and Neon pipette station; store at room temperature; Thermo Scientific, Waltham, MA, USA.

13. Neon Transfection System 100 μL kit: Kits includes resuspension buffer R, electrolytic buffer E2, Neon electroporation tube, and Neon tips; store all buffers at 4 °C, store tubes and tips at room temperature; Thermo Scientific.

2.2 Opsonizing Particles with IgG

2.2.1 Polystyrene Beads

1. Phosphate buffered saline (PBS): 137 mM NaCl, 2.7 mM KCl, 10 mM Na_2HPO_4 and 1.8 mM KH_2PO_4, pH 7.4.

2. Polystyrene beads: 3.87 μm polystyrene beads (see **Note 3**); store at 4 °C (see **Note 4**); Bangs Laboratories, Fishers, IN, USA.

3. Human Immunoglobulin G (human IgG): prepare 10 mg/mL solution of human IgG with 150 mM NaCl in sterile ultrapure water; aliquot and store at –20 °C (see **Note 5**).

4. Microcentrifuge tube.

5. Microcentrifuge.

6. Nutator.

7. Fluorescent secondary anti-human IgG: Prepare 0.75 mg/mL solutions of Cy2-, Cy3- or Cy5-conjugated donkey anti-human IgG (see **Note 6**) by adding sterile 50 % glycerol in sterile ultrapure water; aliquot and store at –20 °C; Jackson ImmunoResearch, West Grove, PA, USA.

2.2.2 Escherichia coli

1. PBS.

2. *E. coli*: DH5α or other lab strain (see **Note 7**).

3. Anti-*E. coli* antibody: Rabbit anti-*E. coli* antibody (see **Note 8**); store at 4 °C.

4. Microcentrifuge tube.

5. Microcentrifuge.

6. Nutator.

7. Fluorescent secondary anti-rabbit IgG: DyLight 488-, DyLight 550-, or Dylight 650-conjugated donkey anti-rabbit IgG (see **Note 6**); store at 4 °C; Bethyl Laboratories, Montgomery, TX, USA.

2.2.3 Sheep Red Blood Cells

1. PBS.

2. Red blood cells (RBCs): sheep RBCs; store at 4 °C for up to 2 month; MP Biomedicals, Santa Ana, CA, USA.

3. Anti-sheep antibody: Prepare 40 mg/mL solution of rabbit anti-sheep antibody with sterile ultrapure water; store at 4 °C for up to 2 months; MP Biomedicals.

4. Microcentrifuge tube.

5. Microcentrifuge.

6. Nutator.

7. Fluorescent secondary anti-rabbit IgG: DyLight 488-, DyLight 550-, or Dylight 650-conjugated donkey anti-rabbit IgG (see **Note 6**); store at 4 °C; Bethyl Laboratories.

2.3 Inside–Outside Staining of Phagocytosed Particles

1. Swinging bucket centrifuge with microplate rotor.

2. PBS.

3. 4% paraformaldehyde (PFA) (*see* **Note 9**): Prepare 4% PFA solution by diluting 16% PFA stock solutions with PBS; store at 4 °C for up to 1 week.

4. 100 mM glycine: Prepare 100 mM glycine by dissolving powered glycine in PBS; store at 4 °C for up to 1 week.

5. 0.1% Bovine serum albumin (BSA): Prepare 0.1% (w/v) BSA solution by dissolving lyophilized BSA in PBS; store at 4 °C for up to 1 week.

6. 0.1% Triton (*see* **Note 9**): Prepare 0.1% (v/v) Triton X-100 by diluting stock Triton X-100 with in PBS; store at 4 °C for up to 1 week.

7. Fluorescent mounting media; store at 4 °C.

8. Glass slide.

2.4 Microscopy and Quantification Phagocytosis

1. Confocal fluorescent microscope (*see* **Note 10**) equipped with the following, or related, lasers and band pass emission filters with 40 nm bandwidths: 491 nm laser with 515 nm emission filter for green fluorescent fluorophores (Cy2 and DyLight 488), 561 nm laser with 594 nm emission filter for red fluorescent fluorophores (Cy3 and DyLight 550), and 642 nm laser with 670 nm emission filter for far red fluorescent fluorophores (Cy5 and DyLight 650).

2. ImageJ or proprietary image analysis software.

3   Methods

3.1 Cell Culture of RAW 264.7 Macrophage

1. Grow RAW macrophages in T-25 flasks with 10 mL of full growth media in the presence of 5% CO_2 at 37 °C. Allow cells to grow to 80% confluence (*see* **Note 11**), replacing media every 2–3 days.

2. Replace old medium with 10 mL of fresh, prewarmed full growth media. Gently lift cells using a cell scraper and resuspend cells by pipetting.

3. For sub-culturing, retain 1 mL of cells and discard remaining cells. Top up media to 10 mL with fresh full growth media and check confluency in 2–3 days.

4. For plating, add sterilized glass coverslips to each well of a 6-well plate (*see* **Note 12**). Add 1 mL of cells (~5×10^5 cells) to each well and top up media to 2 mL with fresh full growth media.

5. If plasmid transfection is required, prepare 6-well plate with coverslips. Add 0.5 mL of cells (~2.5×10^5 cells) to each well and top up media to 2 mL with fresh full growth media. On the following day, prepare DNA-transfection mix in a centrifuge tube by diluting plasmid DNA to a final concentration of 0.02 μg/μL

in 51 µL of serum-free DMEM. Next, add 3.9 µL of FuGene HD transfection reagent to the diluted DNA and pipette to mix. Incubate DNA-transfection mix for 10 min at room temperature, then add 50 µL of the DNA-transfection mix to each well. Replace media containing the DNA-transfection mix 4–6 h post-transfection with fresh full growth media and perform experiments 24 h after transfection (*see* **Note 13**).

6. If siRNA knockdown is required, prepare 6-well plate with coverslips and incubate 2 mL of full growth medium in the CO_2 incubator while setting up electroporation (*see* **Note 14**). Prepare one additional well for each siRNA oligonucleotide without coverslips with 2 mL of full growth medium (*see* **Note 15**). Pellet 10 mL of cells (~5×10^6 cells) in a conical tube at $500 \times g$ for 5 min (*see* **Note 16**). Aspirate full growth media and wash cells with 10 mL of PBS. Pellet cells again and resuspend the cells with 600 µL of resuspension buffer R. Separate the 600 µL of cells into two 300 µL aliquots in separate microcentrifuge tubes and add control, nontargetting siRNA or siRNA against your gene of interest to a final concentration of 200 nM. Meanwhile, add 3 mL of electrolytic buffer E2 to the Neon electroporation tube and set the loaded tube into the Neon pipette station. Load the Neon pipette with a Neon tip and draw up 100 µL of cells containing nontargetting siRNA. Insert the loaded pipette into the electroporation tube and initiate a protocol with the following parameters: voltage of 1730 V, pulse width of 20 ms, and 1 pulse. After electroporation is complete, dispense electroporated cells into a single 6-well containing pre-incubated full growth media. Repeat electroporation for remaining cells. With a new tip, repeat electroporation for cells containing siRNA against the gene of interest. Perform experiments 24–48 h post-electroporation (*see* **Note 17**).

3.2 Opsonization of Particles with IgG

3.2.1 Opsonization of Polystyrene Beads with IgG

1. Mix 40 µL of human IgG with 20 µL of polystyrene beads and 100 µL of PBS in a microcentrifuge tube. Rotate bead mixture for 1 h at room temperature (*see* **Note 18**).

2. Pellet beads at $500 \times g$ for 1 min and resuspend in 100 µL of PBS. Repeat once (*see* **Note 19**).

3.2.2 Opsonization of E. coli with IgG

1. Streak DH5α or related *E. coli* from −80 °C glycerol stock onto LB agar plate. Grow *E. coli* overnight (12–16 h) at 37 °C.

2. Grow an overnight culture of *E. coli* in 1 mL of liquid LB media in a 10 mL round bottom tube (*see* **Note 20**). Incubate overnight culture at 37 °C with shaking.

3. Subculture the overnight culture at a 1:200 ratio in 10 mL of liquid LB and grow cells to mid log phase (OD_{600} 0.5–0.7) (*see* **Note 21**).

4. Aliquot *E. coli* cell mass ($OD_{600} \times$ volume) of 0.6 in a microfuge tube.

5. Pellet cells at $4000 \times g$ for 5 min and resuspend in 100 µL of PBS. Repeat once.

6. Opsonize 100 µL of *E. coli* with 1 µL of anti-*E. coli* antibody and rotate for 30 min at room temperature (*see* **Note 22**).

7. Pellet cells at $4000 \times g$ for 5 min and resuspend in 100 µL of PBS. Repeat once (*see* **Note 19**).

3.2.3 Opsonization of Sheep RBCs with IgG

1. Mix 25 µL of RBC with 1 mL of PBS.

2. Pellet RBCs at $500 \times g$ for 1 min and resuspend in 50 µL of PBS. Repeat once.

3. Add 1 µL of rabbit anti-sheep RBC antibody to 50 µL of washed RBCs and rotate for 30–60 min at room temperature.

4. Pellet RBCs at $500 \times g$ for 1 min and resuspend in 50 µL of PBS. Repeat once (*see* **Note 19**).

3.3 Phagocytosis and Inside-Out Staining of Phago cytosed Particles

1. If cells require stimulation with drugs, cytokines, etc., pretreatment may be done prior to the addition of opsonized particles.

2. Add 100 µL of IgG-opsonized beads or IgG-opsonized *E. coli* or 50 µL of IgG-opsonized RBCs to each well with cells.

3. Synchronize bead or RBC binding to macrophages by spinning 6-well plates at $500 \times g$ for 1 min or for 5 min for *E. coli* (*see* **Note 23**).

4. Wash off unbound particles with 2 mL of PBS and incubate cells in 2 mL of prewarmed, full growth media (*see* **Note 24**). Allow for internalization and trafficking for the desired time.

5. Fix cells with 2 mL of 4% PFA for 10–15 min (*see* **Note 25**).

6. Quench fixation reaction with 2 mL of 100 mM glycine for 5 min.

7. Wash cells three times with 2 mL of 0.5% BSA for 5 min each.

8. Incubate cells with 1 mL of far-red fluorescent secondary antibodies (anti-human for beads or anti-rabbit for RBCs and *E. coli* at 1:400 dilution in 0.5% BSA) for 30 min at room temperature (*see* **Note 26**). Protect the plate from light by wrapping the plate in foil or using a dark room or dark cupboard (*see* **Note 27**).

9. Wash cells 3 times with 2 mL of 0.5% BSA for 5 min each.

10. Permeabilize cells with 2 mL of 0.1% Triton solution for 5 min. Wash cells 3 times with 2 mL of 0.5% BSA for 5 min each.

11. Incubate cells with 1 mL of green fluorescent secondary antibodies (anti-human for beads or anti-rabbit for RBCs and *E. coli* at 1:400 dilution in 0.5% BSA) for 30 min at room temperature

(*see* **Note 26**). Avoid using the same colour fluorescent secondary antibody as in **step 8**. Protect the plate from light by wrapping the plate in foil or using a dark room or dark cupboard (*see* **Note 28**).

12. Wash cells 3 times with 2 mL of 0.5% BSA for 5 min each. Then wash cells 2 times with 2 mL of PBS for 5 min each.

13. Add one to two drops of fluorescent mounting media onto a glass slides. Remove excess PBS by touching coverslip to a tissue and mount coverslips face down onto slides. Allow slides to cure overnight in the dark. The edges of the coverslip may be sealed with clear nail polish to prevent drying. Slides may be stored for up to 1 month at 4 °C.

3.4 Quantification of Phagocytosis

1. Image cured slides using a confocal microscope. Acquire a single slice through the middle of the cell (*see* **Note 29**) in the differential interference contrast (DIC) channel (*see* **Note 30**), far-red channel (**step 8** of Subheading 3.3) and green channel (**step 11** of Subheading 3.3). Acquire enough fields to count at least 100 cells (*see* **Note 31**).

2. To determine phagocytic index, count (*see* **Note 32**) the number of internalized particles (*see* **Note 33**) and divide by the total number of cells counted (Fig. 1).

3. To determine phagocytic efficiency, determine the number of cells with at least one internalized particle and divide by the total number of cells counted, expressed as a percentage (Fig. 1).

4 Notes

1. RAW 264.7 can be substituted with primary phagocytes or with other cell line macrophages, such as the murine J774A.1 or human U937.

2. If 12-well format is desired, use ½ amounts indicated.

3. Beads may contain divinylbenzene (DVB) to improve stability of the microsphere.

4. Particles of interest will settle out of solution over time. Gently agitate the solution to homogenize the suspension before use.

5. Avoid repeated freeze–thaw cycles of IgG.

6. Other fluorophore-conjugated antibodies are also acceptable. Follow manufacturer instructions for recommended dilutions and storage conditions.

7. Some particles may not require opsonization for phagocytosis, such as *E. coli* which will be phagocytosed via other phagocytic receptors. Particles may also be opsonized with other opsonins, such as C3b and phagocytosed via the complement receptor type 3.

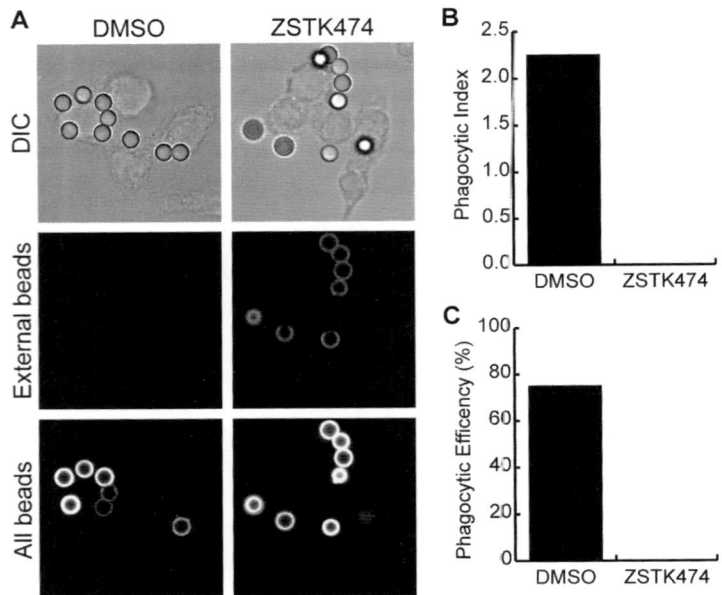

Fig. 1 Determination of phagocytic index and phagocytic efficiency. (**a**) RAW 264.7 macrophages were pretreated with either the class I PI3K inhibitor ZSTK474 or its vehicle DMSO. Cells were allowed to phagocytose human IgG-opsonized beads for 60 min, followed by staining with secondary antibodies against human IgG before (*external beads*) and after (*all beads*) permeabilization. Thus, internalized beads are single-stained, while external beads are double-stained. (**b**) Using the figure shown as an example, DMSO-treated control cells internalized nine beads across four cells, giving a phagocytic index of 2.25. In cells treated with ZSTK474, beads are bound to the cells but have not been internalized. There are zero internalized beads between the five cells, representing a phagocytic index of 0. (**c**) In control cells, three of the four cells contained phagocytosed beads, giving a phagocytic efficiency of 75 %. In ZSTK474-treated cells, none have internalized beads; therefore, the phagocytic efficiency is 0 %

8. There are various anti-*E. coli* antibodies. Monoclonal *E. coli* antibodies are not recommended; polyclonal *E. coli* antibodies will target multiple surface antigens and provide greater opsonization with IgG that enhances uptake and labeling.

9. Although most antibodies will use 4 % PFA for fixation and 0.1 % Triton X-100 for permeabilization, some antibodies may require different conditions such as 100 % methanol. Check manufacturer's instructions for recommended fixation and permeabilization conditions.

10. Wide field fluorescence microscopy can substitute confocal microscopy for *E. coli* and RBCs, but not polystyrene beads. The beads will refract light causing halos around the beads that make it difficult to distinguish if the particle is internal or external.

11. Do not allow cells to reach 100% confluence as they will rapidly metabolize nutrients and undergo starvation. Use cells with less than 20 passages.

12. Round 25 mm coverslips or square 18×18 mm coverslips can be used in 6-well plates.

13. Plasmid DNA may be transfected with other available transfection reagents, such as Lipofectamine. Use transfection conditions as recommended by the manufacturer. Due to low transfection efficiency of RAW cells, plasmid DNA may also be transfected by electroporation if necessary.

14. Other electroporation systems may also be used, such as the Amaxa Nucleofector Kit. siRNA may also be introduced by transfection reagents. However, due to the low transfection efficiency, it is not recommended unless the user has access to a reliable way to identify cells that have transfected siRNA.

15. Electroporation of an additional well of cells allows the user to check for knockdown efficiency via Western Blot.

16. All plasticware, including pipette tips and microcentrifuge tubes, should be RNase-free while working with siRNA.

17. The length of knockdown depends on the turnover rate of the protein of interest. Each siRNA knockdown should be checked for efficiency and optimized for maximal knockdown.

18. Opsonization of polystyrene beads can be done overnight at 4 °C.

19. Opsonized particles may be stored at 4 °C for up to 2 h prior to use. Longer incubation may lead to clumping.

20. *E. coli* should be grown with round bottom tubes or flasks that are 5 times the volume of the culture to ensure proper aeration. If tubes with screw caps are used, do not seal tubes completely to prevent cultures from undergoing anaerobic respiration.

21. Dilution of overnight culture directly to OD_{600} of 0.5–0.7 will contain too many nonviable bacterial cells.

22. Opsonization of live *E. coli* overnight is not recommended as they will continue to divide and cells may dilute the available IgG and/or degrade. Alternatively, bacteria may be heat-killed at 65 °C for 15 min prior to opsonization.

23. If swinging bucket centrifuge with microplate rotor is unavailable, synchronize particle binding by allowing particles to settle by gravity for 10–20 min on ice. However, incubation in the cold may disrupt microtubules.

24. RBCs may be burst in the presence of sterile water for 30 s. However, lysed erythrocytes may still be phagocytosed as RBC ghosts.

25. Avoid prolonged exposure of cells to 4% PFA. Over-fixation can cause cracks in the plasma membrane and will allow antibodies to leak into the cell before intentional permeabilization.

26. Far-red and green fluorescent secondary antibodies are used here as a generic place holder and will be used for the rest of the chapter as an example. Fluorescent secondary antibodies are interchangeable between green fluorescence (Cy2 and DyLight 488), red fluorescence (Cy3 and DyLight 550), and far-red fluorescence (Cy5 and DyLight 650) based on preference and/or microscope compatibility. If cells have been transfected prior to phagocytosis, be careful not to choose a secondary antibody with the same colour fluorescence as the transfected protein (i.e. GFP and Cy2).

27. The immunolabeling of particles with secondary antibody before cell permeabilization will label particles that have not been internalized by macrophages.

28. The immunolabeling of particles with secondary antibody after cell permeabilization will label all particles.

29. Acquiring a single slice through the middle of the cell offers a quick way to look at the number of particles internalized. However, particles above and below the plane will be missed and would not be considered in the final quantification of phagocytosis. The acquisition of Z-stacks through the entire cell may be used to determine the total number of particles internalized, though this method is slower.

30. White light acquisition of cells may be done by bright field, differential interference contrast (DIC), or phase-contrast microscopy.

31. If fluorescently tagged proteins are transfected into the cell, also acquire the channel corresponding to its fluorescence. This will require three-channel acquisition (transfected protein, outside particles, and all particles).

32. Counting the number of phagocytosed particles may be done manually with the help of useful plugins on ImageJ, such as the Cell Counter plugin. Automatic particle counting may be done with the "Analyze particle..." plugin, after images have been thresholded, made binary, and watershed.

33. Internalized particles will be single stained (with green secondary antibody after permeabilization) while non-internalized particles will be dual stained (with far red and green secondary antibody).

Acknowledgements

C.C. is supported through an Ontario Graduate Scholarship. R.J.B. is supported through a Canada Research Chair and Early Researcher Awards and grants from the Canadian Institutes of Health Research and the Natural Sciences and Engineering Research Council of Canada.

References

1. Gordon S (2016) Phagocytosis: an immunobiologic process. Immunity 44:463–475

2. Flannagan RS, Jaumouillé V, Grinstein S (2012) The cell biology of phagocytosis. Annu Rev Pathol 7:61–98

3. Freeman SA, Grinstein S (2014) Phagocytosis: receptors, signal integration, and the cytoskeleton. Immunol Rev 262:193–215

4. Suzuki T, Kono H, Hirose N et al (2000) Differential involvement of Src family kinases in Fc receptor-mediated phagocytosis. J Immunol 165:473–482

5. Kiefer F, Brumell J, Al-Alawi N et al (1998) The Syk protein tyrosine kinase is essential for Fcgamma receptor signaling in macrophages and neutrophils. Mol Cell Biol 18:4209–4220

6. Hoppe AD, Swanson JA (2004) Cdc42, Rac1, and Rac2 display distinct patterns of activation during phagocytosis. Mol Biol Cell 15:3509–3519

7. Botelho RJ, Teruel M, Dierckman R et al (2000) Localized biphasic changes in phosphatidylinositol-4,5-bisphosphate at sites of phagocytosis. J Cell Biol 151:1353–1368

8. Marshall JG, Booth JW, Stambolic V et al (2001) Restricted accumulation of phosphatidylinositol 3-kinase products in a plasmalemmal subdomain during Fc gamma receptor-mediated phagocytosis. J Cell Biol 153:1369–1380

9. Shalaby MR, Aggarwal BB, Rinderknecht E et al (1985) Activation of human polymorphonuclear neutrophil functions by interferon-gamma and tumor necrosis factors. J Immunol 135:2069–2073

10. Wan M, van der Does AM, Tang X et al (2014) Antimicrobial peptide LL-37 promotes bacterial phagocytosis by human macrophages. J Leukoc Biol 95:971–981

11. Martinez FO, Gordon S (2014) The M1 and M2 paradigm of macrophage activation: time for reassessment. F1000Prime Rep 6:13

12. Hölzl MA, Hofer J, Steinberger P et al (2008) Host antimicrobial proteins as endogenous immunomodulators. Immunol Lett 119:4–11. doi:10.1016/j.imlet.2008.05.003

13. Ogle JD, Noel JG, Sramkoski RM et al (1990) Effects of chemotactic peptide f-Met-Leu-Phe (FMLP) on C3b receptor (CR1) expression and phagocytosis of microspheres by human neutrophils. Inflammation 14:337–353

14. Nimmerjahn F, Ravetch JV (2008) Fcγ receptors as regulators of immune responses. Nat Rev Immunol 8:34–47. doi:10.1038/nri2206

Single Cell Analysis of Phagocytosis, Phagosome Maturation, Phagolysosomal Leakage, and Cell Death Following Exposure of Macrophages to Silica Particles

Gaurav N. Joshi*, Renée M. Gilberti*, and David A. Knecht

Abstract

Chronic inhalation of silica in various occupational settings results in the development of silicosis, a disease characterized by lung fibrosis. Uptake of silica particles by alveolar macrophages results in cell death and this is one of the contributing factors to the development of silicosis. We have characterized the uncoated or protein-coated (non-opsonized) and Fc receptor-mediated (antibody-opsonized) routes of silica phagocytosis and toxicity. Numerous microscopy techniques and fluorescent probes are outlined in this chapter to carefully measure particle uptake, by macrophages, phagosome maturation, phagosomal reactive oxygen species generation, phagolysosomal leakage, and cell death.

Key words Silica, Phagocytosis, Phagosome, Endosome, Lysosome, Phagolysosomal leakage, ROS, Cell death, Fluorescence, Microscopy

1 Introduction

1.1 Particle Phagocytosis and Phagosome Maturation

Silica particles up to 5 μm in size can reach the alveolar spaces of the lungs [1]. Macrophages are professional phagocytes that are able to recognize various ligands on the surface of pathogens and clear them. Inorganic particulate matter such as silica (toxic) and latex (nontoxic) particles are also phagocytosed even though they lack these ligands [2]. In vivo, inhaled silica particles might be coated with protein or antibodies [3] resulting in their recognition by a non-opsonized or opsonized particle phagocytosis pathway [2, 4].

Regardless of the uptake pathway, a series of signaling events take place within a cell after particle interaction, resulting in membrane extension around the particle facilitated by actin polymerization. This actin-rich phagocytic cup quickly closes and the engulfed particle resides in a structure known as a phagosome. A series of

---

*Author contributed equally with all other contributors.

Roberto Botelho (ed.), *Phagocytosis and Phagosomes: Methods and Protocols*, Methods in Molecular Biology, vol. 1519, DOI 10.1007/978-1-4939-6581-6_5, © Springer Science+Business Media New York 2017

phagosome maturation steps then occur, including fusion with endosomes and lysosomes. Conversion of membrane lipids from one form of phosphoinositide (PI) to another is important for phagosome maturation [5, 6]. A phagosome that has fused with an endolysosome is referred to as a phagolysosome throughout the text. When phagocytosed by a macrophage, bacteria are usually degraded by a concerted action of reactive oxygen species (ROS) generated by NADPH oxidase-2 that is recruited to a phagosome and also by lysosomal proteases [6–8]. However, phagolysosomal leakage and cell death is unique to silica particle phagocytosis [9, 10, 15]. In contrast, phagocytosis of a latex particle does not result in phagolysosomal leakage or cell death [10, 11]. Leakage of proteases such as cathepsins from phagolysosomes can cause inflammasome activation, which results in the release of pro-inflammatory cytokines [12, 13] and cell death [14]. Figure 1 summarizes the processes leading up to silica induced phagolysosomal leakage (Fig. 1).

The internalization of both crystalline and porous, spherical amorphous silica results in phagolysosomal leakage and cell death [2, 11, 15]. The spherical shape, porosity, and an even size distribution of the amorphous silica particles help with tracking of these particles via microscopy [10]. This series of protocols will aid in the

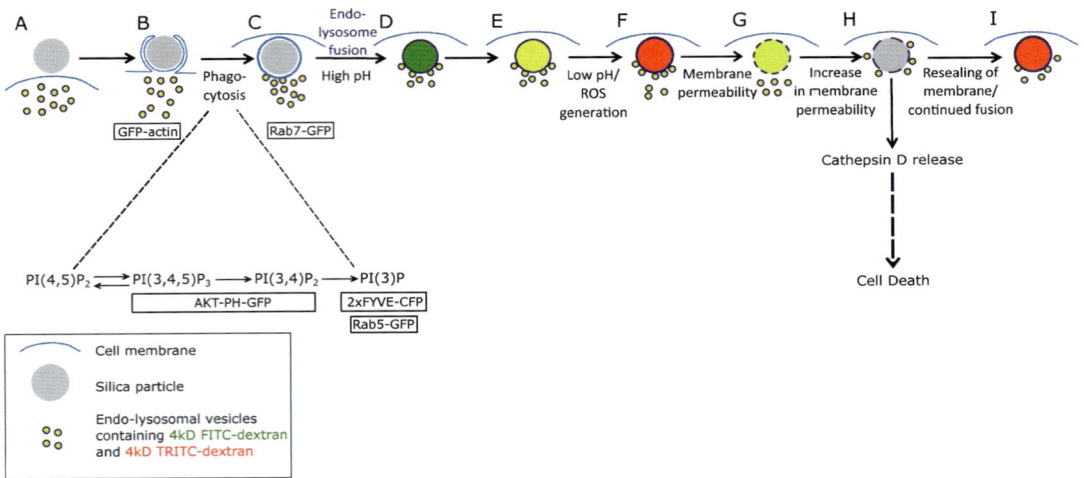

Fig. 1 A schematic of events of 3 μm amorphous silica particle uptake through cell death. (a) Dextran will localize to endolysosomal vesicles giving a punctate fluorescence pattern. (b) Following particle–cell interaction, the cell membrane extends around the particle to form a phagocytic cup. (c) A phagocytic cup transitions into a complete phagosome that is mediated in part by PI dynamics. Endolysosomes fuse with the phagosome. (d) Since FITC-dextran moves to a higher pH compartment, there is an increase in FITC-dextran fluorescence. (e) As more vesicles fuse with the phagosome, there is an increase in TRTIC-dextran fluorescence. The phagosome is therefore both *red* and *green*. (f) pH decreases in the phagosome due to vATPase activity resulting in a decrease in FITC-dextran fluorescence. (g) A transient increase in FITC-dextran fluorescence is observed in the phagosome due to an increase in pH, suggestive of a limited membrane permeabilization. (h) With an increase in permeability, both dextrans leak into the cytoplasm. (i) This phagolysosomal leakage is transient and is resolved by fusion of new endolysosomes as evidenced by an increase in TRITC-dextran fluorescence

preparation of non-opsonized and opsonized particles as well as the quantification of internalized particles at different time points. Immunofluorescence assays can measure uptake at increasing time points by fixing the cells and staining the particles with antibodies specific to their protein coating.

In order to visualize the dynamic processes of phagocytosis and phagosome maturation, macrophages were transfected with DNA from a vector encoding a fluorescent protein of interest. These include, GFP-actin or mRFP-ruby-Lifeact for visualizing actin dynamics (Fig. 1b), AKT-PH-GFP for visualizing $PI(3,4,5)P_3$ and $PI(3,4)P_2$ dynamics, 2xFYVE-CFP for visualizing $PI(3)P$ dynamics, and Rab5-GFP for fusion of early endosomes to a phagosome. These probes appear before complete closure of the phagosome [5, 6, 16]. Rab7-GFP, a marker for fusion of late endosomes to a phagosome, appears following closure of the phagosome (Fig. 1c). Alternatively, phagosome maturation can be studied by monitoring delivery of the endolysosomal contents to the phagosomes. Cells can be loaded with fluorescent-dextrans that are delivered to phagolysosomes by fusion of endosomes and phagosomes. This assay is also useful for quantifying phagolysosomal leakage [10].

1.2 Generation of Phagosomal Reactive Oxygen Species (ROS)

Phagosomal ROS is generated when NADPH oxidase-2 (NOX2) is recruited to a phagosome. NOX2 results in a single electron reduction of oxygen within a phagosome and leads to the generation of phagosomal superoxide, which undergoes self or enzymatic dismutation to form hydrogen peroxide (H_2O_2). Usually, H_2O_2 is converted to water by various enzymes [17]. It has been hypothesized that in the presence of silica, H_2O_2 is converted into more harmful hydroxyl radicals, which cause lipid peroxidation of the phagolysosomal membrane and subsequent leakage [18, 19]. Linking an oxidation sensitive dye such as H_2HFF to the protein bovine serum albumin (BSA), coating the particle, allows for detection of phagosomal ROS [10, 20]. The H_2HFF fluorescence increases upon oxidation by phagosomal ROS and as the protein becomes proteolyzed inside a phagolysosome, H_2HFF becomes free and during phagolysosomal leakage, is released into the cytoplasm.

1.3 Detection of Cell Death

Lysosomal damage and the subsequent release of cathepsin-D have been proposed to initiate the intrinsic cell death pathway leading to apoptosis [14, 21–24]. Apoptosis can be distinguished from necrotic cell death using Annexin-V conjugated to a fluorophore [15]. Annexin-V binds phosphatidylserine (PS) which becomes exposed on the external face of the plasma membrane during the apoptotic process but not during necrosis [25]. Cell death can also be measured using propidium iodide, which passes through the cell membrane of dead cells and binds to nucleic acids. Since the nucleus is the major site of DNA and RNA, propidium iodide is detected as a bright nuclear signal. Another methods paper from our research group details techniques to investigate the intrinsic cell death pathway using silica and other inducers [26].

It should be noted that the assays in this chapter can be multiplexed to simultaneously image various processes as long as spectrally separable fluorescent probes are used.

2 Materials

2.1 Cell Culture

1. MH-S murine alveolar macrophage cell line (ATCC: CRL-2019).

2. Complete medium (for MH-S cells): 500 ml RPMI-1640 medium supplemented with 2 mM L-glutamine, 10 mM HEPES, heat inactivated 10% fetal bovine serum, 100 μg/ml ampicillin, and 100 μg/ml dihydrostreptomycin sulfate.

3. 25 mm round glass coverslips or 22 mm square #1.5 glass coverslips.

4. 35 mm tissue culture grade petri dishes.

5. 35 mm Delta T (also known as Bioptechs) glass-bottom culture dish.

6. 35 mm cover glass bottom dish (see **Note 1**).

7. 8-Well Thermo Scientific Nunc Lab-Tek II chamber cover glass slide (see **Note 1**).

8. CO_2-Independent Medium (CIM) (ThermoFisher Scientific) (see **Note 1**).

2.2 Particle Preparation

1. Min-U-Sil 5 alpha-quartz crystalline silica (US Silica Corporation) (see **Note 1**).

2. Alltech Allsphere amorphous, 3 μm spherical porous silica (Alltech allsphere, now a part of Grace-Davison) (see **Note 1**).

3. 25 mg/ml, 3 μm plain latex beads. Store at 4 °C (Polysciences, Inc.) (see **Note 1**).

4. Particle coating buffer (PCB): 1.8 mM Na_2CO_3, 3.2 mM $NaHCO_3$, 135 mM NaCl, pH 9.5, in a 1.5 ml centrifuge tube.

5. Particle coating-chicken egg ovalbumin (PC-CEA): 10 mg/ml solution of chicken egg albumin (Ovalbumin) in PCB. Either use fresh or store at −20 °C.

6. Particle coating-bovine serum albumin (PC-BSA): 10 mg/ml solution of bovine serum albumin (BSA) in PCB. Either use fresh or store at −20 °C.

7. 1 mg/ml Rabbit anti-chicken egg albumin (Rabbit anti-ovalbumin) in PBS, pH 7.2. Store at 4 °C.

8. 1 mg/ml Rabbit anti-bovine serum albumin (Rabbit anti-BSA) in PBS, pH 7.2. Store at 4 °C.

9. 0.75 mg/ml FITC-conjugated goat anti-rabbit antibody prepared in 1:1 sterile dH_2O:glycerol and stored at −20 °C.

10. 10 mg/ml Tetramethyl-rhodamine isothiocyanate or Rhodamine B stock made up in DMSO.

11. A tumbler for rotation of micro-centrifuge tubes.

2.3 Inhibitors

The following inhibitors were all dissolved in DMSO to prepare stock solutions as follows:

1. 16.5 mM PP2, a tyrosine kinase phosphorylation inhibtior.

2. 32.5 mM LY294002 or 5.8 mM Wortmannin, phosphatidyl inositol 3-kinase inhibitors.

3. 0.24 mM Latrunculin-A, 5 mM Latrunculin-B, or 20 mM Cytochalasin-D, actin polymerization inhibitors.

4. 300 μM Nocodazole, a microtubule polymerization inhibitor.

2.4 Fixed Cell Phagocytosis assay

1. 4% Formaldehyde: Prepared by diluting EM grade, methanol-free 16% formaldehyde solution in 1× PBS.

2. PBS: 137 mM NaCl, 2.7 mM KCl, 10 mM Na_2HPO_4, 1.8 mM KH_2PO_4, pH 7.4.

3. 100 mM NH_4Cl prepared in 1× PBS, pH 7.25.

4. Mounting medium.

5. Slides.

6. Coverslips.

7. Tweezers (preferably curved with a fine tip).

2.5 Fluorescent Probes for Phagosome Maturation

1. GFP-actin.

2. mRFP-ruby-Lifeact.

3. AKT-PH-GFP.

4. 2xFYVE-CFP.

5. Rab5-GFP.

6. Rab7-GFP.

7. FuGene HD transfection reagent (Promega) (*see* **Note 2**).

8. 100 mg/ml 4 kDa FITC-dextran and 4 kDa TRITC-dextran in sterile H_2O.

2.6 Phagosomal ROS and Cell Death Assays

1. 1 mg dihydro hexafluorofluorescein-BSA (H_2HFF-BSA) per vial, stored at –20 °C. Dissolve in particle coating buffer when necessary.

2. Nitrogen gas tank.

3. 16 μg/ml AnnexinV-FITC in Tris buffer (15 mM Tris, 80 mM NaCl, 1 mM EDTA, 0.2% BSA). Store at 4 °C.

4. 1 mg/ml Propidium Iodide prepared in sterile deionized water and stored at 4 °C.

2.7 Microscopy and Data Quantification

1. Epifluorescence microscope.
2. Confocal microscope.
3. Spinning disk microscope.
4. Automated image acquisition software.
5. Motorized and computer-controlled XY stage and Z focus drive.
6. Environmental chamber for live cell imaging.
7. FIJI (Image J) software.

3 Methods

3.1 Tissue Culture and Plating of Macrophage Cell Line

1. Maintain murine macrophage MH-S cell line in complete medium at 37 °C in a 5% CO_2 incubator in a 60 mm dish or 100 mm tissue culture grade dish in a total of either 5 ml or 8 ml of complete medium, respectively.
2. Harvest cells by gently triturating them using a pipette or a 1 ml micropipettor. These cells are not tightly adherent to plastic surfaces, so they are relatively easy to dislodge (see **Note 3**).
3. Plate 1.5×10^5 macrophage cells in 1 ml complete medium in a 35 mm glass bottom dish. Alternatively, plate 3×10^4 cells/well in an 8-well chamber slide in a total of 500 µl complete medium per well (see **Notes 4–6**).
4. Let cells adhere and grow for 12 h at 37 °C in a 5% CO_2 incubator.
5. After 12 h, check cells to confirm they are about 80% confluent and remove any dead cells or debris by aspirating the medium and replacing it with fresh complete medium that is warmed to 37 °C.

3.2 Particle Preparation for Immunofluorescence Assays

3.2.1 Generating Protein-Coated (Non-opsonized) or Antibody-Coated (Opsonized) Latex or Silica Particles

1. Before using silica particles, make them endotoxin-free by adding silica particles to a glass container, cover it with aluminum foil and bake in an oven at 83 °C for 18 h.
2. In a 1.5 ml centrifuge tube, add 1 ml of PC-BSA or PC-CEA and 2 mg of either crystalline or spherical silica particles or 100 µl (2.5 mg) of plain 3 µm latex beads.
3. Incubate the centrifuge tube for 90 min at 37 °C while rotating on the tumbler.
4. Centrifuge the particles at $8000 \times g$ for 1 min, aspirate the supernatant, and replace with 1 ml PBS. Vortex the tube to resuspend the particles (see **Note 7**).
5. Repeat the PBS wash twice.
6. Resuspend the particles in 1 ml PBS, aliquot 500 µl of these protein-coated particles into a new 1.5 ml centrifuge tubes and

store them at 4 °C. These are the protein-coated, non-opsonized particles. The remaining 500 µl non-opsonized particles will be used for opsonization (antibody labeling).

7. Opsonization is accomplished by adding 5 µl of the stock solution of antibody (rabbit anti-ovalbumin or rabbit anti-BSA) to 500 µl of protein-coated particles. Incubate the centrifuge tube for 90 min at 37 °C while rotating on the tumbler.

8. Add 500 µl PBS and spin the particles at 8000×g for 1 min, aspirate the supernatant, and replace with 1 ml PBS. Vortex the tube to resuspend the particles. Repeat this wash two times (*see* **Note 7**) and then resuspend the particles in 500 µl of PBS and store the particles at 4 °C.

3.2.2 Confirmation of Particle Opsonization

Particle opsonization in protocol 3.2.1 can be confirmed by the steps outlined below.

1. Add 1 µl of secondary antibody conjugated to a fluorophore (such as, goat anti-rabbit FITC) to 149 µl PBS for 150 µl total volume in a centrifuge tube.

2. Vortex opsonized particles to resuspend and add 10 µl to the diluted antibody.

3. Incubate for 30 min at room temperature.

4. Centrifuge at 8000×g for 1 min to spin down the particles and aspirate supernatant.

5. Add 100 µl PBS and vortex. Add 50 µl to a glass bottom 8-well chamber slide or a 100 µl to the center of 35 mm WillCo well (*see* **Note 8**). Image the particles with a fluorescence microscope (*see* **Note 9**). As a negative control, have particles that were not opsonized but incubated with antibody. Opsonized particles conjugated with secondary antibody will be fluorescent whereas the control particles will not be fluorescent.

6. Capture images of the fluorescent particles with defined camera and illumination settings and then use those same settings each time particles are prepared to confirm that the coating is consistent.

3.2.3 Labeling of Silica with TRITC-R

1. Add 2 mg of crystalline silica to 1 ml of PCB containing 50 µg/ml of TRITC or Rhodamine B in a 1.5 ml centrifuge tube.

2. Incubate the tube, wrapped in aluminum foil, for 90 min at 37 °C while tumbling.

3. Centrifuge the particles at 8000×g for 1 min and aspirate the supernatant.

4. Stop the reaction by resuspending the pellet in 50 mM NH$_4$Cl, for 2 h at 37 °C while tumbling.

5. Centrifuge the particles at $8000 \times g$ for 1 min, aspirate the supernatant, and replace with 1 ml PBS. Vortex the tube to resuspend the particles. Repeat the wash five times.

6. Resuspend the particles in 1 ml PBS and store at 4 °C.

3.3 Temporal Analysis of Silica Phagocytosis by Macrophage Cells

The protocols in this section detail the use of both fixed and live-cell imaging methods to visualize and quantify particle uptake in macrophages. Data from both methods complement each other for a thorough understanding of particle uptake.

3.3.1 Fixed Cell Phagocytosis Assay

1. Sterilize 22 mm round glass or square #1.5 glass coverslips by placing them in a beaker with 100 % ethanol. Coverslips can be stored in the ethanol. When needed, remove a coverslip with a fine tip tweezers and place it in a well of 35 mm tissue culture dish. Allow the ethanol to evaporate in a tissue culture hood to prevent contamination.

2. Plan on the number of coverslips depending on the time-points for the study. Plate cells for each time point as described in Subheading 3.1 in a 35 mm dishes containing the coverslip.

3. Before the experiment, replace medium with 1 ml of CIM and incubate for 15 min in a 37 °C ambient atmosphere incubator to equilibrate cells with the new medium.

4. If it is desired to the study the role of various uptake mechanisms [4, 27], replace the medium with CIM containing either a tyrosine kinase phosphorylation inhibitor (20 µM PP2), phosphatidylinositol 3-kinase inhibitors (20 µM LY294002 or 20 nM wortmannin), actin polymerization inhibitors (500 nM latrunculin-A, 500 nM latrunculin-B, or 500 nM cytochalasin-D), or microtubule polymerization inhibitor (800 nM nocodazole) and incubate for 45 min at 37 °C in an ambient atmosphere incubator (see **Note 10**).

5. Chill the dishes on ice for 10 min.

6. Add 15 µg/cm^2 of either non-opsonized or opsonized silica or 10.5 µg/cm^2 non-opsonized or opsonized latex particles to the cells while on ice, swirl the dish to disperse the particles and then centrifuge at $300 \times g$ for 5 min at 4 °C (see **Note 11**).

7. Immediately following centrifugation remove the 0 time control, aspirate the media, and add 1 ml 4 % formaldehyde for 6 min to fix the cells.

8. Place the remaining dishes in the 37 °C atmospheric incubator and incubate for various times before fixing. Aspirate and add 1 ml 4 % formaldehyde for 6 min.

9. After fixation, aspirate the medium and add 1 ml of 50 mM NH$_4$Cl, for 3 min to quench the fixative and then aspirate and add 1 ml of PBS. Caution: **Steps 10** and **11** are only for cells exposed to non-opsonized particles. For cells exposed to opsonized particles, proceed to **step 12** after **step 9**.

10. Aspirate the PBS from the dish and add 200 µl of 1:800 rabbit anti-ovalbumin or rabbit anti-BSA in PBS directly to the coverslip and incubate for 35 min at room temperature.

11. Aspirate the primary antibody and wash the plate twice with 1 ml of PBS.

12. Aspirate the PBS and carefully add 200 µl of 1:150 FITC-conjugated goat anti-rabbit secondary antibody in PBS directly to the coverslip and incubate for 35 min.

13. Aspirate the secondary antibody and wash the coverslips twice with 1 ml of PBS followed by one wash with 1 ml of distilled water to prevent the drying of salt crystals on the coverslip surface.

14. To mount the coverslip on a slide, pipet 10 µl of mounting medium onto the center of a microscope slide.

15. Lift the coverslip from the 35 mm dish with tweezers keeping track of the surface of the coverslip onto which cells are plated. Gently dab the sides of the coverslip on a tissue to remove excess water and place the coverslip onto the slide at a 45° angle such that one of the edges of the coverslip touches the mounting medium.

16. Make sure that the surface of the coverslip on which the cells are plated is facing the mounting medium. Gently release the coverslip from the tweezers in a controlled manner such that the mounting medium spreads evenly between the coverslip and slide without leaving any air pockets.

17. Seal the edges of coverslips with clear nail polish to prevent drying of the mounting medium and shifting of the coverslip (*see* **Note 12**).

18. Image the cells on an epifluorescence or confocal microscope.

19. Quantify phagocytosis and represent it as either "percent phagocytosis" (**step 18**) or "phagocytic index" (**step 19**).

20. To calculate "percent phagocytosis," count the total number of cell-associated particles in the DIC/phase contrast image and the number of cell associated particles in the corresponding fluorescent image (representative of secondary antibody) for 100 cells (**Fig. 2**). To calculate the total number of internalized particles, subtract the number of fluorescent particles from the total particles. Express this as percent phagocytosis, by dividing the number of internalized particles by the total number of particles and multiplying by 100. Perform this quantification for every time point in the study (*see* **Notes 13** and **15**).

21. To calculate "phagocytic index," count total number of cell associated particles in DIC/phase contrast image and in the corresponding fluorescence image (representative of particles that were not taken up) for each cell (**Fig. 2**). To get the total

A) Fixed cell internalization assay for non-opsonized particles

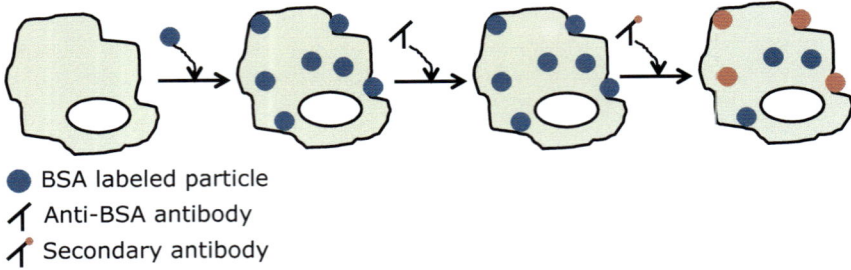

● BSA labeled particle
🡅 Anti-BSA antibody
🡅 Secondary antibody

B) Fixed cell internalization assay for opsonized particles

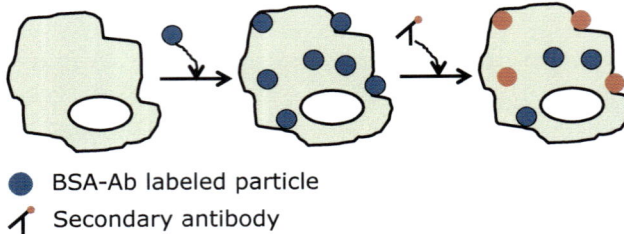

● BSA-Ab labeled particle
🡅 Secondary antibody

Fig. 2 Fixed cell internalization assay for non-opsonized and opsonized particles. (**a**) Cells exposed to non-opsonized particles are fixed, followed by exposure to an antibody against the protein that binds to particles on the outside of the cell. To visualize this, a secondary antibody conjugated to a fluorophore (depicted here in *red*) is added. Particles on the outside appear *red*, whereas the internalized particles are not stained. (**b**) Cells exposed to opsonized particles are fixed, followed by exposure to a secondary antibody conjugated to a fluorophore with a similar outcome to that of non-opsonized particles. As an example, the protein BSA is shown in this schematic

number of internalized particles for each cell, subtract the number of fluorescent particles from the particles observed in DIC/phase contrast image. Repeat this for one hundred cells and calculate the mean value (*see* **Notes 14** and **15**).

3.3.2 Assay for Phagocytosis of Particles That Cannot Be Coated or Opsonized

This assay is particularly helpful to identify the cytoplasmic location of uncoated particles.

1. Plate cells in a 35 mm glass bottom dish as described in Subheading 3.1.

2. Replace the complete medium with CIM and let cells equilibrate at 37 °C in this medium in an ambient air incubator for 15 min.

3. Replace CIM with 1 ml CIM containing 50 μM CMFDA and incubate for 30 min at 37 °C in an ambient atmosphere incubator.

4. At the end of the incubation period, wash two times with CIM.

5. Expose cells to 15 µg/cm² crystalline or amorphous silica particles labeled with TRITC-R and incubate for 1 h at 37 °C in an ambient air incubator.

6. Bring the plate to a confocal microscope and acquire z-stacks 0.5 µm apart in channels appropriate for DIC, CMFDA (GFP filter set), and TRITC-R (TRITC filter set).

7. Reconstruct the z-stacks into a volume view either using the software on the confocal microscope or using FIJI [28].

8. Enable the orthogonal view feature and focus on the particle. If the particle is inside the cell, it will appear red within the green volume of CMFDA in the cytoplasm.

3.4 Live-Cell Imaging of Phagocytosis and Phagosome Maturation

3.4.1 FugeneHD DNA Transfection

1. Plate 1.5×10^5 cells in each of two 35 mm glass bottom dishes in 1 ml of complete medium. Incubate the dishes overnight at 37 °C in a 5 % CO_2 incubator.

2. The next morning, aspirate the medium and add 875 µl of fresh complete medium.

3. Mix 2 µg of DNA with 100 µl RPMI-1640 medium (serum free) that only contains 10 mM HEPES in a 5 ml sterile polystyrene tube (*see* **Notes 2** and **16**). Prepare a second tube with no DNA.

4. Add 6 µl of room temperature FuGeneHD transfection reagent to the center of the tube, being careful to only allow the FuGeneHD to mix with the DNA-medium complex and not directly touch the tube.

5. After 15 min of incubation at room temperature, add the contents of the polystyrene tube to the cells in a drop-wise fashion and adjust the total volume to 1 ml of complete medium.

6. Incubate the dishes in a CO_2 incubator for up to 12 h. Depending on the cells and vector, expression can be seen as early as 6 h after transfection, but usually the cells are incubated overnight.

7. Check for cellular fluorescence using a fluorescence microscope and appropriate filter sets. The cells with transient protein expression can be used for experiments or stable cell lines can be selected by adding the appropriate antibiotic followed by several rounds of cloning fluorescent cell colonies (*see* **Note 17**).

3.4.2 Live Cell Imaging

1. Set the environmental chamber surrounding the microscope and the stage top incubator to the appropriate temperature (37 °C) and humidity setting and, once it has reached these settings, place the plate of cells in the microscope stage insert (*see* **Note 18**).

2. Allow cells to equilibrate for about 10 min.

3. Select fluorescent cells in either a single field or multiple fields of view (*see* **Notes 19** and **20**).

4. Set appropriate imaging parameters for the fluorescent cells (DIC and fluorescence channels) using a 63× oil immersion objective (*see* **Note 21**).

5. Set time intervals 15–30 s apart and an appropriate z-slice range for the cells (2–3 μm apart) (*see* **Note 22**) for a total of 30 min on a laser scanning confocal or spinning disk confocal microscope (*see* **Note 23**).

6. Initiate time-lapse imaging and then expose cells to 15 μg/cm^2 silica or 10.5 μg/cm^2 latex particles to the cells. To expose cells to particles, add 500 μl of CIM to a 1.5 ml centrifuge tube and add appropriate amount of particles and vortex. Bring this tube to the microscope stage. Carefully remove 500 μl of medium from the plate such that the plate is not disturbed. Add the mixture of particles in CIM to the plate (*see* **Note 24**). These concentrations yield approximately 5 particles per cell and do not crowd up in the cytoplasm. This concentration of silica is also toxic to macrophages, but cell death occurs on a much longer time scale than phagocytosis.

3.4.3 Quantification of the Timing of Association of Probes with Phagosomes

1. Open the data set using FIJI.

2. Draw a circular region of interest (ROI) around the particle at the initial location of particle–cell interaction in the transmitted light channel and record the ROI position using the ROI Manager. Switch to the fluorescence channel and measure the average fluorescence intensity in the ROI in each image to determine the timing of actin, PI(3,4,5)P$_3$, and PI(3)P association with and subsequent dissociation from the phagosome.

3. The particle's position may shift following uptake so it may be necessary to move the ROI over time without changing the ROI area in order to track the phagosome.

3.5 Comparison of the Timing Between GFP-Actin Association and Endolysosomal Fusion with Phagosomes

1. Transfect cells with a probe in which actin is fused to a fluorescent protein as described in Subheading 3.4.1 or, if there are macrophages stably expressing this probe, plate cells in a glass bottom 35 mm dish in complete medium overnight as described in Subheading 3.1.

2. The next morning, replace the medium with fresh complete medium containing 1 mg/ml 4 kDa TRITC dextran and incubate for 2 h at 37 °C with 5 % CO$_2$ to load the internal vesicle compartments with dextran.

3. Wash cells twice with complete medium to remove extracellular dextran. Between each wash, place cells in an ambient air incubator at 37 °C for 5 min before the next wash.

4. Replace the medium in the dish with CIM and bring the plate of cells to the microscope stage and set up live cell imaging as described in Subheading 3.4.2.

5. The increase in phagosomal dextran fluorescence can be quantified by drawing an ROI around the phagosome and measuring mean fluorescence over time as described in Subheading 3.4.3 (*see* **Note 25**).

3.6 Detecting Fusion of Phagosome with Endolysosomes and Phagolysosomal Leakage

This is an alternative to the method described in Subheading 3.5. Cells are loaded with dual dextran (both FITC-dextran and TRITC-dextran, 4 kDa in size). This allows for a detailed visualization and quantification of the dynamics of phagosome maturation, pH changes, and phagolysosomal leakage (Fig. 3) [10].

1. Plate cells as described earlier in Subheading 3.1 in a 35 mm glass bottom dish or 8-well chamber slide.

2. Incubate cells with 0.5 mg/ml each of 4 kDa FITC-dextran and 4 kDa TRITC-dextran, respectively in complete medium for 2.5 h at 37 °C (*see* **Notes 26** and **27**). Alternatively, cells can also be incubated with either 1 mg/ml of 4 kDa FITC-dextran or 1 mg/ml of 4 kDa TRITC-dextran in complete medium for 2.5 h at 37 °C (*see* **Note 28**).

3. Wash cells twice with complete medium to remove excess dextran. Between each wash, place cells in an ambient air incubator at 37 °C for 5 min before the next wash. Replace the medium with 1 ml of CIM for imaging.

4. Bring the plate to a microscope fitted with an appropriate environmental chamber and image using appropriate settings for FITC and TRITC. Also include the DIC channel (*see* **Note 29**).

5. Add 15 μg/cm² silica particles (toxic) or 10.5 μg/cm² latex particles as a control and image 30 s apart for at least 1 h (*see* **Note 30**).

6. If imaging is performed using a confocal microscope, acquire images in multiple z-planes (3 μm apart) so that a particle can be found in a different focal plane due to movement of the phagosome.

7. Quantify phagosome–endolysosomal fusion and phagolysosomal leakage by drawing an ROI over the phagosome and measuring mean pixel intensity for every time point in FIJI. Phagolysosomal leakage can also be quantified by drawing an ROI over the nuclear area and by measuring mean pixel intensity for every time point.

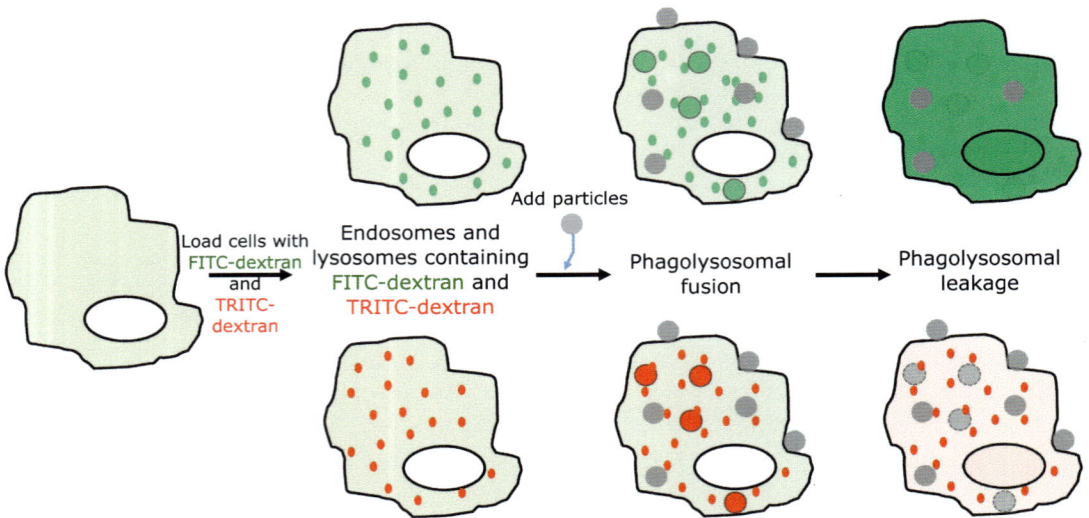

Fig. 3 Investigation of phagosome maturation and phagolysosomal leakage in cells loaded with FITC-dextran and TRITC-dextran. Cells were loaded with 4 kDa FITC-dextran and 4 kDa TRITC dextran that co-localize within endolysosomal vesicles. For an ease of understanding of the processes, cells are shown to contain FITC-dextran and TRITC-dextran separately. Following exposure to 3 μm amorphous silica particles, uptake, and phagolysosomal fusion, varying levels of dextran fluorescence is observed in a phagolysosome. FITC-dextran fluorescence is quenched in a phagolysosome due to low pH where as the TRITC-dextran fluorescence is not affected by pH. Upon phagolysosomal leakage, there is an increase in FITC-dextran fluorescence in the cytoplasm and nuclear area since the FITC-dextran fluorescence is no longer quenched at a higher pH. TRITC-dextran also leaks into the cytoplasm and nucleus but the quanta of fluorescence is weak compared to FITC-dextran. The nuclear area can be used for quantification of dextran release

3.7 Detection of Phagosomal Reactive Oxygen Species (ROS) by Using Particles Labeled with H₂HFF-BSA

3.7.1 Particle Preparation

1. Deoxygenate the particle coating buffer (PCB) by bubbling nitrogen gas through the solution for 30 min. This is achieved by filling up a 50 ml tube with 20–25 ml of PCB. Two layers of parafilm are kept on the mouth of the tube and a sterile glass pasteur pipette is inserted through the center of the parafilm. Holes are made on the periphery of the parafilm by puncturing it with 200 μl pipette tip to allow gas to pass through during the bubbling process. The nitrogen gas tank regulator is set at a pressure low enough for continuous bubbling of the gas through the buffer. The tube coming out of the nitrogen tank regulator is attached with a male luer coupler. The tubing on the other side of the coupler is inserted into the pasteur pipette and nitrogen gas is passed continuously for 30 min.

2. After 30 min, the pasteur pipette is removed and the cap is placed immediately on the mouth of the 50 ml tube without removing the parafilm to minimize the mixing of atmospheric oxygen into the tube.

3. Add 1 ml of deoxygenated PCB to a screw cap (with O-ring) vial containing 1 mg of H₂HFF-BSA. Purge the space between the 1 ml liquid and the cap area with nitrogen gas to remove oxygen and tighten the cap. Vortex to mix the contents.

4. Add 2 mg silica particles or 100 μl latex particles to the vial in **step 2**. Purge the space with nitrogen gas to remove oxygen and tighten the cap. Vortex to mix contents and incubate for 90 min at 37 °C while rotating the tube on the tumbler.

5. Meanwhile, deoxygenate 25 ml of PBS by bubbling nitrogen gas for 30 min as described in **step 1**.

6. Upon completion of the incubation period in **step 3**, transfer the contents to a 1.5 ml microcentrifuge tube. Spin the particles at $8000 \times g$ for 2 min, aspirate the medium, and replace with 1 ml deoxygenated PBS. Vortex the tube to resuspend the particles. Repeat the deoxygenated PBS wash two more times. Perform these steps quickly to minimize the exposure of particles to oxygen and to avoid auto-oxidation of the dye.

7. Aliquot 500 μl of H_2HFF-BSA labeled (non-opsonized) particles to a screw cap tube with O-ring. Purge the space between the liquid and cap area with nitrogen gas before capping and store at 4 °C.

8. To opsonize particles, add 10 μg of rabbit anti-BSA antibody to the remaining 500 μl of non-opsonized particles. Rotate for 90 min at 37 °C on the tumbler. Repeat washing steps in **step 5**. At the last step, add 500 μl of deoxygenated PBS and vortex the particles. Transfer it to a screw cap tube with O-ring. Purge the space between the liquid and cap area with nitrogen gas before capping and store at 4 °C.

3.7.2 Determining Particle Labeling Using H_2O_2

Labeling of particles with H_2HFF-BSA can be determined by checking the oxidation of H_2HFF.

1. Suspend 10 μg of labeled particles in 500 μl PBS and add aliquots to two well of a multiwell plate.

2. To one well, add 0.5 M H_2O_2 and mix the contents well by aspirating several times using a micropipette (handle H_2O_2 by wearing gloves).

3. Compare the fluorescence of the H_2O_2 treated particles to the untreated using GFP/FITC filter set on the same microscope that will be used to image cells exposed to particles. The difference between the treated and untreated particles is a measure of the dynamic range of detectable oxidation. The same microscope should be used to image cells upon exposure to H_2HFF-BSA labeled particles so that the maximum dynamic range of the change in fluorescence can be measured.

4. Use the H_2HFF labeled particles within 48 h of preparation as the particles become oxidized with time reducing their dynamic range. Store at 4 °C and protect from light.

3.7.3 Determining Phagosomal ROS

1. Use the microscope settings as determined in Subheading 3.7.2.

2. Bring the plate of cells to the microscope fitted with an appropriate environmental chamber (*see* **Note 31**).

3. Expose cells to H$_2$HFF-BSA labeled particles or further opsonized 15 μg/cm^2 of 3 μm amorphous silica particles or 10.5 μg/cm^2 latex particles and image 1 min apart (*see* **Note 32**).

4. Quantify an increase in H$_2$HFF fluorescence by drawing an ROI around the phagosome for every time point and measure the mean pixel intensity using FIJI.

3.8 Detecting Phosphatidylserine Externalization

1. Plate cells in a 35 mm glass bottom dish overnight as described in Subheading 3.1.

2. Next day, replace the complete medium from the plate with CIM and let cells equilibrate at 37 °C in this medium in an ambient air incubator for 15 min and replace with fresh CIM.

3. Bring the dish to a microscope fitted with an appropriate environmental chamber.

4. Replace the medium in the plate with 500 μl of CIM containing 40 ng AnnexinV-FITC.

5. Add 15 μg/cm^2 silica particles.

6. Acquire images using FITC settings at appropriate time intervals (*see* **Notes 33** and **34**).

7. Alternatively, remove the dishes from the incubator at appropriate intervals (several hours) and capture images of each dish.

3.9 Detecting Cell Death Using Propidium Iodide

1. Plate cells overnight in a 35 mm glass bottom dish as described in Subheading 3.1.

2. Replace the complete medium from the plate with CIM and let cells equilibrate at 37 °C in this medium in an ambient air incubator for 15 min and replace with fresh CIM.

3. Bring the dish to microscope fitted with an appropriate environmental chamber.

4. Replace the medium in the plate with 1 ml of CIM containing 1 μg/ml propidium iodide (*see* **Notes 35** and **36**).

5. Add 15 μg/cm^2 silica particles.

6. Capture images every 15 min in a red channel set up for detecting propidium iodide fluorescence. It is useful to also collect DIC or phase contrast images in parallel to visualize the cellular morphology.

7. Alternatively, remove the dishes from the incubator at appropriate intervals (several hours) and capture images of each dish.

4   Notes

1. 35 mm cover glass bottom dishes can be purchased from either WillCo wells, MatTek or Nunc. Use of 8-well Thermo Scientific Nunc Lab-Tek II chamber cover glass slide is recommended as

they have more surface area that allows for more cells to be plated and a greater well-depth allowing it to hold more media compared to other dishes. This additional well-depth compared to other dishes has been found to be extremely useful during washing steps following incubation of cells with a dye or any chemical compound. CO_2-independent medium (CIM) is a proprietary formulation of ThermoFisher Scientific. The recipe of this product is not available from the company. US Silica Corporation has been a preferred vendor for the purchase of crystalline silica particles in various studies that investigate silicosis. Porous amorphous silica particles were purchased from Alltech, which is now a part of Grace Davison. Polystyrene latex beads can be purchased from several vendors. However, we have always bought them from Polysciences Inc.

2. Many transfection reagents can likely be used to transfect cells with DNA. We prefer FuGeneHD because it was less toxic and had higher transfection efficiency compared to other reagents we have tested on macrophage cell lines. Also, cells can be left in the FuGene/DNA medium overnight eliminated the need for another media change.

3. While this protocol is focused on macrophages, other macrophage cell types and other cell lines can also be used. Many cell lines will phagocytose non-opsonized particles, albeit more slowly than macrophage cells. If the cells are adherent (for example, epithelial cells), then they may require trypsinization to remove the cells from the plastic dish surface.

4. Some cell lines may also require coating glass with extracellular matrix molecules like fibronectin, laminin, or collagen to enable them to adhere. Coat the glass surface of a 35 mm dish or 8-well chamber slide with appropriate protein based on the routine protocol before plating the cells.

5. If the goal is to image using an immersion objective, then one should use dishes with coverglass thick bottoms such as WillCo wells, Bioptechs dishes, Nunc, or MatTek dishes. Alternatively, if one seeks to do a semi-throughput imaging experiment, an 8-well chamber slide with a glass bottom will be helpful. If semi-throughput imaging is performed using an immersion objective, then in addition to putting a drop of oil on the objective, the entire surface of the coverglass facing the objective should be coated with oil. Care should be taken that bubbles are not formed when the entire coverglass surface is covered with oil, as this will affect imaging.

6. These dishes and plates are extremely useful for live cell imaging. However, specialized environmental chamber systems are required to manage the temperature, humidity, and CO_2 during the experiments. If the environmental chamber does not have a port for gas, a specially formulated CO_2-independent

medium can be used during live cell imaging. This does not require extra CO_2 to maintain physiological pH of the medium unlike complete medium. The WillCo wells and 8-well chamber slide fit on a stage insert with a round or rectangular hollow space, respectively. A specialized Bioptechs hardware and stage insert is required for Bioptechs dishes with a Bioptechs temperature control system.

7. Some types of particles become aggregated after centrifugation. Instead of adding 1 ml of PBS and vortexing, we suggest adding 100–200 μl PBS using a micropipettor with a narrow diameter tip to pipette up and down until a homogenous particle mixture is observed. Bring the volume to a total of 1 ml.

8. It may take a while for the particles to settle down to the surface of the plate. Silica particles will settle quickly, within a few minutes. Latex particles may take 5 min or longer to settle due to their lower density. To expedite the settling process, the plates can be centrifuged at 4 °C using a rotor with an attachment that is capable of holding plates.

9. Antibodies cannot penetrate the membrane of fixed cells that have not been permeabilized and hence do not have access to the cytoplasm, so only external particles become fluorescently labeled.

10. It is important to add the inhibitors in this way since they are often dissolved in DMSO, and if added directly the DMSO will sink to the bottom of the dish and expose the cells to a higher than desired concentration of the compound. Also, these concentrations are lower than what is sometimes utilized in the literature. We have optimized to determine the lowest concentrations that disrupt the actin and microtubule cytoskeletons in order to minimize secondary effects on cellular morphology and physiology. These concentrations are not toxic to cells (minimal propidium iodide-positive nuclei) over the maximum time period (8–10 h) used in these experiments [4].

11. When particles are added to the dishes, they settle slowly onto the cells so different particles contact the cell at different times. In order to better synchronize the timing of phagocytosis, cells are chilled to 4 °C to stop uptake. Particles are added and then dishes are centrifuged at 4 °C to spin the particles onto the cells. When the plates are shifted to 37 °C, uptake of particles begins at the same time for all cells.

12. One can either use a commercially available mounting media such as Prolong Gold or use a homemade preparation such as DABCO-MOWIOL. The former hardens after a day, whereas DABCO-MOWIOL does not. With a mounting media such as Prolong Gold, slides can be imaged after the mounting media has hardened. If one wishes to image the slide with a nonhardening media then the edges of the coverslip can be sealed with clear

nail-polish to prevent any slipping of the coverslip during imaging. Sealing of the coverslip also prevents drying or cracking of the medium during storage. Slides should be stored at 4 °C.

13. Percent phagocytosis" is used to measure the proportion of particles that contact a cell and become internalized over time.

14. "Phagocytic index" is used to determine the total number of particles that are internalized per 100 cells. Phagocytic index is used because the same number of cells (100) is being compared for each variable (particle type, inhibitor, etc.).

15. It is easy to count the total number of particles associated with cell and distinguish the ones that are outside (i.e. fluorescent) by overlaying the DIC/phase contrast image with the fluorescence image in FIJI.

16. This protocol describes use of a single vector to transfect cells. Two different types of DNA can be used to transfect cells simultaneously. Not all cells will express both proteins of interests so fields of view with cells of this type need to be identified first.

17. The use of antibiotic will depend on the resistance gene in the vector. A standard killing curve should be performed with nontransfected cells to check for an appropriate concentration of the drug. A concentration should be used that kills 100% of the cells between 24 and 48 h after the drug addition. After 48 h of drug treatment, dead cells should be removed by gently rocking the plate to suspend the dead cells and then aspirate the medium. Add back fresh medium and antibiotic. Monitor for growth of colonies that are fluorescent using a low magnification long working distance objective that can image through the plastic dish. Colonies containing a high proportion of fluorescent cells can be picked manually using a 200 μl pipetter, set to 50 μl and a 35 mm dish with 5 ml complete medium adjacent to the microscope station. Once a fluorescent colony is in focus, gently tip back the condenser arm and bring the tip of the micropipettor onto the colony. In the fluorescence mode, the edge of pipette tip will be easily visible when it is in the region of interest. Make sure that the piston of the pipettor is depressed while dipping into the plate. Bring the pipette tip to the surface of the dish near the colony and gently scrape the cells while withdrawing the plunger to suck up some of the fluorescent cells. Transfer the cells into the new dish containing complete medium. Once a stable expressing cell line is established (can take several weeks), they should be frozen, as the cells will generally lose expression with time.

18. The plate of cells should be properly secured on the plate insert on the microscope stage. This is usually performed by securing metal spring tabs onto the plate. To better secure the plate or if these tabs are missing, apply double-sided tape on the edges

of the plate and then put the plate onto the stage insert. This is to ensure that the plate does not move while imaging with an oil immersion objective, or while adding particles during which medium is removed or added.

19. If the microscope stage is motorized, and the software allows selection of multiple XY fields of view, select fluorescent cells in multiple XY fields of view so that more data can be acquired from a single experiment.

20. Cells with varying levels of fluorescence will be observed. Generally, cells that are not bright expressers should be used. A proper signal of the probe fusing with the phagosome will depend on instrument settings such as the intensity of laser and photomultiplier gain on a laser scanning confocal microscope or on the camera exposure settings and the intensity of the fluorescence light source for an epifluorescence microscope. Parameters should be selected such that an appropriate signal is obtained without causing photobleaching or phototoxicity.

21. Do not image cells in the same or an adjacent field of view that is used to set imaging parameters of microscope. Usually cells are exposed to a relatively large amount of fluorescent light during this process. We have observed a loss or delay in phagocytic ability of cells exposed to continuous or high levels of light. Move to a new field of view once the settings are determined.

22. Since particles are taken up in various z planes of the cell and they may move to a different z-plane after internalization, it is useful to image in several different z-planes so that a single particle can be tracked over time and more events can be imaged in each cell. Given the large size of the particles, it is not necessary to try to image at optimal z resolution. Usually, it is sufficient to collect z planes at 3 μm intervals to capture all phagocytosis events. However, this increases the time it takes to collect images and the exposure of the cells to fluorescent light, so this parameter may have to be traded off against other considerations to obtain optimal data.

23. We prefer the spinning disk confocal because it acquires images faster and causes less cellular phototoxicity.

24. Particle addition to the cells can also be done in the following manner. Carefully pipette 500 μl of CIM from the plate and transfer it to the 1.5 ml centrifuge tube. Add the appropriate amount of particles to this tube, vortex and add this volume back to the plate. It may take up to 5 min for particles to settle down, so imaging can be paused for a few minutes after particle addition to prevent any unnecessary exposure of cells to laser or fluorescence source.

25. Because of the porosity of amorphous silica particles, the fluorescent-dextran occupies the entire volume of the phagosome.

An increase in endolysosomal fusion with the phagosome results in an increase in phagosomal fluorescence. In contrast, latex particles are not porous and the phagosomal fluorescence only appears around the periphery of the particle [10].

26. Following phagolysosomal leakage, dextran is released into the cytoplasm and if it 20 kDa or below is able to enter the nucleus. FITC-dextran is advantageous for this assay. Vesicular FITC-dextran fluorescence is quenched by the low pH of the endosomal and phagosomal compartments. There is an increase in the leaked FITC-dextran fluorescence due to unquenching of FITC in the pH neutral environment of the cytoplasm. It is easier to quantify the leakage in the nuclear area, as there is no background from out of plane vesicles or phagosomes. The cytoplasmic leakage can be measured with either FITC or TRITC, but FITC-dextran release is easier to detect because the signal-to-noise ratio is much higher due to the quenching of the nonreleased signal, leading to lower overall background.

27. During phagolysosomal leakage, the release of proteases such as cathepsins (~50 kDa) into the cytoplasm should be confirmed by loading cells with FITC-dextran that is larger than 50 kDa. Upon release of FITC-dextran into the cytoplasm, there will be an increase in cytoplasmic fluorescence but not the nuclear fluorescence because of the diffusion limit of the nuclear pore complex.

28. The maturation of phagosomes is visualized in a greater detail with dual-dextran experiments. A continued phagosomal maturation is not seen when only FITC-dextran is loaded as FITC fluorescence is very low due to quenching. Having TRITC-dextran allows phagosomes to be tracked over time because its fluorescence is not pH dependent.

29. At the start of the experiment, set the imaging parameters to only weakly detect the FITC-dextran signal so you have dynamic range for when it is unquenched during phagolysosomal leakage.

30. Porous amorphous 3 μm silica particles should be used if the goal is to detect dynamics of phagosome–endolysosomal fusion and phagolysosomal leakage. If crystalline silica particles are used, then only phagolysosomal leakage can effectively be detected because they are not porous and not uniform in shape and size.

31. Both phagosomal ROS generation and phagosome maturation can be tracked simultaneously if cells are pre-loaded with 4 kDa TRITC-dextran before the addition of H_2HFF-BSA labeled particles.

32. There is batch-to-batch variation in the coating of particles making it difficult to relate any increase in fluorescence between experiments when particles from different batches are used.

33. Since there is no signal initially, it is important to establish settings that visualize the binding before running the actual exper-

iment. Because cells are autofluorescent, which can be visualized in the GFP channel, the fluorescence intensity and exposure should be set so that weak cellular background fluorescence is observed. Upon AnnexinV-FITC binding to the cells, there will be an increase in the signal above the background.

34. Because FITC is sensitive to photobleaching, we do not recommend frequent imaging or using excitation light any higher than necessary to image the probe. It is therefore advisable to image at intervals of 5 min or longer. If one would like to obtain higher temporal resolution, then determine the approximate time of PS flipping and start acquiring images for shorter duration at time points prior to PS flipping.

35. Cells become propidium iodide positive whether they have gone through apoptosis or necrosis. Hence, this method does not allow an unequivocal determination of the mechanism of cell death. Apoptotic cells become propidium iodide positive once they transition to secondary necrosis and hence there is a temporal delay in detecting cell death.

36. Determine the maximum propidium iodide fluorescence settings before the start of the experiment by treating macrophages with propidium iodide containing medium with adding 0.5 M H_2O_2 (handle H_2O_2 by wearing gloves) in CIM containing propidium iodide. Macrophages will die within 10 min yielding propidium iodide stained nuclei.

Acknowledgements

We thank Sergio Grinstein at the Hospital for Sick Children (Toronto, Canada) for providing the AKT-PH-GFP and 2xFYVE-CFP constructs, and Roland Wedlich-Soldner at the Max Planck Institute of Biochemistry (Munich, Germany) for the mRFP-Ruby Lifeact construct. Carol Norris at the University of Connecticut Confocal Microscopy Facility for microscopy assistance.

References

1. Oberdörster G, Maynard A, Donaldson K et al (2005) Principles for characterizing the potential human health effects from exposure to nanomaterials: elements of a screening strategy. Part Fibre Toxicol 2:8. doi:10.1186/1743-8977-2-8

2. Gilberti RM, Joshi GN, Knecht DA (2008) The phagocytosis of crystalline silica particles by macrophages. Am J Respir Cell Mol Biol 39:619–627. doi:10.1165/rcmb.2008-0046OC

3. Huang SH, Hubbs AF, Stanley CF et al (2001) Immunoglobulin responses to experimental silicosis. Toxicol Sci 59:108–117. doi:10.1093/toxsci/59.1.108

4. Gilberti RM, Knecht DA (2015) Macrophages phagocytose nonopsonized silica particles using a unique microtubule-dependent pathway. Mol Biol Cell 26:518–529. doi:10.1091/mbc.E14-08-1301

5. Vieira OV, Botelho RJ, Grinstein S (2002) Phagosome maturation: aging gracefully. Biochem J 366:689–704. doi:10.1042/BJ20020691

6. Swanson JA, Hoppe AD (2004) The coordination of signaling during Fc receptor-mediated phagocytosis. J Leukoc Biol 76:1093–1103. doi:10.1189/jlb.0804439

7. Halliwell B (2006) Phagocyte-derived reactive species: salvation or suicide? Trends Biochem Sci 31:509–515. doi:10.1016/j.tibs.2006.07.005

8. Tian W, Li XJ, Stull ND et al (2008) Fc R-stimulated activation of the NADPH oxidase: phosphoinositide-binding protein p40phox regulates NADPH oxidase activity after enzyme assembly on the phagosome. Blood 112:3867–3877. doi:10.1182/blood-2007-11-126029

9. Thibodeau MS, Giardina C, Knecht DA et al (2004) Silica-induced apoptosis in mouse alveolar macrophages is initiated by lysosomal enzyme activity. Toxicol Sci 80:34–48. doi:10.1093/toxsci/kfh121

10. Joshi GN, Goetjen AM, Knecht DA (2015) Silica particles cause NADPH oxidase independent ROS generation and transient phagolysosomal leakage. Mol Biol Cell 26:3150–3164. doi:10.1091/mbc.E15-03-0126

11. Costantini LM, Gilberti RM, Knecht DA (2011) The phagocytosis and toxicity of amorphous silica. PLoS One 6:e14647. doi:10.1371/journal.pone.0014647

12. Hornung V, Bauernfeind F, Halle A et al (2008) Silica crystals and aluminum salts activate the NALP3 inflammasome through phagosomal destabilization. Nat Immunol 9:847–856. doi:10.1038/ni.1631

13. Cassel SL, Eisenbarth SC, Iyer SS et al (2008) The Nalp3 inflammasome is essential for the development of silicosis. Proc Natl Acad Sci U S A 105:9035–9040. doi:10.1073/pnas.0803933105

14. Thibodeau M, Giardina C, Hubbard AK (2003) Silica-induced caspase activation in mouse alveolar macrophages is dependent upon mitochondrial integrity and aspartic proteolysis. Toxicol Sci 76:91–101. doi:10.1093/toxsci/kfg178

15. Joshi GN, Knecht DA (2013) Silica phagocytosis causes apoptosis and necrosis by different temporal and molecular pathways in alveolar macrophages. Apoptosis 18:271–285. doi:10.1007/s10495-012-0798-y

16. Flannagan RS, Jaumouillé V, Grinstein S (2012) The cell biology of phagocytosis. Annu Rev Pathol 7:61–98. doi:10.1146/annurev-pathol-011811-132445

17. Kalyanaraman B (2013) Teaching the basics of redox biology to medical and graduate students: Oxidants, antioxidants and disease mechanisms. Redox Biol 1:244–257. doi:10.1016/j.redox.2013.01.014

18. Fubini B, Hubbard A (2003) Reactive oxygen species (ROS) and reactive nitrogen species (RNS) generation by silica in inflammation and fibrosis. Free Radic Biol Med 34:1507–1516. doi:10.1016/S0891-5849(03)00149-7

19. Persson HL (2005) Iron-dependent lysosomal destabilization initiates silica-induced apoptosis in murine macrophages. Toxicol Lett 159:124–133. doi:10.1016/j.toxlet.2005.05.002

20. VanderVen BC, Yates RM, Russell DG (2009) Intraphagosomal measurement of the magnitude and duration of the oxidative burst. Traffic 10:372–378. doi:10.1111/j.1600-0854.2009.00877.x

21. Boya P, Kroemer G (2008) Lysosomal membrane permeabilization in cell death. Oncogene 27:6434–6451. doi:10.1038/onc.2008.310

22. Repnik U, Turk B (2010) Lysosomal-mitochondrial cross-talk during cell death. Mitochondrion 10:662–669. doi:10.1016/j.mito.2010.07.008

23. Aits S, Jäättelä M (2013) Lysosomal cell death at a glance. J Cell Sci 126:1905–1912. doi:10.1242/jcs.091181

24. Bidère N, Lorenzo HK, Carmona S et al (2003) Cathepsin D triggers Bax activation, resulting in selective apoptosis-inducing factor (AIF) relocation in T lymphocytes entering the early commitment phase to apoptosis. J Biol Chem 278:31401–31411. doi:10.1074/jbc.M301911200

25. Segawa K, Kurata S, Yanagihashi Y et al (2014) Caspase-mediated cleavage of phospholipid flippase for apoptotic phosphatidylserine exposure. Science 344:1164–1168. doi:10.1126/science.1252809

26. Joshi GN, Knecht DA (2013) Multi-parametric analysis of cell death pathways using live-cell microscopy. Curr Protoc Toxicol 58:Unit 4.40. doi: 10.1002/0471140856.tx0440s58

27. Flannagan RS, Harrison RE, Yip CM et al (2010) Dynamic macrophage "probing" is required for the efficient capture of phagocytic targets. J Cell Biol 191:1205–1218. doi:10.1083/jcb.201007056

28. Schneider CA, Rasband WS, Eliceiri KW (2012) NIH image to ImageJ: 25 years of image analysis. Nat Methods 9:671–675. doi:10.1038/nmeth.2089

Chapter 6

Quantitative Live-Cell Fluorescence Microscopy During Phagocytosis

Stella M. Lu, Sergio Grinstein, and Gregory D. Fairn

Abstract

Phagocytosis is a receptor-mediated process whereby professional phagocytes internalize invading pathogens and apoptotic bodies into an intracellular vacuole or phagosome, leading to their degradation. During the formation and maturation of the phagosome, several lipids undergo changes and effector proteins are recruited on the nascent phagosome in a concerted manner. These highly localized, dynamic, and transient processes can only be studied by methods capable of high spatial and temporal resolution. The use of genetically encoded chimeric constructs coupled with fluorescence confocal microscopy enables the continuous, noninvasive analysis of the distribution and metabolism of lipids and effector proteins during phagocytosis. Here, we describe a method where the mouse macrophage cell line, RAW 264.7, and primary macrophages are transiently transfected with fluorescent chimeric probes to analyze and quantify phagocytosis of immunoglobulin-opsonized particles, using confocal microscopy.

Key words Phagocytosis, Fluorescent probes, Macrophage, Confocal microscopy

1 Introduction

Phagocytosis plays an essential role in both the innate and adaptive immune responses by first internalizing particles that are ≥ 0.5 µm in diameter and subsequently presenting antigens to other immune cells. Professional phagocytes, such as neutrophils and macrophages, can internalize bacteria and apoptotic bodies to resolve the infection and maintain tissue homeostasis [1]. This process is mediated by receptors that recognize intrinsic ligands on the surface of the target particles, or serum factors (opsonins) that coat the surface of unwanted materials and microbes. The engagement of these receptors induces a downstream signaling cascade that leads to extensive actin and membrane remodeling [2]. The receptor-initiated metabolism of phospholipids and recruitment of effector proteins drive pseudopod formation, leading to the internalization of particles into a membrane-bound vacuole known as the phagosome.

Roberto Botelho (ed.), *Phagocytosis and Phagosomes: Methods and Protocols*, Methods in Molecular Biology, vol. 1519,
DOI 10.1007/978-1-4939-6581-6_6, © Springer Science+Business Media New York 2017

Phagocytosis can be subdivided into two phases, phagosome formation and maturation. Following formation, the nascent phagosome undergoes maturation via fusion and fission with endosomes, and later lysosomes, to create a microbicidal environment that can degrade the sequestered prey [3]. Both of these processes involve complex signaling cascades that combine to coordinate the uptake and degradation of the phagosomal contents. Phagosome formation is initiated when cells recognize specific ligands on the surface of particles that mark them for elimination; immunoglobulins bound to surface antigens are one such ligand. The interaction between immunoglobulins and their cognate receptors, FcγR, has been studied in most detail. Clustering of FcγR on the surface of phagocytes induces the activation of Src-family kinases, which phosphorylate immunoreceptor tyrosine-based activation motifs (ITAMs) on the cytosolic tail of the receptor [4]. The resulting phospho-ITAMs then recruit enzymes such as Syk tyrosine kinase that further activates downstream enzymes such as phosphatidylinositol 3-kinase (PI3K) and phospholipase Cγ (PLCγ), leading to dynamic and localized changes in phosphoinositide metabolism [5].

Phosphoinositide dynamics are critical not only to maintain the basal actin cytoskeleton but also to promote the stimulation of actin assembly used to drive pseudopods around the particle, and the disassembly of actin that is required for completion of the internalization process. Actin dynamics are controlled partly by phosphatidylinositol-4,5-*bis*phosphate (PI(4, 5)P$_2$), which undergoes biphasic changes during phagocytosis [6]. PI(4, 5)P$_2$ initially increases modestly at sites of phagocytosis, then disappears as the phagosome seals. The loss of PI(4, 5)P$_2$ is mediated in part by the conversion to PI(3–5)P$_3$ [5, 7] and by the recruitment of PLCγ which hydrolyzes the phosphoinositide generating diacylglycerol (DAG) and inositol *tris*phosphate (IP$_3$) [6]. Once the phagosome undergoes closure and scission from the membrane, actin and PI(4, 5)P$_2$ are lost from its membrane [8].

As phagosome maturation begins, phosphatidylinositol 3-phosphate (PI3P), Rab5, and early endosomal antigen 1 (EEA1) are acquired. Subsequently, after fusion with late endosomes, late phagosomes acquire Rab7 and lysosomal-associated membrane protein 1 (LAMP1) [9, 10].

Phagocytosis is a localized, transient, and asynchronous process; for these reasons, it cannot be studied in sufficient detail using most classical biochemical approaches, which are population-based, end point measurements. Ideally, it should be analyzed using continuous, noninvasive measurements with high spatial resolution. This can be accomplished using genetically encoded fluorescent proteins, which have revolutionized the way information can be obtained in live cells. This approach uses specific probes attached to a fluorescent protein (Fig. 1) such as green, red, cyan, or yellow fluorescent protein (GFP, RFP, CFP, YFP, respectively).

A

B

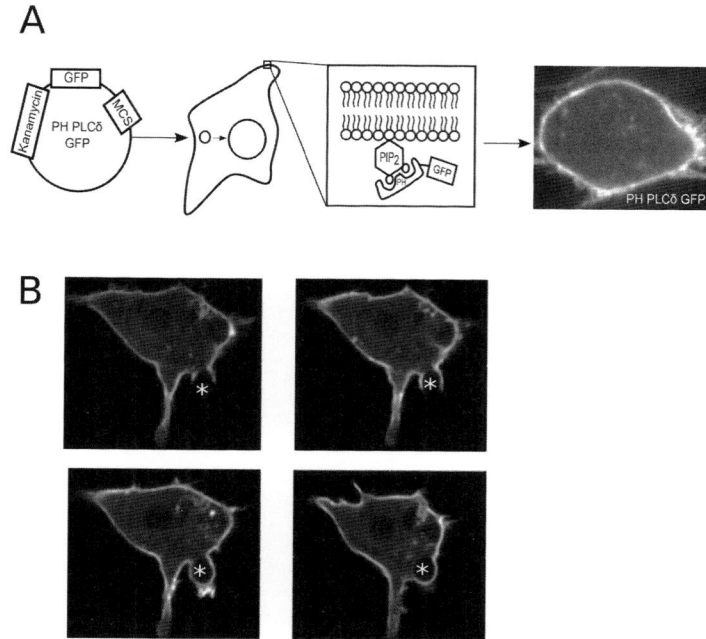

Fig. 1 Genetically encoded chimeric proteins can be transiently transfected into cells through electroporation or lipofection transfection. Once the plasmid is introduced into the cell and expressed, the recognition domain, e.g. the PH of PLCδ, will bind to its target, PI(4, 5)P$_2$. The fluorescent protein attached to the recognition domain will allow for the visualization of the target lipid (**a**). Murine macrophages were transiently transfected with Lifeact-GFP and the distribution of polymerized actin during phagocytosis of an IgG opsonized particle is shown (**b**)

Single-cell microscopy and digital imaging can be used to quantify alterations in the degree of association of a given probe with the phagosome. The distinct spectral properties of the individual fluorescent proteins allow for the study of several probes simultaneously. Several proteins, lipids, and small molecules are known to associate with the phagosome during its formation and maturation, and with the use of these chimeric constructs one can monitor their spatial and temporal distribution in real time during phagocytosis [11]. Processes such as phagosome sealing, membrane remodeling, and cytoskeletal dynamics can be monitored [10, 12, 13] using laser scanning or spinning disc confocal microscopy or even epifluorescence microscopy. Some of the probes and fusion proteins commonly used to monitor the progression of phagocytosis are listed in Table 1.

Using transient or stable transfection, constructs of interest can be expressed in the cells of choice. We routinely transfect the mouse macrophage cell line RAW 264.7 cells using lipofection, and primary human or mouse macrophages using electroporation.

Table 1
Probes that can be used in live-cell imaging to monitor lipids during phagocytosis

Ligand	Binding domain	Protein source	Reference
PI3P	FYVE	EEA1	[14]
		Hrs	[15]
PI4P	PH	OSH2	[16]
		FAPP1	[17]
	P4M	SidM from *Legionella pneumophila*	[18]
PI(4, 5)P$_2$	PH	PLCδ	[19]
PI(3–5)P$_3$	PH	Btk	[20]
		Gab2	[21]
PI(3, 4)P$_2$/PI(3–5)P$_3$	PH	Akt	[22]
PS	C2	Lactadherin	[23]
DAG	C1	Protein Kinase D	[24]
	C1	Protein Kinase Cδ	[25]
Cholesterol	Recombinant D4	Perfringolysin O toxin	[26]

Cell lines are commonly used in transfection studies, as they are readily available and more easily transfected than primary cells. Conversely, primary cells are more reactive and biologically relevant, but are more difficult to transfect. Here, we describe the use of genetically encoded chimeric constructs in combination with confocal fluorescence microscopy to analyze and quantify events during phagocytosis.

2   Materials

2.1   Cell Culture

1. RAW 264.7 murine macrophage cell line.

2. RPMI 1640 medium with l-glutamine and sodium bicarbonate supplemented with 10% heat-inactivated fetal bovine serum (FBS).

3. Phosphate-Buffered Saline (PBS) without calcium and magnesium: 0.144 g/L KH_2PO_4, 9 g/L, 0.795 g/L Na_2HPO_4, pH 7.4.

4. 1.7 cm cell scrapers.

5. 15 mL conical tubes.

6. T-25 tissue culture flasks.

7. 12 Well tissue culture plates.

8. 18 mm round glass coverslips.

9. 37 °C incubator with 5% CO_2.

2.2 Primary Cell Culture

1. Heparin.

2. Density gradient separation medium for isolating lymphocytes and monocytes from human peripheral blood such as Lympholyte-H (Cedarlane) or Ficoll Paque (GE Healthcare).

3. PBS.

4. Pasteur pipettes.

5. Antibiotic/antimycotic: 10,000 IU/mL penicillin sodium salt, 10,000 IU/mL streptomycin sulfate, 25 μg/mL amphotericin B, pH 7.9.

6. 10 cm tissue culture dishes.

7. 50 mL conical tubes.

8. 10 μg/mL Granulocyte Macrophage Colony-Stimulating factor (GM-CSF) in PBS and 0.1% BSA, aliquot and store at −80 °C.

9. 10 μg/mL Macrophage Colony-Stimulating Factor (M-CSF) in PBS and 0.1% BSA, aliquot and store at −80 °C.

10. 0.1 mg/mL Lipopolysaccharide (LPS) in PBS, aliquot and store at −80 °C.

11. 10 μg/mL Interferon gamma (INF-γ) in PBS and 0.1% BSA, aliquot and store at −80 °C.

12. 10 μg/mL Interleukin 4 (IL-4) in PBS and 0.1% BSA, aliquot and store at −80 °C.

2.3 Lipofection-Mediated Transient Transfection

1. Transfection reagent such as Fugene HD or Lipofectamine 3000. Our lab routinely uses Fugene HD for transient transfection of RAW macrophages as this gives a higher transfection efficiency and is less toxic to cells.

2. Dulbecco's Modified Eagle's Medium (DMEM) with sodium pyruvate, sodium bicarbonate, glucose, and l-glutamine supplemented with 10% heat-inactivated FBS.

3. Serum-free DMEM.

4. DNA plasmids isolated from *E. coli* transformed with the plasmid of interest and using an endotoxin-free maxiprep kit.

2.4 Electroporation-Mediated Transient Transfection

1. Electroporation system such as Neon (ThermoFisher Scientific), Nucleofector (Lonza) or Gene Pulse Xcell (Bio-Rad). Our lab routinely uses the Neon transfection system for electroporation of primary human macrophages and macrophage cell lines. This system uses an electronic pipette as the transfection chamber and is compatible with several cell types giving a high transfection efficiency and viability of cells. Using the Neon transfection system, the tips, pipettes, tubes, and pipette station equipment are required in addition to the Resuspension Buffer R and Electrolytic Buffer E2 reagent.

2. 0.4% Trypan Blue Solution in 0.81% sodium chloride and 0.06% potassium phosphate.

3. Accutase Enzyme Cell Detachment Solution (BD Biosciences) in Dulbecco's PBS containing 0.5 mM EDTA and phenol red.

2.5 Phagocytosis Assay

1. 3.87 μm polystyrene beads with 2% (v/v) divinylbenzene (DVB).

2. 50 mg/mL human IgG in PBS; aliquot and store at –20 °C.

3. Fluorophore conjugated anti-human IgG.

4. PBS.

5. Hank's Buffered Salt Solution with calcium and magnesium without phenol red.

2.6 Microscopy

1. A spinning-disc confocal microscope with a 63x, 1.4 NA objective, light source, filter set that is appropriate for the fluorescent protein of choice, objective heater, and heated stage. The setup in our laboratory consists of a spinning-disc microscopy system with an Axiovert 200 M microscope (Zeiss) equipped with diode-pumped solid-state lasers emitting at 440, 491, 561, 638, 655 nm (Spectral Applied Research) and a motorized X-Y stage (API). Images are captured using a back-thinned cooled EM-CCD camera (Hamamatsu).

2. Live cell-imaging chamber for 18 mm coverslips. Our lab uses the Chamlide CMB magnetic chamber (Quorum Technologies) for live cell imaging. The chamber holds a single round coverslip and has a silicon O-ring and magnetic body that prevents the leaking of the medium. Alternative suppliers of coverslip holders/chambers include Bioscience Tools and ThermoFisher Scientific.

3. Image analysis software such as Volocity (Perkin Elmer) or ImageJ.

3 Methods

3.1 Cell Culture

1. RAW 264.7 macrophage cells are grown in tissue culture flasks in RPMI supplemented with 10% FBS at 37 °C under 5% CO_2. To passage cells, aspirate medium and wash cells with 10 mL sterile prewarmed PBS (see **Note 1**).

2. Aspirate PBS and gently scrape cells with a sterile cell scraper (see **Note 2**). Add 10 mL sterile prewarmed RPMI + 10% FBS and pipette gently up and down to suspend the cells into a homogenous suspension.

3. Transfer cells into a 15 mL conical tube and spin down at $500 \times g$ for 5 min.

4. Gently aspirate the medium without disturbing the pellet.

5. Resuspend the pellet in RPMI + 10 % FBS. Pipette the cell suspension up and down to break up cell clumps.

6. To passage cells, add the resuspended cells into a new sterile tissue culture flask with RPMI + 10 % FBS to achieve the desired dilution. A 1 in 10 dilution will need to be split every 2–3 days.

7. To seed tissue culture plates for phagocytosis assays, add a few drops of the cell suspension from **step 5** to a 12-well tissue culture plate containing sterile 18 mm round glass coverslips and RPMI + 10 % FBS (*see* **Note 3**).

8. Incubate the plate or flasks in a tissue culture incubator at 37 °C under 5 % CO_2.

3.2 Primary Cell Isolation and Cell Culture

As an alternative to RAW cells, one can use primary human macrophages as follows:

1. Collect blood from healthy donors in a 30 mL syringe with 150 µL heparin (*see* **Note 4**).

2. Isolate human peripheral blood mononuclear cells from the blood using a density gradient separation medium such as Lympholyte-H or Ficoll Paque. Dilute the blood with an equal volume of PBS.

3. Overlay 6 mL diluted blood onto 3 mL of Lympholyte-H. Lympholyte-H has a greater density than the diluted blood, so two distinct layers should form.

4. Centrifuge for 20 min at $800 \times g$ without imposed deceleration, to prevent turbulence and mixing of layers.

5. A layer of cells will have accumulated between PBS and Lympholyte-H after centrifugation. Carefully transfer this layer into a new tube using sterile Pasteur pipettes.

6. Wash cells with PBS, by adding PBS and spinning the tubes at $800 \times g$ for 10 min without imposed deceleration.

7. Aspirate PBS and resuspend the cell pellet in RPMI + 10 % FBS + 1 % antibiotic/antimycotic. Plate the monocytes onto sterile tissue culture dishes and let the monocytes adhere for 1 h in a tissue culture incubator at 37 °C under 5 % CO_2.

8. After cells have adhered, change the media and culture the monocytes for 5 days in RPMI + 10 % FBS + 1 % antibiotic/antimycotic and 25 ng/mL GM-CSF for M1 macrophages or 25 ng/mL M-CSF for M2 macrophages, changing the medium every 2 days. After 5 days, treat macrophages with 500 ng/mL LPS and 10 ng/mL IFN-γ for M1 macrophages, or 25 ng/mL IL-4 for M2 macrophages, in addition to the respective growth factors for another 2 days (*see* **Note 5**).

3.3 Lipofection of RAW Macrophages

1. Transfection of cells is performed using Fugene HD. One μg of plasmid DNA and 3 μL of Fugene HD are used in addition to 100 μL of sterile prewarmed serum-free DMEM per transfection. Mix serum-free DMEM, plasmid DNA, and Fugene HD in an Eppendorf tube (*see* **Notes 6–8**).

2. Let stand at room temperature for 15 min.

3. Add 50 μL Fugene HD/DNA mixture into each well of a 12-well plate in a drop-wise manner.

4. Incubate cells at 37 °C under 5 % CO_2 overnight.

3.4 Electroporation of RAW Macrophages

1. Wash the cells with 10 mL PBS and gently scrape with a sterile cell scraper. Aliquot the cell suspension into an Eppendorf tube and add trypan blue to count the number of cells (*see* **Note 9**).

2. Aliquot 1×10^6 cells into a 15 mL conical tube and spin down the cells at $500 \times g$ for 5 min.

3. Resuspend the cell pellet in 100 μL Resuspension Buffer R and add 5 μg DNA (*see* **Note 10**).

4. Add 3 mL Electrolytic Buffer into the Neon Tube and place it into the Neon Pipette Station.

5. Insert the Neon Tip into the Neon Pipette and take up 100 μL of the resuspended cells (*see* **Note 11**).

6. Set up the electroporation parameters (1680 V, 20 ms pulse width, 1 pulse).

7. Start electroporation and aliquot 100 μL of the cell suspension onto 18 mm round glass coverslips in a 12-well plate with medium without antibiotics.

3.5 Electroporation of Primary Macrophages

1. Wash the dish of cells three times with PBS and add 3 mL Accutase Cell Detachment Solution. Incubate cells at 37 °C for 20 min (*see* **Note 12**).

2. Add 3 mL of HBSS to the dish and mix.

3. Spin down the cells at $500 \times g$ for 10 min.

4. Resuspend the pellet in Resuspension Buffer R and add 5 ng of plasmid cDNA.

5. Add 3 mL of Electrolytic Buffer into the Neon Tube and place it into the Neon Pipette Station.

6. Insert the Neon Tip into the Neon Pipette and take up 100 μL of the resuspended cells.

7. Set up the electroporation parameters. For M1 macrophages: 1475 V, 20 ms pulse width, 1 pulse. For M2 macrophages: 1100 V, 30 ms pulse width, 2 pulses.

8. Start electroporation and aliquot 100 μL of the cell suspension onto 18 mm round glass coverslips in a 12-well plate with media without antibiotics.

9. Incubate cells at 37 °C for 1 h and add cytokines to activate cells. For M1 macrophages: 500 ng/mL LPS, 10 ng/mL recombinant human IFN-γ. For M2 macrophages: 25 ng/mL IL-4.

3.6 Preparation of Phagocytic Targets

1. Mix 400 μL of PBS, 15 μL of human IgG and 30 μL of the 3.87 μm polystyrene bead slurry in an Eppendorf tube (*see* **Note 13**).

2. Incubate at 37 °C shaking for 1 h.

3. Wash the opsonized beads two times with PBS by spinning them down at $15,000 \times g$ for 30 s, aspirating the supernatant and resuspending in 400 μL PBS.

4. Add 4 μL DyLight 405-conjugated anti-human IgG to the opsonized beads and incubate for 15 min on a rotator at room temperature. Any fluorescently conjugated anti-human IgG antibody can be used to label and visualize the opsonized beads.

5. Wash the labeled beads with PBS and resuspend in 400 μL PBS; *see* **step 3**.

6. 12 μL of the opsonized beads are added per well of a 12-well plate.

3.7 Phagocytosis Assay and Microscopy

1. Turn on the spinning disc confocal microscope, light source, and lasers for the correct wavelength, and the heater for the heated stage and objective. Allow time for the heated stage to reach 37 °C.

2. Aspirate the medium in the wells after the overnight transfection and replace it with DMEM without serum.

3. Using forceps, transfer one coverslip into the base of the magnetic Chamlide chamber (Quorum technologies) and replace the top portion. Pipette 700 μL of HBSS into the chamber (*see* **Notes 14** and **15**).

4. Add 12 μL of opsonized and labeled polystyrene beads to the imaging chamber and force the beads onto cells by spinning in a table-top centrifuge at $300 \times g$ for 1 min. Alternatively, the beads can be parachuted onto the cell monolayer to initiate phagocytosis.

5. Place the imaging chamber on the heated stage and find the desired focal plane. For time-lapse imaging, the focal plane is usually set in the middle of the cell. To acquire Z-stacks set the bottom and top of the cell and set a distance of 0.4–0.5 μm between optical slices.

6. Set up the acquisition parameters for the experiment using Volocity software (*see* **Notes 16–18**).

3.8 Image Analysis

1. Open the images using an image analysis software program such as ImageJ or Volocity.

2. Select a region of interest (ROI) and measure the fluorescence intensity (mean fluorescence intensity per pixel). For example, the ROI can be a region at the plasma membrane where a particle is engaged or a region of the phagosomal membrane.

3. Select a region outside the cell that exhibits no fluorescence and subtract this background reading from the experimental ROI.

4. Select a region where the fluorescence intensity stays constant and measure the fluorescence intensity, this value can be used to normalize the experimental ROI.

5. To normalize the fluorescence intensity of probes that partition between the membrane and cytosol, determine the ratio of the fluorescence intensities of the experimental ROI to a cytosolic ROI by using the calculation: (plasma membrane-cytosol)/cytosol. The ratios obtained from multiple cells can be compiled for statistical analysis. Typically, we analyze 15–20 cells per experimental condition from 3 to 5 separate experiments.

4   Notes

1. RAW 264.7 macrophages are ready to split when they reach 80-90% confluence. A 1 in 10 dilution of cells will grow for about 2–3 days before they need to be passaged again.

2. An alternative approach to passaging cells using a centrifuge is to resuspend the cells in sterile prewarmed RPMI + 10% FBS and to dilute them directly into new sterile tissue culture flasks.

3. Allow cells to adhere overnight onto the coverslip before transfection. Cells should reach 50–60% confluence before transfection. Do not let cells reach high (80–90%) confluence before transfection.

4. Human blood is potentially hazardous biological agent and isolated cells can be a reservoir of infectious agents. Donor consent and any Institutional specific procedures need to be followed when using primary human samples.

5. Macrophages are tissue resident specific cells that differentiate from monocyte precursors when they are recruited to sites of infection or tissue damage. Monocytes can differentiate into the classically activated macrophages, designated as M1, or the alternatively activated macrophages, designed at M2. In general, M1 macrophages are pro-inflammatory and fight infection and induce tissue damage while M2 macrophages are anti-inflammatory, that resolve inflammation and promote tissue repair.

6. Changing the medium to DMEM + 10% FBS before transfecting cells increases the transfection efficiency. Aspirate the medium and replace with sterile prewarmed DMEM + 10% FBS.

7. Mix the components for the transfection in the same order (serum free DMEM, plasmid DNA, Fugene HD) as listed. Add Fugene HD directly into the medium and prevent it from touching the sides of the Eppendorf tube.

8. When transfecting multiple plasmids, split the amount of each plasmid DNA, so the total plasmid DNA added is 1 μg per 2 wells of a 12-well plate.

9. Using a hemocytometer and trypan blue, count the number of viable cells. 1×10^6 cells are required per transfection.

10. Resuspend pellet in 100 μL Resuspension Buffer R per transfection.

11. Avoid air bubbles during pipetting of the cell suspension into the pipette tip used for electroporation, it will cause a spark and interfere with the electroporation.

12. Electroporate macrophages the same day as activation with cytokines (Day 5).

13. Polystyrene beads come in varying sizes from 0.5 to 25 μm. Using polystyrene beads containing 2 % divinylbenzene opsonizes better with IgG.

14. Ensure that the coverslip is tightly sealed in the imaging chamber. Add HBSS to the imaging chamber with the coverslip to ensure that no liquid is leaking out before adding the particles for phagocytosis. An alternative to the Chamlide CMB imaging chamber is the Attofluor cell chamber (ThermoFisher Scientific) or the coverslip chamber.

15. It is very important to regulate environmental factors such as temperature throughout the imaging process to ensure that cells are growing and functioning in their normal conditions. Using HBSS or a HEPES buffered solution will help maintain the pH of the medium during imaging. Medium without phenol red will help reduce background noise and improve the signal-to-noise ratio.

16. When setting up the acquisition parameters, the exposure, intensity and laser power should be set as to minimize photobleaching and phototoxicity. For time-lapse imaging of phagosomal maturation, acquiring images every 15–20 s for 20–30 min is ideal.

17. Fluorescent proteins with different spectral properties are available allowing for multiple acquisitions across the various wavelengths. For example, yellow, green, red, and cyan fluorescent proteins are commonly used fluorophores. Having several color options available allows for the detection of several proteins or lipids simultaneously.

18. Caution must be exercised when using heterologously expressed fluorescent constructs. Autofluorescence or background fluorescence can contaminate the measurements. Moreover, the probes

may interfere with the normal biological activity of their cognate ligand, particularly when expressed in excess. Overexpression of the fluorescent protein may scavenge a significant fraction of the ligand it is intended to sense, interfering with its biological function, e.g. signaling. All of these caveats must be kept in mind when analyzing data and drawing conclusions.

Acknowledgements

S.M.L. is a recipient of the Alexander Graham Bell Scholarship from the Natural Science and Engineering Research Council of Canada. G.D.F. is a recipient of a New Investigator Award from Canadian Institutes of Health Research (CIHR) and an Early Researcher Award from the Government of Ontario. Original work in the authors' laboratories is supported by a St. Michael's Hospital New Investigator Start-up Fund (to G.D.F), and by grants FDN-143202 and MOP-126069, and MOP-133656 awarded by the Canadian Institutes of Health Research (to S.G. and G.D.F., respectively).

References

1. Hochreiter-Hufford A, Ravichandran KS (2013) Clearing the dead: apoptotic cell sensing, recognition, engulfment, and digestion. Cold Spring Harb Perspect Biol 5:a008748

2. Niedergang F, Chavrier P (2004) Signaling and membrane dynamics during phagocytosis: many roads lead to the phagos(R)ome. Curr Opin Cell Biol 16:422–428

3. Viera OV, Botelho RJ, Grinstein S (2002) Phagosome maturation: aging gracefully. Biochem J 15:689–704

4. Crowley MT, Costello PS, Fitzer-Attas CJ et al (1997) A critical role for Syk in signal transduction and phagocytosis mediated by Fcgamma receptors on macrophages. J Exp Med 186:1027–1039

5. Marshall JG, Booth JW, Stambolic V et al (2001) Restricted accumulation of phosphatidylinositol 3-kinase products in a plasmalemmal subdomain during Fc gamma receptor-mediated phagocytosis. J Cell Biol 153:1369–1380

6. Botelho RJ, Teruel M, Dierckman R et al (2000) Localized biphasic changes in phosphatidylinositol-4,5-bisphosphate at sites of phagocytosis. J Cell Biol 151:1353–1368

7. Vieira OV, Botelho RJ, Rameh L et al (2001) Distinct roles of class I and class III phosphatidylinositol 3-kinases in phagosome formation and maturation. J Cell Biol 155:19–25

8. Scott CC, Dobson W, Botelho RJ et al (2005) Phosphatidylinositol-4,5, bisphosphate hydrolysis directs actin remodeling during phagocytosis. J Cell Biol 169:139–149

9. Ellson CD, Anderson KE, Morgan G et al (2001) Phosphatidylinositol 3-phosphate is generated in phagosomal membranes. Curr Biol 11:1631–1635

10. Vieira OV, Bucci C, Harrison RE et al (2003) Modulation of Rab5 and Rab7 recruitment to phagosomes by phosphatidylinositol 3-kinase. Mol Cell Biol 23:2501–2514

11. Swanson JA, Hoppe AD (2004) The coordination of signaling during Fc receptor-mediated phagocytosis. J Leukoc Biol 76:1093–1103

12. Lee WL, Mason D, Schreiber AD et al (2007) Quantitative analysis of membrane remodeling at the phagocytic cup. Mol Biol Cell 8:2883–2892

13. Riedl J, Creveenna AH, Kessenbrock K et al (2008) Lifeact: a versatile marker to visualize F-actin. Nat Methods 5:605–607

14. Gillooly DJ, Morrow IC, Lindsay M et al (2000) Localization of phosphatidylinositol 3-phosphate in yeast and mammalian cells. EMBO J 19:4577–4588

15. Raiborg C, Bremnes B, Mehlum A et al (2001) FYVE and coiled-coil domains determine the specific localisation of Hrs to early endosomes. J Cell Sci 114:2255–2263

16. Roy A, Levine TP (2004) Multiple pools of phosphatidylinositol 4-phosphate detected using the pleckstrin homology domain of Osh2p. J Biol Chem 279:44683–44689

17. Godi A, Di Campli A, Konstantakopoulos A et al (2004) FAPPs control Golgi-to-cell surface membrane traffic by binding to ARF and PtdIns(4)P. Nat Cell Biol 6:393–404

18. Hammond GR, Machner MP, Balla T (2014) A novel probe for phosphatidylinositol 4-phosphate reveals multiple pools beyond the Golgi. J Cell Biol 205:113–126

19. Stauffer TP, Ahn S, Meyer T (1998) Receptor-induced transient reduction in plasma membrane PtdIns(4,5)P2 concentration monitored in living cells. Curr Biol 8:343–346

20. Varnai P, Rother K I, Balla T (1999) Phosphatidylinositol 3-kinasedependent membrane association of the Bruton's tyrosine kinase pleckstrin homology domain visualized in single living cells. J Biol Chem 274:10983-10989

21. Gu H, Botelho RJ, Yu M et al (2003) Critical role for scaffolding adapter Gab2 in Fc gamma R-mediated phagocytosis. J Cell Biol 161:1151–1161

22. Servant G, Weiner OD, Herzmark P et al (2000) Polarization of chemoattractant receptor signaling during neutrophil chemotaxis. Science 287:1037–1040

23. Yeung T, Gilbert GE, Shi J et al (2008) Membrane phosphatidylserine regulates surface charge and protein localization. Science 319:210–213

24. Chen J, Deng F, Wang QJ (2008) Selective binding of phorbol esters and diacylglycerol by individual C1 domains of the PKD family. Biochem J 411:333–342

25. Colon-Gonzalez F, Kazanietz MG (2006) C1 domains exposed: from diacylglycerol binding to protein-protein interactions. Biochim Biophys Acta 1761:827–837

26. Ohno-Iwashita Y, Shimada Y, Waheed AA et al (2004) Perfringolysin O, a cholesterol-binding cytolysin, as a probe for lipids rafts. Anaerobe 10:125–134

Chapter 7

Intracellular Manipulation of Phagosomal Transport and Maturation Using Magnetic Tweezers

Shashank Shekhar, Vinod Subramaniam, and Johannes S. Kanger

Abstract

Phagocytosis is an important process of the immune system by which pathogens are internalized and eliminated by phagocytic cells. Upon internalization, the phagosome matures and acidifies while being transported in a centripetal fashion. In this chapter, we describe protocols for simultaneous imaging of phagosomal acidification as well as their spatial manipulation by magnetic tweezers. First, we describe the protocols for functionalization of magnetic microbeads with pH-sensitive dyes and pH calibration of these particles. We also describe the preparation of magnetic tweezers and the calibration of forces that can be generated by these tweezers. We provide details of the design of the custom electrical and optical setup used for simultaneous imaging of phagosomal pH and phagosome's location. Finally, we provide a detailed description of the data analysis methodology.

Key words Phagocytosis, Magnetic tweezers, Intracellular manipulation, Acidification, pH-sensitive fluorescent dyes, Quantitative biology

1 Introduction

Phagocytosis is a process by which immune cells like neutrophils and macrophages capture and eliminate pathogens [1, 2]. Via this process, phagocytic cells internalize pathogens in an organelle called the phagosome. The pathogen-containing phagosome then undergoes phagosomal maturation. During the maturation process, both the phagosome's membrane and its lumen evolve as a result of intricately controlled membrane fusion and fission events with lysosomes. Phagosome maturation is marked by a gradual acidification of the phagosomal lumen as a result of the V-ATPase proton pumps in the phagosomal membrane. During a few minutes, the phagosomal pH drops from near neutral to a pH of about 4.5–5.0 over the duration of a few minutes [1].

During the maturation process, the phagosome gets attached to the cytoskeletal network and is transported in a centripetal fashion from the cell periphery to the perinuclear region of the cell [3, 4].

Roberto Botelho (ed.), *Phagocytosis and Phagosomes: Methods and Protocols*, Methods in Molecular Biology, vol. 1519, DOI 10.1007/978-1-4939-6581-6_7, © Springer Science+Business Media New York 2017

Although phagosomes have been demonstrated to exhibit bidirectional motion consisting of both dynein-mediated centripetal motion (towards the minus end of microtubules) and kinesin-mediated centrifugal motion (towards the plus-end of microtubules), the centripetal motion however appears to be preferred [5]. While the centripetal motion and localization is a well-established phenomenon, the reason behind this transport and its relationship to phagosomal maturation remains poorly understood. Centripetal transport has mainly been linked to the high concentration of lysosomes in the perinuclear region. A number of methods have been utilized to visualize lysosome localization including their direct visualization using GFP-tagged Lamp1 and Lamp2 proteins [6]. Lysosomes have also been labeled using acidotropic fluorescent probes like LysoSensor [7] and LysoTracker [8]. All these techniques have demonstrated that although the lysosomes are distributed throughout the cell, they are present in substantially higher numbers in the nuclear periphery. This has led to the expectation that phagosomes travel to the perinuclear region to promote phagosome-lysosome fusion [6], which in turn leads phagosomal maturation and eventual pathogen elimination.

To examine this reasoning, a number of approaches have been used for manipulating phagosomal location to study the relationship between transport and maturation. Most studies either involve disrupting the cytoskeleton–actin and microtubular tracks on which the phagosomes are transported—using drugs like cytochalasin and nocodazole or by disrupting various proteins that are part of the machinery involved in the transport process e.g. molecular motors [9, 10]. Although these drug-based methods inhibit centripetal transport, biochemical intervention also affects a myriad of other cellular processes that require an intact cytoskeleton. For example while microtubular disruption by nocodazole prevents phagosomal transport, it also disrupts all long-range intracellular transport and affects basic cellular processes like cell-division. It is therefore imperative to design an experiment that allows manipulation of phagosome location with minimal side effects on other cellular processes.

We addressed this problem using a combination of magnetic tweezers and magnetic microbead-containing phagosomes [11]. While phagosomal maturation is followed by measuring the pH of the acidifying phagosomal compartment using pH-sensing magnetic microbeads, magnetic tweezers are used for spatially manipulating microbead-containing phagosomes.

Phagosomal pH has been traditionally measured using ratiometric approaches. This approach requires a combination of two fluorescent dyes, one of which is pH-sensitive while the other is pH-insensitive. The ratio of intensities of these two fluorophores is a measure of pH. While Oregon Green and Fluorescein are the two commonly used pH-sensitive dyes, Texas Red and

Carboxyltetramethyl Rhodamine are the common pH-insensitive dyes used in ratiometric studies [12]. In this study however, SNARF-4F fluorescent dye has been used. The emission spectrum of this dye consists of two pH-dependent peaks which exhibit pH-dependent changes in intensity. The ratio of these two intensity peaks is a measure of pH. This eliminates the requirement of a second pH-insensitive dye. In this study, we delivered the pH-sensing dye to the phagosomes by attaching the dye on microbeads and allowing the functionalized microbeads to be phagocytosed [12]. A wide variety of microbeads can be used for studying phagocytosis. The majority of studies so far have used polystyrene/latex microbeads [13]. The main advantage of these microbeads is that they are by themselves chemically inert and they can easily be functionalized with surface-reactive chemical groups which can be used for immobilizing other ligands through cross-linking reactions. A pH-sensing fluorophore can thus be attached to the microbeads. The microbeads' size and surface chemical reactivity can also be easily varied depending on the phagocytic assay of interest. In case of quantitative imaging assays, the microbead autofluorescence should also be considered (*see* **Note 1**).

Additionally, given the requirement for magnetic manipulation, microbeads also had to be magnetic in nature. Instead of latex microbeads, superparamagnetic microbeads were therefore used as they get magnetized (i.e. exhibit magnetic moment) only in the presence of an external magnetic field. Once magnetized, they experience a magnetic force when placed in an external field gradient and can therefore be physically manipulated (*see* **Note 2**). In addition to spatial manipulation, the translational and rotational movement of the magnetic particle in a magnetic field can also be used to study micromechanical properties of phagocytosis [11, 14].

In order to spatially manipulate magnetic microparticles in the intracellular milieu, a single-pole magnetic tweezers setup was developed. Magnetic tweezers have an important advantage over other approaches: the magnetic tweezers are very selective to magnetic materials only and as a result the force is selectively applied only on the organelle containing the magnetic microbead (i.e. phagosomes) [15]. Magnetic tweezers are based on the principle that a magnetic microbead placed in a magnetic field gradient experiences a translational force in the direction of increasing magnetic flux. It is possible to exert forces on the order of several nN on micron-sized magnetic microbeads placed in magnetic field gradients [16]. A magnetic microbead of moment $\vec{m}$ placed in an external magnetic field of flux density $\vec{B}$ experiences a force. $\vec{F}_{mag} = \nabla(\vec{m} \cdot \vec{B})$ The amplitude of the magnetic moment $\vec{m}$ of a microbead scales with its volume ($|\vec{m}| \propto d^3$, with d being the microbead diameter). To be able to apply high enough forces on micron-sized small microbeads, it is therefore essential to generate magnetic fields with large enough gradients [17].

For generating a magnetic flux gradient, the magnetic tweezers setup consisted of three functional parts: generation of the magnetic flux, transporting the flux to the area of interest and producing the required gradient in the flux density [18]. Although magnetic flux can be generated using permanent magnets, this approach suffers from the limitation that it can neither be changed in amplitude nor be rapidly turned on and off. The only way to change or switch off the magnetic flux of a permanent magnet is to physically (re)move the magnet, which can be rather complicated and can also lead to disturbances in the experiment. For this reason, magnetic flux is generated electromagnetically by passing a current through a copper coil. Since the magnetic flux is controlled by the electric current passing through the coil, not only can the magnitude of the generated flux be varied but it can also be rapidly switched on or off by turning the current on or off. The generated magnetic flux is then transported via an iron rod with a micromachined sharp tip. The generated flux depends, among others, on the geometry of the tip.

Below, we describe the detailed protocol for preparation and calibration of pH-sensitive magnetic microbeads as well as a single-pole magnetic tweezers setup [11, 12]. We also provide details of the design of the electrical and optical setup used as well as a description of the data acquisition and analysis methodology.

2   Materials

2.1   Cell Culture

1. The experiments described here were performed with murine macrophage cell line RAW 264.7 (ATCC catalogue number: TIB-71). These cell lines were transfected with Kmyr-GFP.

2. DMEM culture medium without Phenol Red and supplemented with 10% Fetal Bovine Serum, 2 mM L-Glutamine and Pen-Strep (*see* **Note 3**).

3. Standard tissue culture flasks.

4. Chambers (petri dish) with a cover glass bottom (Cellview, Greiner Bio-One) for microscopy experiments.

5. Cell scrapers.

2.2   Bead Functionalization

1. Dynal M270 microbeads with amine surface functionalization (Thermo Fisher, USA). Store at 4 °C.

2. SNARF-4F 5-(and-6)-Carboxylic Acid Fluorescent dye (Thermo Fisher, USA). Dissolve in DMSO to 2mM concentration (final volume 1ml). Divide in 20 equal parts in microcentrifuge tubes. Let DMSO evaporate in a desiccator, store tubes at −20 °C (*see* **Note 4**).

3. Phosphate-buffered saline (PBS): 140 mM NaCl, 5 mM KCl, 8 mM NaH_2PO_4, and 2 mM KH_2PO_4, adjusted to pH 7.4 with 1 M NaOH. Store at 4 °C for up to 6 months.

4. 2-(N-morpholino) ethanesulfonic (MES) buffer: 50 mM MES in water with pH adjusted to 5.7 (Sigma Aldrich, USA). Store at 4 °C for up to 6 months once freshly prepared.

5. 1-Ethyl-3-(3-dimethylaminopropyl) carbodiimide hydrochloride (EDC) (Sigma Aldrich, USA). Store at –20 °C.

6. N-Hydroxysuccinimide (NHS) (Sigma Aldrich, USA). Store at –20 °C.

7. 30 mg/ml Human IgG (Jackson Labs, USA). Store at 4 °C for up to 6 months.

8. pH Buffers for pH calibration. These are commercially available. Alternatively, citric acid ($C_6H_8O_7$) and disodium hydrogen phosphate (Na_2HPO_4) can be mixed to prepare pH-buffers in the range pH 3–pH 8 (*see* **Note 5**). Store at 4 °C for up to 6 months.

2.3 Magnetic Tweezers

1. Soft Iron sheets for magnetic pole preparation. These should be available at any metal supplier.

2. Copper coil (5 mm × 5 mm, rectangular cross-section) with 250 turns to be wrapped around a 5 mm × 5 mm iron core.

3. Three-axis Oil Hydraulic Micromanipulator (Narishige, Japan).

4. Home-built amplifier with four PA39 power amplifiers (Apex Microtechnology Corp.).

5. Glycerol for force calibration.

6. A computer for controlling the amplifier.

2.4 Optical Setup

1. A widefield fluorescence microscope.

2. LED for fluorophore excitation (Lumileds, 520–550 nm).

3. Near infra-red (NIR) long-pass filter (>750 nm) (Chroma Technology Corp.).

4. Band-pass filters (570–615 nm, 655–740 nm) (Semrock).

5. Dichroics and mirrors (Semrock).

6. Filter-wheel with 6-positions (Thorlabs).

7. Video camera (WOTEC 902H).

8. An intensified CCD camera (Princeton Instruments Roper Scientific PentaMAX 512 FT).

9. Temperature-controlled microscope incubator (Solent Scientific, UK).

2.5 Software

1. ImageJ (National Institute of Health) for image processing.
2. MATLAB (MathWorks) for data processing.
3. LabVIEW (National Instruments) for hardware control as well as data processing.

3 Methods

3.1 pH-Sensitive Magnetic Microbeads

3.1.1 Bead Functionalization and Opsonization

Dynal M270 magnetic microbeads (*see* **Note 2**) were functionalized with SNARF-4F fluorescent dye and opsonized with IgG as follows:

1. Resuspend 20 μl of stock (30 mg/ml) Dynal M270 microbeads solution in 250 μl of MES buffer at pH 5.7.
2. Wash twice with 250 μl MES buffer to ensure removal of traces of storage buffer that the beads were supplied in. To do this, add 250 μl MES buffer to the beads and hold a permanent magnet close to the bottom of the microcentrifuge tube. Once the beads get collected at the site close to the magnet, the supernatant can be removed. Resuspend the beads in fresh buffer.
3. Remove the buffer and resuspend the beads in MES buffer supplemented with 50 mM N-Hydroxysuccinimide (NHS) (*see* **Note 6**).
4. Mix SNARF-4F dye in MES buffer to a final volume of 25 μl and final dye concentration of 4 mM in a separate microcentrifuge tube.
5. Combine the solutions from **steps 3** and **4** and mix well. Since the carboxylic group on the fluorophore is inert, no reaction occurs at this stage.
6. Add 20 mg EDC to the above bead solution to activate the carboxylic groups on the fluorophores and enable the formation of the amide bond between the fluorophores and the beads. Mix well by pipetting and then incubate for 2 h at room temperature in the dark, with gentle mixing on a rotator.
7. Wash microbeads five times using magnetic separation to remove excess dye and resuspend them in PBS buffer at pH 7.4.
8. Opsonize the dye-functionalized microbeads via nonspecific binding of human IgG by resuspending microbeads in 240 μl of PBS and adding 10 μl of antibody (1.2 mg/ml final concentration).
9. Incubate further for 1 h in the cold room (4 °C) with gentle mixing.

10. Wash microbeads three times with PBS to remove free IgG and then resuspend in 250 μl PBS. These beads can be kept at 4 °C for up to 1 week (*see* **Note 7**). Microbeads thus prepared are referred to as "M270-SNARF" microbeads in the rest of the text.

3.1.2 M270-SNARF Beads pH Calibration

The emission spectrum of SNARF-4F dye consists of two pH-dependent peaks (570–615 and 655–740 nm) which exhibit pH-dependent changes in intensity. The ratio of these two intensity peaks is used as a measure of pH. Prior to using these microbeads for measuring phagosomal pH, their pH-dependence was calibrated against buffers of known pH (*see* Fig. 1) as follows:

1. Prepare pH-buffers in the range pH 3–pH 8 (*see* **Note 5**).

2. Add 1 ml of pH 3 buffer in a petri dish with a cover glass bottom. To this, add 1 μl of SNARF-M270 beads. Allow the beads to settle at the bottom of the chamber on the coverslip.

3. Use any widefield fluorescence microscope to illuminate the beads with a light source at ~543 nm. Acquire an image each in the two pH-dependent peaks (Channel 1: 570–615 nm and Channel 2: 655–740 nm) using appropriate band-pass filters (*see* Fig. 7b, c).

4. A custom-written macro script in ImageJ can be used for further image analysis. Select one microbead and measure its integrated intensity over its entire surface in the two channels (see Image Analysis section for detailed methodology).

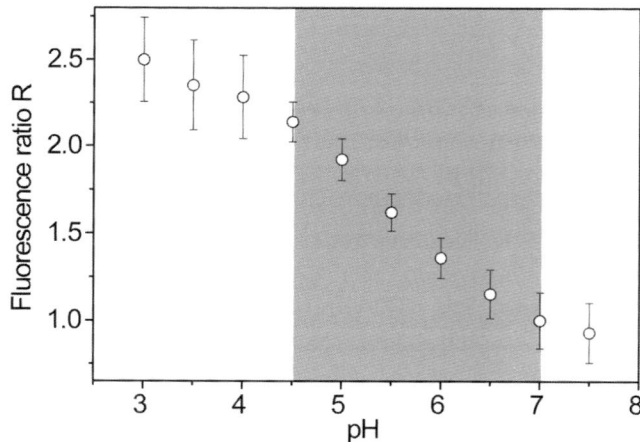

Fig. 1 Calibration of pH-sensitive magnetic beads against buffers of known pH. Mean fluorescence ratio *R* of SNARF-M270 microbeads in different pH-buffers (error bars are standard deviations). The grayed area indicates the relevant physiological pH range for phagocytosis (pH 4.5–7.0). This calibration curve can be used for ratiometric imaging of phagosomal pH. The fluorescence ratios *R* are normalized such that ratio *R*=1 at pH 7. Adapted from Shekhar et al. [12] with permission from Elsevier

5. Determine the fluorescence ratio R for the microbead by dividing the integrated intensity of the microbead in Channel 1 by the integrated intensity of the microbead in Channel 2.

6. Repeat **steps 4** and **5** for at least 50 more beads over a few fields of view.

7. Repeat **steps 2–6** for buffers with varying pH up to pH 8. Calculate the mean ratio R for each pH value.

8. Draw a calibration curve of fluorescence ratio R as a function pH (*see* Fig. 1). This calibration curve can be used to convert ratio R in to phagosomal pH.

3.2 Magnetic Tweezers: Design and Development

The magnetic tweezers used here consists of following physical parts: a magnetic pole, an electric coil to magnetize the pole and an amplifier or current generator.

3.2.1 Magnetic Tweezers Pole

A single pole magnetic tweezers consisting of an iron rod with a sharp tip was used. It was prepared as follows:

1. Cut rectangular rods with dimensions of 5 mm × 5 mm × 150 mm from commercially available soft iron plates.

2. Micro-machine one end of the rod with a lathe in the workshop to an approximate parabolic shape with a feature size of about 100 μm (*see* Fig. 2).

3. Prepare multiple tips of varying curvature in order to generate a range of magnetic field gradients (*see* **Notes 8** and **9**).

4. To allow very precise movement of the tweezers tip across the microscope field of view, attach the magnetic tweezers to a micromanipulator.

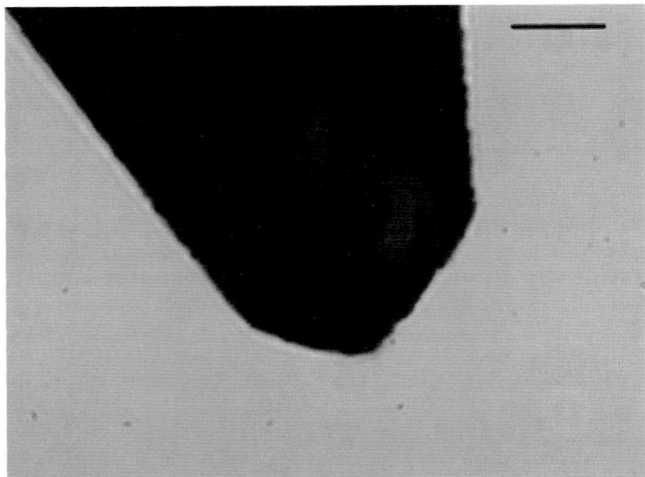

Fig. 2 Magnetic tweezers pole tip. Microscopic image of the micro-machined iron tip of the magnetic tweezers pole. Scale bar: 50 μm

3.2.2 Coils

1. Acquire or prepare a hollow copper coil ($5 \text{ mm} \times 5 \text{ mm}$) with 250 turns for generating a magnetic field. It yields a maximum of 500 A-turns at 2 A (the maximum allowed current of the amplifier used).

2. Mount the coil on the magnetic pole.

3. Current carrying coils can generate substantial heat (Joule Heating). Take appropriate steps to minimize heat generated (*see* **Note 10**).

3.2.3 Amplifier

1. A home-built amplifier can be used for driving the magnetic tweezers coil. The amplifier was built with four PA39 power amplifiers (Apex Microtechnology Corp). This amplifier can be controlled via a computer which allows very accurate control over the magnitude and nature of the current generated (step function, sine wave etc.) which is used for driving the coils.

2. Use the same amplifier also for controlling the LED illumination for fluorescence excitation.

3. Alternatively, commercially available, computer-controlled (or manually controlled), DC power sources can also be used. However, separate manual control of each part may hinder the synchronization of the entire setup via the computer.

3.3 Magnetic Tweezers Calibration

The magnetic forces can be increased or decreased by varying the driving current in the coils. The calibration is carried out via the viscous drag method as follows (*see* **Note 11**):

1. Suspend Dynal M270 magnetic microbeads in a high-viscosity liquid like Glycerol ($\eta = 1.5$ Pa s) in a chamber with a cover glass bottom. Glycerol is used since its viscosity is close to the viscosities expected in the intracellular environment.

2. Allow microbeads to settle down at the bottom of the chamber (~30 min).

3. Place the chamber with the microbeads on the microscope. The microbeads can be imaged in brightfield.

4. Dip the magnetic tweezers pole into the sample and move the tip in the field of view using the micromanipulator (*see* **Note 12**).

5. Turn on the current in the coil; microbeads should be seen to rapidly move towards the magnetic pole. Record microbeads' movement via a video camera connected to the microscope (*see* Fig. 6).

6. Track microbead movement under force over the entire video and calculate microbead velocity (see section on particle tracking).

7. Using the velocity and the known viscosity of the solution, calculate the force experienced by the microbeads from Stokes' law $\left(F_{drag} = 6\pi\eta rv \right)$ as a function of the distance from the pole tip (*see* **Note 11**).

8. Draw a force calibration curve (Fig. 3). The maximum magnetic force that can be exerted on a M270 microbead is in the range of ~1 nN at a distance of 10 μm from the pole tip and ~0.1 nN at 100 μm from the pole tip (*see* Fig. 3a). The magnetic force at a given distance from the tip can be increased up to saturation by increasing the current through the coil (*see* Fig. 3b).

3.4 Intracellular Magnetic Manipulation

For intracellular manipulation of phagosomes, the following steps were followed:

1. Incubate RAW 264.7 cells expressing Kmyr-GFP (*see* **Note 13**) overnight in a petri dish with cover glass bottom at 37 °C and 5% CO_2 atmosphere in DMEM culture medium with 10% Fetal Bovine Serum, 2 mM L-Glutamine and Pen-Strep.

2. Prior to the experiment the next morning, replace the culture medium with ice-cold DMEM medium (without FCS and l-glutamine but with 25 mM HEPES to maintain the pH) for 10 min in order to synchronize microbead internalization. Keep the cells on ice.

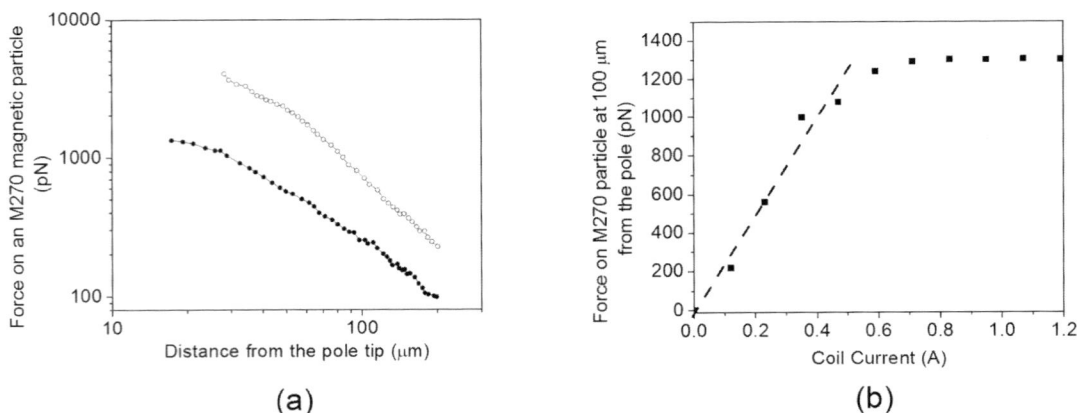

(a)   (b)

Fig. 3 Calibration of the single-pole magnetic tweezers. (**a**) Force applied on a Dynal M270 magnetic microbead as a function of distance between the pole and the microbead is shown for two different coil currents : 0.12A(*solid symbols*) and 0.23A(*open symbols*). The force experienced by the microbead increases as the distance between the microbead and the pole decreases. The force can also increases with increasing coil current (up to pole saturation). (**b**) The force applied on an M270 microbead by the magnetic pole for varying coil currents measured at a distance of 100 μm from the pole (the *dashed line* shows a linear fit for data-points prior to saturation). Reproduced from Shekhar et al. [11] with permission from Elsevier

3. Prepare a suspension of M270-SNARF microbeads in ice-cold medium (final concentration approx. 4×10^5 microbeads/ml of medium).

4. Add these microbeads to the cells while continuing to keep the cells on ice.

5. Allow beads to settle down and get attached to cells (~10 min).

6. Place the chamber on the prewarmed-microscope stage and allow the medium to warm up to 37 °C (temperature of the microscope incubator box).

7. Select a field of view with enough cells and microbeads. Also ensure that the field is not over-crowded with cells.

8. As the sample warms up, place the magnetic tweezers pole in the sample and carefully move the pole tip to the field of interest using the micromanipulator.

9. As the cells warm up, cells will start phagocytosing magnetic microbeads attached to them. Follow microbead internalization by observing the closure of the phagocytic cup using Kmyr-GFP (*see* **Note 13**) (Fig. 4).

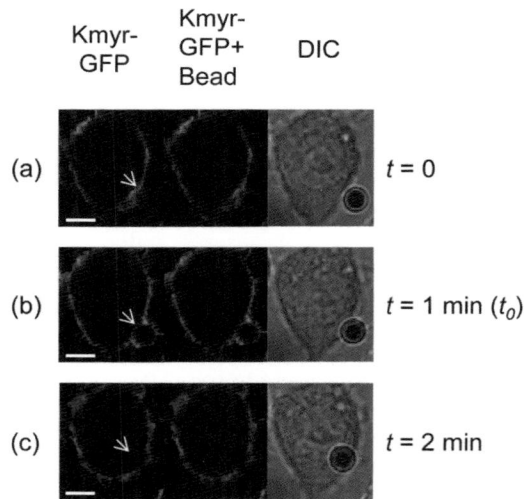

Fig. 4 Defining the time-point of microbead internalization (t_0). Time-lapse images of a RAW 264.7 cell expressing Kmyr-GFP (*green*) internalizing a M270-SNARF magnetic microbead (*red*). Successive images are a minute apart from each other. Figure shows (**a**) beginning of formation of the phagocytic cup (**b**) completion of the phagocytic cup at t_0 (**c**) the microbead moving inwards after internalization. At t_0, the Kmyr-GFP ring forms a uniform ring around the phagosome, indicating the closure of the phagocytic cup. Scale bar: 5 μm. Adapted from Shekhar et al. [11] with permission from Elsevier

$t = 0$ min $t = 10$ min

No force

In presence of force

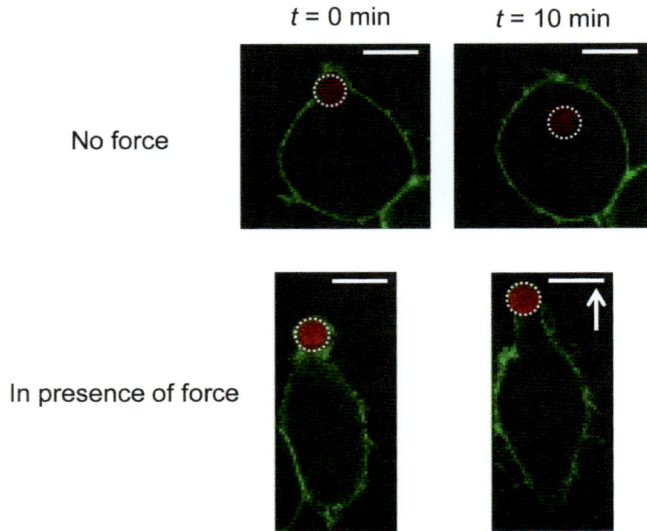

Fig. 5 Effect of force on a magnetic microbead containing phagosome. Time-lapse images showing the location of the magnetic microbead (*red, white dotted circle*) with respect to the plasma membrane of the cell (*green*) in absence (*top*) and presence (*bottom*) of an external magnetic force. The two frames are 10 min apart. In the absence of force (*top*), the microbead is transported inwards towards the cell nucleus. However, in the presence of force the microbead remains at the plasma membrane throughout the experiment (*bottom*). The white arrow denotes the direction in which the microbead experiences the magnetic force. Scale bar—5 μm. Reproduced from Shekhar et al. [11] with permission from Elsevier

10. As soon as the phagocytic cup gets closed, turn on the magnetic tweezers by switching on the current in the copper coil around the magnetic pole. The microbeads which are unattached (to cells) should immediately get attracted to the magnetic pole. The internalized magnetic microbeads should remain inside the cells (Fig. 5).

11. Vary the magnitude of the current in the coils to stall the centripetal transport of the magnetic microbead containing phagosome (Fig. 5). The phagosome transport through the cell can also be accelerated by changing the direction of the magnetic force by physical moving the tip to the opposite side with the micromanipulator.

3.5 Data Acquisition   In order to study the effect of intracellular manipulation of phagosomes on their acidification, the location of the phagosome (using brightfield imaging) and its pH (by fluorescence imaging) need to be simultaneously recorded. This requires simultaneous brightfield and widefield imaging, which is impractical to do in commercial microscopes. Since our experiment required fluorescence acquisition

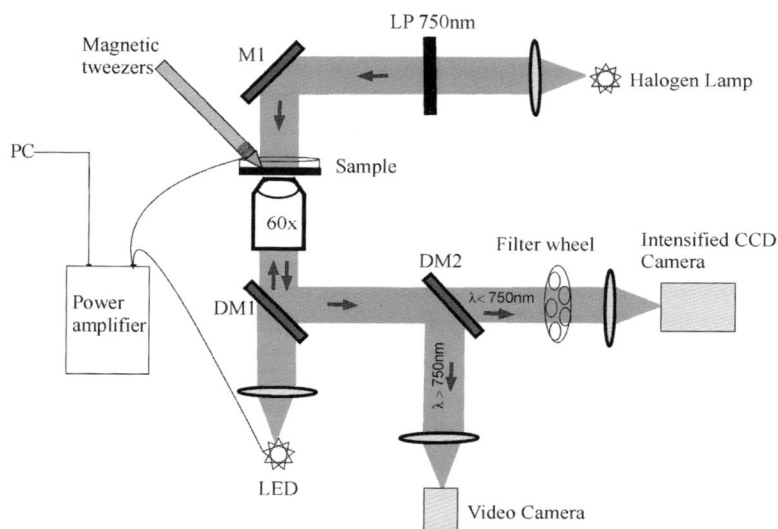

Fig. 6 Schematic representation of the customized optical setup as well the computer-controlled magnetic tweezers. The optical setup was assembled with a Nikon Eclipse TE 2000 microscope as the base. The setup is designed for simultaneous brightfield (in Near Infra-Red wavelength range) and fluorescence imaging. Fluorescence excitation is carried out by a computer-controlled LED. The filter wheel contains the appropriate band pass filters for ratiometric pH measurements. DM1 and DM2 are two dichroic mirrors, M1 is a mirror and LP 750 is a long pass 750 nm filter. A single-pole computer-controlled magnetic tweezers can also be seen placed on the microscope stage. Reproduced from Shekhar et al. [11] with permission from Elsevier

for pH-measurement, video acquisition would have to be temporarily stopped to allow fluorescence acquisition to prevent polluting the fluorescence image with white light (visible wavelengths) and overexposing the intensified CCD camera. To overcome these limitations, a home-built optical setup was designed.

1. Any commercially available widefield microscope can be used as a base (we used a Nikon Eclipse TE 2000). Figure 6 shows a schematic representation of the setup.

2. For fluorescence excitation, use an LED (wavelength 520–550 nm) instead of the mercury lamp. This also eliminates the need of an excitation filter in the filter cube. Remove the filter cube from the microscope.

3. The LED illumination is computer-controlled such that the LED is turned on only during acquisition, thus preventing needless illumination (and bleaching) of the fluorophores in between successive pH measurements (*see* **Note 14**).

4. Since the emission spectrum of the dye contains two pH-dependent peaks, pH-measurement require acquiring fluorescence images in two channels (Channel 1: 570–615 nm and Channel 2: 655–740 nm). Instead of using emission filters in the filter cube, use a computer-controlled filter-wheel instead. Place the band-pass emission filters for each channel in the filter wheel.

5. Acquire images in the two channels in series with the maximum integration time of 1 s per image using an intensified CCD (ICCD) camera. The thermally induced dark noise of the camera was reduced by internal cooling of the CCD chip to –20 °C.

6. Acquire fluorescence images once every 30 s to track phagosomal pH over time. The ratio of the microbead intensity in the two images is used to determine the phagosomal pH (*see* Subheading 3.6 for details).

7. Carry out brightfield imaging simultaneously with fluorescence imaging (*see* Fig. 7). Brightfield acquisition should be done at video rate (25 images per second) to allow monitoring small microbead displacements.

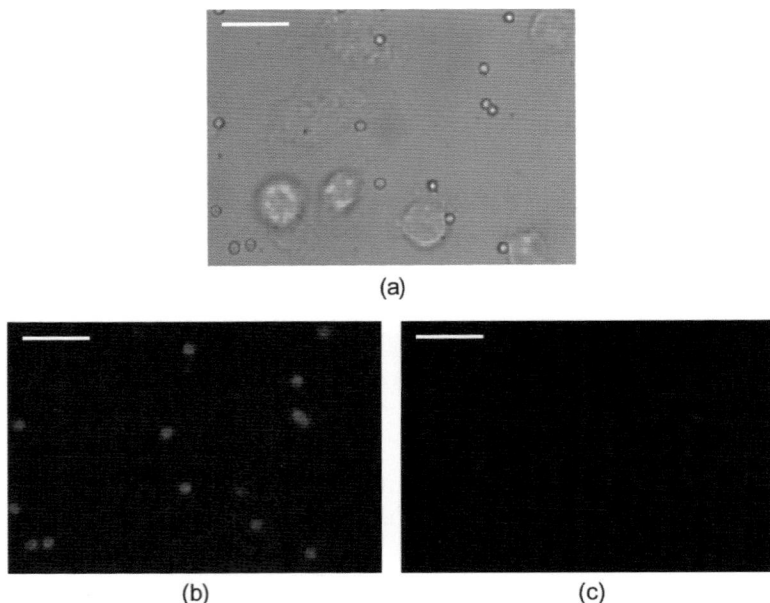

(a)

(b)   (c)

Fig. 7 Simultaneous brightfield and fluorescence imaging of phagocytosis. An example of (**a**) brightfield image acquired in Near-Infrared (NIR) wavelength region along with its complementary simultaneously acquired pH-dependent fluorescence images in (**b**) Channel 1 and (**c**) Channel 2. The ratio of microbead intensity in Channel 1 and Channel 2 is a measure of pH. Scale bar: 15 μm

8. Place a near infrared (NIR) filter (>750 nm) in front of the halogen lamp for brightfield imaging. Use the transmitted light as a light-source for brightfield imaging. Record using a video camera.

9. Use appropriate dichroic mirrors, to simultaneously record brightfield video in NIR wavelength using a video camera (WOTEC 902H) and fluorescently measure pH (once every 30 s) without interruption (*see* Fig. 8).

3.6   Data Analysis: Image Analysis and Particle Tracking

3.6.1   Image Analysis for pH-Measurement

Phagosomal pH is measured from two fluorescence images of beads (Channel 1: 570–615 nm and Channel 2: 655–740 nm). Although we used ImageJ, any other image analysis program can easily be used instead to carry out the following steps.

1. Carry out background correction on individual images in the two channels (Fig. 7b, c) using the rolling ball background subtraction algorithm to improve signal-to-noise.

2. Determine the microbead position in the Channel 1 image by applying an appropriate thresholding and subsequent particle (microbead) detection.

3. Be careful to select only spherical microbeads with an approximate expected size of 2.8μm (microbead size of M270 microbeads) for further analysis. In this way, microbead aggregates and/or other impurities are ignored.

4. Draw a circle of 2.8μm diameter at the center of microbead location and calculate the integrated fluorescence intensity of this circular area in both Channel 1 and Channel 2 images.

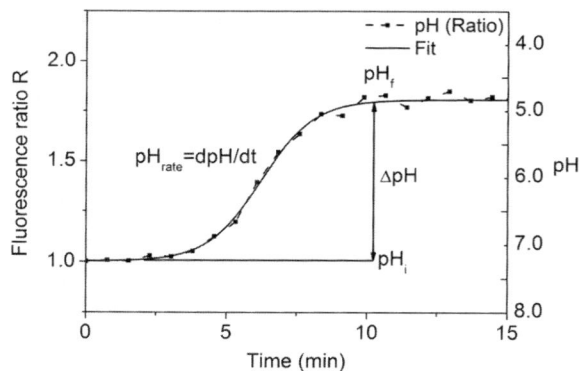

Fig. 8 Phagosomal acidification followed by a phagocytosed M270-SNARF microbead. Ratiometric measurement of phagosomal acidification by an M270-SNARF microbead upon its phagocytic uptake. As the phagosome acidifies, the ratio *R* increases (symbols, *dotted line*) and the pH decreases. The *thick line* represents the sigmoidal fit of the acidification curve (pH vs. time). The phagosome begins acidifying from an initial pH = pH_i at a rate of acidification (pH_{rate}) to a final pH of pH_f at the completion of acidification

5. Determine the fluorescence ratio R for each microbead by dividing the integrated intensity of the microbead in Channel 1 by the integrated intensity of the microbead in Channel 2 i.e.
$$R = \frac{I_{ch1}}{I_{ch2}}.$$

6. Normalize the fluorescence ratios R are such that the Ratio $R = 1$ at pH 7.

7. Determine the pH of the microbead using the calibration curve prepared earlier (*see* Fig. 1).

8. The above steps are repeated over the time-lapsed pH-images taken over the duration of the experiment.

3.6.2 Particle Tracking in Brightfield

A large number of particle tracking algorithms exist in order to accurately track the location of the microbeads. A custom-written program in LabVIEW based on pattern-recognition routines was used for particle tracking. Microbead tracks are also used for determining microbead velocities. Specifically the program functions as:

1. Identify the position of the microbead of interest manually in the first frame of the video.

2. Define a region of interest (ROI) around the microbead in which the program can attempt to find the microbead in the subsequent frame.

3. The ROI selection ensures that the microbead trajectory does not get polluted by a trajectory of any other microbeads in close vicinity.

4. Repeat above steps in an iterative fashion over successive frames to record microbead localization through the full duration of the video.

5. Determine the localization accuracy of your software by measuring the accuracy in tracking a microbead fixed on the glass surface. This will give the sensitivity of specific tracking method used.

4   Notes

1. Some magnetic beads can have appreciable autofluorescence in the visible wavelength range. This autofluorescence can pollute the signal coming from the pH-sensitive dye and can adversely affect ratiometric pH-measurement. It is therefore important to keep this in mind while choosing the beads.

2. Size and magnetic content of the particles (magnetic moment) are the two most important criteria while choosing magnetic microbeads. In this protocol, we have used amine-functionalized Dynal M270 microbeads. These are 2.8 μm in diameter,

adequately sized to follow phagosomal transport. Their magnetic content (volume magnetization = 11.5 kA/m) is also high enough to allow up to 1nN forces when placed in magnetic flux density gradients of a few tens of kT/m (typical gradients generated by single-pole magnetic tweezers).

3. Phenol red, which is the red-colored substance in cell culture media, is autofluorescent and will therefore interfere with the ratiometric pH-measurements. It is therefore important to use medium without phenol red during imaging. We are aware that some labs solve this problem by running their cell cultures in medium with phenol-red and they use non-phenol red medium during imaging. Nevertheless, we prefer to never expose cells to phenol red containing medium.

4. SNARF-4F can be dissolved in DMSO and divided into single-use volumes in microcentrifuge tubes. The dye can be concentrated under vacuum until all the solvent evaporates. The microcentrifuge tube can then be stored at –20 °C for up to 1 year at least.

5. pH-buffers for pH-calibrations can be prepared by mixing together 0.1 M citric acid and 0.2 M disodium hydrogen phosphate as follows [19]:

pH	0.2 M Na_2HPO_4	0.1 M citric acid
3.0	20.55 ml	79.45 ml
4.0	38.55 ml	61.45 ml
5.0	51.50 ml	48.50 ml
6.0	63.15 ml	36.85 ml
7.0	82.35 ml	17.65 ml
8.0	97.25 ml	2.75 ml

6. NHS is needed for the cross-linking reaction required for functionalizing the dye to the microbead surface. NHS increases the lifetime of the active intermediate formed from the EDC-COOH reaction, which reacts with the amine groups on the microbeads [20].

7. M270-SNARF beads that are functionalized only with fluorophores can be used for up to a month once prepared. However, once opsonised with the IgG, the beads should be used within 1 week. This is done to ensure that the antibodies on the beads remain active.

8. Over time, it will be noticed that the magnetic pole continues to stay magnetized even when the current has been turned off. When this occurs, the magnetic pole should be degaussed. This can be done by running a coil current with a decaying sinewave signal.

9. The tip of the magnetic tweezers pole can get rusted over time due to its frequent exposure to humidity. To prevent this, the tip can be kept dipped in edible oil during its storage. Ensure that the oil is completely removed prior to using the pole in cell experiments.

10. Electromagnets with high currents often produce substantial heat due to Joule Heating (Q). This in turn can cause heating of the cell culture medium close to the pole, leading to abnormal cellular behavior and even cell-death. If I is the current and N is the number of turns, $Q \propto I^2 \times N$. Therefore, Joule Heating increases with the square of the current but only linearly with N. Coil current should therefore be kept low to ensure this doesn't happen. In case heating occurs even at low current, adequate passive or active cooling strategies should be employed [15].

11. An object moving through a liquid experiences a retarding force against its motion due to the viscosity of the liquid [15]. This retarding force, called viscous drag, can be used for calibration. The viscous drag on a microbead is given by $\bar{F}_{drag} = \gamma \bar{v}$, where v is the velocity of the microbead and γ is the drag coefficient. For a spherical microbead of radius r, placed in a liquid of known viscosity η, the drag coefficient is given by $\gamma = 6\pi\eta r$. Under equilibrium conditions (~constant velocity), the magnitude of the drag force equals the magnitude of exerted magnetic force ($F_{drag} = F_{mag}$). By measuring the velocity of the microbead using video microscopy the position dependent magnetic force can be determined.

12. For magnetic tweezers calibration, make sure that the magnetic pole is as close as possible to the glass surface and only beads that stay in the same focal plane during their movement are used for calibration. This ensures that microbead velocity is not underestimated. It also helps with the automated tracking of the microbead afterwards.

13. Prior to manipulating a phagocytosed magnetic microbead, it is important to ensure that the microbead has been internalized and is not just sticking to the cell surface. Also, to compare parameters (e.g. beginning of acidification, end of acidification etc.) between different phagocytic events, it is essential to define an absolute time-point relative to which all other parameters can be set. We use the time-point of internalization (t_0), the time at which the microbead is completely enclosed in the phagocytic cup. For this we used a GFP-tagged polycationic probe Kmyr, a K-Ras-derived peptide that binds to anionic lipids [21]. Upon transfection and subsequent

expression, Kmyr-GFP is recruited to the plasma membrane of cells. Clear labeling of the plasma membrane by Kmyr-GFP ensures that formation and the closure of the phagocytic cup can be followed by observing a clear ring of Kmyr-GFP around the microbead (Fig. 4).

14. Photobleaching of SNARF-4F can sometimes cause a change in the ratio R (and therefore the measured pH). It can be partially prevented by using a low-intensity light source and reducing exposure as much as possible. In any case, it is important to always perform proper control experiments to take into account the effect of photobleaching. One way of easily doing this is to expose labeled beads to illumination of varying intensity and duration (at a fixed pH) and following the effect on ratio R.

References

1. Flannagan RS, Jaumouille V, Grinstein S (2012) The cell biology of phagocytosis. Annu Rev Pathol 7:61–98. doi:10.1146/annurev-pathol-011811-132445

2. Heinrich V (2015) Controlled one-on-one encounters between immune cells and microbes reveal mechanisms of phagocytosis. Biophys J 109(3):469–476. doi:10.1016/j.bpj.2015.06.042

3. Toyohara A, Inaba K (1989) Transport of phagosomes in mouse peritoneal macrophages. J Cell Sci 94(Pt 1):143–153

4. Harrison RE, Bucci C, Vieira OV, Schroer TA, Grinstein S (2003) Phagosomes fuse with late endosomes and/or lysosomes by extension of membrane protrusions along microtubules: role of Rab7 and RILP. Mol Cell Biol 23(18):6494–6506

5. Blocker A, Severin FF, Burkhardt JK, Bingham JB, Yu H, Olivo JC, Schroer TA, Hyman AA, Griffiths G (1997) Molecular requirements for bi-directional movement of phagosomes along microtubules. J Cell Biol 137(1):113–129

6. Falcon-Perez JM, Nazarian R, Sabatti C, Dell'Angelica EC (2005) Distribution and dynamics of Lamp1-containing endocytic organelles in fibroblasts deficient in BLOC-3. J Cell Sci 118(Pt 22):5243–5255

7. Diwu Z, Chen CS, Zhang C, Klaubert DH, Haugland RP (1999) A novel acidotropic pH indicator and its potential application in labeling acidic organelles of live cells. Chem Biol 6(7):411–418

8. VonSteyern FV, Josefsson JO, Tagerud S (1996) Rhodamine B, a fluorescent probe for acidic organelles in denervated skeletal muscle. Journal of Histochemistry & Cytochemistry 44(3):267–274

9. Gruenberg J, Griffiths G, Howell KE (1989) Characterization of the early endosome and putative endocytic carrier vesicles in vivo and with an assay of vesicle fusion in vitro. J Cell Biol 108(4):1301–1316

10. Matteoni R, Kreis TE (1987) Translocation and clustering of endosomes and lysosomes depends on microtubules. J Cell Biol 105(3):1253–1265

11. Shekhar S, Cambi A, Figdor CG, Subramaniam V, Kanger JS (2012) A method for spatially resolved local intracellular mechanochemical sensing and organelle manipulation. Biophys J 103(3):395–404. doi:10.1016/j.bpj.2012.06.010

12. Shekhar S, Klaver A, Figdor CG, Subramaniam V, Kanger JS (2010) Spatially resolved local intracellular chemical sensing using magnetic particles. Sensors and Actuators B-Chemical 148(2):531–538

13. Desjardins M, Griffiths G (2003) Phagocytosis: latex leads the way. Curr Opin Cell Biol 15(4):498–503

14. Irmscher M, de Jong AM, Kress H, Prins MW (2013) A method for time-resolved measurements of the mechanics of phagocytic cups. Journal of the Royal Society, Interface / the Royal Society 10(82):20121048. doi:10.1098/rsif.2012.1048

15. Tanase M, Biais N, Sheetz M (2007) Magnetic tweezers in cell biology. Methods Cell Biol 83:473–493

16. Bausch AR, Moller W, Sackmann E (1999) Measurement of local viscoelasticity and forces in living cells by magnetic tweezers. Biophys J 76(1 Pt 1):573–579

17. Kanger JS, Subramaniam V, van Driel R (2008) Intracellular manipulation of chromatin using magnetic nanoparticles. Chromosome Res 16(3): 511–522. doi:10.1007/s10577-008-1239-1

18. de Vries AH (2004) High force magnetic tweezers for molecular manipulation inside living cells. University of Twente, Enschede, The Netherlands

19. Dawson RMC, Elliot DC, Elliot WH, Jones KM (1986) Data for Biochemical Research. 3rd ed. edn. Oxford Science Publ.,

20. Hermanson GT (2008) Bioconjugate techniques. Academic, San Diego

21. Yeung T, Terebiznik M, Yu L, Silvius J, Abidi WM, Philips M, Levine T, Kapus A, Grinstein S (2006) Receptor activation alters inner surface potential during phagocytosis. Science 313(5785):347–351

Chapter 8

Quantitative Immunofluorescence to Study Phagosome Maturation

Roya M. Dayam and Roberto J. Botelho

Abstract

Cells such as macrophages and neutrophils can internalize a diverse set of particulate matter, illustrated by bacteria and apoptotic bodies through the process of phagocytosis. These particles are sequestered into phagosomes, which then fuse with early and late endosomes, and ultimately with lysosomes to mature into phagolysosomes, through a process known as phagosome maturation. As phagosomes change, they acquire and divest proteins that are associated with the various stages of phagosome maturation. These changes can be assessed at the single-phagosome level by using immunofluorescence methods to study phagosome maturation. Typically, we use indirect immunofluorescence methods that rely on primary antibodies against specific molecular markers that track phagosome maturation. Most commonly, phagosome maturation in macrophages can be determined by staining the cells for Lysosomal-Associated Membrane Protein I (LAMPI) and measuring the fluorescence intensity of LAMPI around each phagosome by microscopy or flow cytometry.

Key words Immunofluorescence, Antibodies, Lysosomes, LAMPI, Macrophages, Phagosome maturation

1 Introduction

During phagocytosis, particulate matter such as bacteria and apoptotic bodies are sequestered into phagosomes [1]. Phagosomes then undergo a striking alteration in their properties through phagosome maturation. This process is typically thought to follow a sequence, whereby phagosomes first fuse with early and late endosomes, and ultimately with lysosomes, transforming into phagolysosomes [2]. Thus, maturation converts phagosomes from inert to highly acidic and degradative organelles that catalyze the decomposition of the internalized particle. Through this process, phagosomes acquire and divest various proteins and lipids associated with each stage of phagosome maturation: from a plasma-membrane-like to a lysosome-like stage [3]. For example, phagosomes lose phosphatidylinositol-4,5-bisphosphate and actin filaments associated with the plasma membrane [4], then acquire

Roberto Botelho (ed.), *Phagocytosis and Phagosomes: Methods and Protocols*, Methods in Molecular Biology, vol. 1519, DOI 10.1007/978-1-4939-6581-6_8, © Springer Science+Business Media New York 2017

and rapidly eliminate EEA1 and phosphatidylinositol-3-phosphate associated with early endosomes [5], then gain and divest of lyso-bisphosphatidic acid (LBPA) and mannose-6-phosphate receptors, which are typically associated with late endosomes, and ultimately acquire lysosomal proteins such as LAMP1 [6], V-ATPase complex and the Rab7 GTPase [7]. While this is the canonical phagosome maturation, many infectious agents have evolved strategies that circumvent phagosome maturation. For example, *Mycobacterium tuberculosis* prevents fusion of phagosomes with lysosomes after they are engulfed by macrophages, thereby protecting themselves from lysosomal degradation [8].

To understand phagosome maturation and how pathogens usurp this machinery, we need methods to track maturation progress. Immunofluorescence techniques are commonly used to determine and visualize phagosome maturation by staining for well-characterized markers of phagosome maturation, in particular EEA1 for early phagosomes and LAMPI for phagolysosomes. Typically, we employ indirect immunofluorescence, whereby fixed cells are exposed to primary antibodies against a specific marker, followed by a fluorescently labeled secondary antibody that binds to the primary antibody [9]. Cells can then be semi-quantitatively or quantitatively analyzed by confocal or epifluorescence microscopy [10], or by flow cytometry. Flow cytometry can measure the total intensity of the fluorescent tag per particle (could be a cell or isolated phagosomes) for thousands of particles, but does not provide any information about the localization of the target protein [11]. In contrast, microscopy can provide us with the location and relative fluorescence signal, but is typically low-throughput [10].

Here, we detailed methodology to quantify phagolysosome biogenesis in macrophage cell lines using immunofluorescence against LAMP1 of whole cells or isolated phagosomes. This method can be adopted to stain other markers, including for double-staining.

2 Materials

1. RAW 264.7 macrophages (ATCC) (or other cell type).

2. Cell culture medium (complete DMEM medium): Dulbecco's Modified Eagle Medium (DEMEM) supplemented with 10% Fetal Bovine Serum (FBS).

3. Fresh 1× phosphate buffered saline (PBS).

4. Fixation solution: Freshly made 4% paraformaldehyde (PFA) in 1× PBS.

5. 100 mM Glycine in 1× PBS, store at 4 °C.

6. Permeabilization solution: 100% ice-cold methanol.

7. Primary antibodies: Rat anti-mouse LAMP1 monoclonal antibodies (Developmental Hybridoma, *see* **Note 1**).

8. Secondary antibodies: Fluorescently conjugated secondary antibodies (Donkey anti-rat, used at 1:1000).

9. Mounting medium.

10. Blocking solution: 0.5% Albumin from bovine serum (BSA) in 1× PBS.

11. Polystyrene plain (hydrophobic) microspheres beads with diameter of 3.87 μm.

12. 10 mg/mL immunoglobulin G (IgG) from human serum, reconstituted in 150 mM NaCl.

13. Microscope slides.

14. Glass cover slips.

15. Homogenization buffer for phagosome isolation (*see* **Note 2**): 20 mM Tris (pH 7.4), 2.5 μL/mL protease inhibitor cocktail, 1 mM ABSF, 1 mM $MgCl_2$, 1 mM $CaCl_2$, 1 mg RNase, 1 mg DNase.

16. Sucrose gradient for phagosome isolation: 60% sucrose in 1× PBS.

17. Microscope (Spinning Disk Confocal).

18. Ultracentrifuge.

19. Ultracentrifuge tubes.

20. Nutator.

21. Flow cytometer: Fluorescence-activated cell sorting (FACS) Calibur.

3 Methods

3.1 Immunofluorescence and Microscopy

3.1.1 Cell Culture

1. Seed RAW 264.7 macrophages onto T25 flask in complete DMEM medium and grow at 37 °C and 5% CO_2 until cells reach 70–80% confluency (*see* **Note 3**).

2. Next day, wash cells with 5 mL PBS.

3. Then add 10 mL new DMEM medium and scrape cells gently.

4. Take 1 mL of cell suspension and seed cells onto 12-well plate (*see* **Note 4**) containing glass coverslips to obtain 50–60% confluency using complete DMEM (*see* **Note 5**).

5. Grow cells at 37 °C and 5% CO_2.

3.1.2 Phagocytosis and Phagosome Maturation

1. Opsonization: Add 50 μL of PBS, 10 μL of polymer beads, and 10 mg/mL of human IgG into a microcentrifuge tube and use a nutator to mix it for 30 min at room temperature (*see* **Note 6**).

2. Wash excess IgG with 1× PBS and centrifuge at $1180 \times g$ for 1 min. Remove the supernatant and repeat this step three times (*see* **Note 7**).

3. Resuspend beads in PBS (*see* **Note 8**).

4. Add opsonized beads to the media containing cells and incubate at 37 °C and 5 % CO_2 for 20 min (*see* **Note 9**).

5. After 20 min pulse, remove the unbound beads from the media by washing the cells with PBS three times.

6. Replace the PBS with prewarmed media and incubate the cells (chase) at 37 °C and 5 % CO_2 for 60 min (*see* **Note 10**).

7. After 60 min chase, fix cells with 4 % PFA for 20 min at room temperature.

8. Replace PFA with 100 mM glycine and incubate for another 20 min at room temperature to quench the remaining PFA.

9. Remove glycine, add 100 % ice-cold methanol, and incubate for 3–5 min on ice to permeabilize the cells (*see* **Note 11**).

10. Wash the cells three times with 0.5 % BSA.

3.1.3 Fixation and Antibody Staining

1. Add primary antibody (rat anti-LAMPl antibody with a dilution of 1:200 in PBS), and incubate for 1 h at room temperature (*see* **Note 12**).

2. Wash the cells three times with 0.5 % BSA every 5 min for 15 min to remove the unbound primary antibodies.

3. Add secondary antibodies (with a dilution of 1:1000 in 1× PBS) and incubate for 1 h at room temperature in the dark (*see* **Note 13**).

4. Wash the excess secondary antibodies three times with 0.5 % BSA every 5 min for 15 min.

5. Mount the cover slips on a microscope slide using mounting medium and store the slide in a microscope slide box at 4 °C until microscopy.

3.1.4 Microscopy and Analysis of Matured Phagosomes in Whole Cells

1. Acquire images (*see* **Note 14**) of 10–15 fields of view per condition using a confocal microscope (*see* **Note 15**).

2. Open the images using Image J and split the channels into DIC and fluorescence (Image J: Image → color → split channels).

3. Convert to 8-bit if necessary.

4. Use the pseudo-color processing (*see* **Note 16**) to help determine the intensity of LAMP1 around the phagosome (Image J: Image → look-up tables → fire).

5. Count the total number of phagosomes in each field of view and assign each phagosome as being LAMPl-positive, partial, or negative based on the intensity of LAMPl around each phagosome (Fig. 1a, c) (*see* **Note 17**).

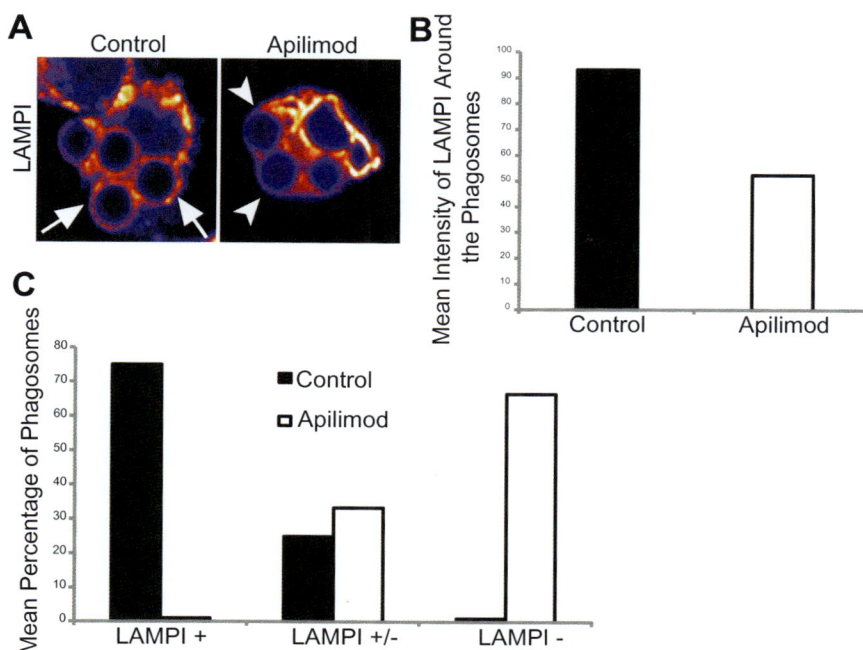

Fig. 1 Whole-cell immunofluorescence. (**a**) Phagosomal acquisition of LAMPI in control and PIKfyve inhibited (Apilimod) RAW macrophages: Cells were fixed and stained with LAMPI antibodies, *arrows* point to LAMPI-positive phagosomes, and *arrow heads* point to LAMPI-negative phagosomes. (**b**) Quantitative analysis of phagosome maturation in whole cell: mean intensity of LAMPI was measured around each phagosome in control and Apilimod-treated cells. (**c**) Semi-quantitative analysis of phagosome maturation in RAW macrophages: applying pseudo-color to the images and assigning each phagosome as LAMPI-positive, partial or negative. *White-yellow* color indicates high intensity of LAMP1 with grayscale intensities of 225-180 and was assigned LAMP1-positive, *orange* to *red* indicates the partial intensity of LAMP1 (grayscale intensities of 180-80) and *purple* to *blue* indicates the absence of LAMP1 (grayscale intensities of 80-1) around the phagosomes, which was assigned LAMP1-negative

3.2 Immunofluore-scence and Flow cytometry Analysis of Isolated Phagosomes

3.2.1 Cell culture for Phagosome Isolation

1. Grow RAW macrophages to 80 % confluency in T-25 culture flasks with complete DMEM (one flask for each condition, e.g. Control, drug treatment, etc.).

2. Next day, wash the cells with 5 mL of 1× PBS.

3. Add 3 mL of new DMEM medium into each flask.

3.2.2 Phagosome Isolation

1. Sucrose gradient preparation: Pipette 1 mL of 60 % sucrose into 1 mL ultracentrifuge tube and centrifuge at $21,000 \times g$ for 1 h at 4 °C (*see* **Note 18**). After 1 h place the tubes on ice without disturbing.

2. To opsonize beads, add 60 μL of beads, 120 μL of 10 mg/mL of human IgG in 300 μL of PBS and rotate for 30 min at room temperature. Wash beads 3× in 1 mL of PBS by centrifuging.

3. Add 300 μL of opsonized beads into each flask containing cells and incubate for 30 min at 37 °C and 5% CO_2 (*see* **Note 19**).

4. Wash cells three times with cold PBS to remove the unbound beads.

5. Replace the PBS with prewarmed media and incubate the cells at 37 °C and 5% CO_2 for 60 min (chase) (*see* **Note 20**).

6. After 60 min chase, remove the media and add 10 mL of cold homogenization buffer to the T-25 flask and gently scrape the cells.

7. Transfer the buffer containing the scraped cells into a 15 mL conical tube and centrifuge at $500 \times g$, 4 °C, for 5 min.

8. To disrupt cells, after centrifugation the pellet is resuspended in 1 mL of homogenization buffer and passed through a syringe with 22 gauge needle, 5–10 times (*see* **Note 21**) and centrifuge for 5 min, 4 °C, at $1180 \times g$ (*see* **Note 22**).

9. The pellet is then resuspended in 200 μL of PBS and transferred into a 1 mL ultra centrifuge tube containing the sucrose gradient (*see* **Note 23**).

10. Beads are separated from the sample by centrifuging the sucrose gradient at $21,000 \times g$ for 10 min at 4 °C.

11. Beads are withdrawn from the sucrose gradient using a syringe with 22 gauge needle, transferred into a fresh microcentrifuge tube and washed with ice-cold PBS (*see* **Note 24**).

12. Fix the beads with 4% PFA for 20 min at room temperature.

13. Replace PFA with 100 mM glycine to quench the remaining PFA and incubate for 20 min at room temperature (*see* **Note 25**).

3.2.3 LAMPI Staining of Isolated Phagosomes

1 Add primary antibodies (anti-LAMPI antibodies with a dilution of 1:200 in 1× PBS), and incubate for 1 h at room temperature (*see* **Note 26**).

2 To remove the unbound primary antibodies, spin down the isolated phagosomes at $1180 \times g$ for 1 min and remove the supernatant (excess primary antibodies).

3 Wash phagosomes with 500 μL of 0.5% BSA and centrifuge at $1180 \times g$ for 1 min, repeat this step three times every 5 min for 15 min.

4 Add secondary antibodies (with a dilution of 1:1000 in PBS) and incubate for 1 h at room temperature in the dark.

5 Wash the excess secondary antibodies three times with 0.5% BSA every 5 min for 15 min by spinning down the phagosomes after each wash using a centrifuge at $1180 \times g$ for 1 min.

6 Either resuspend the phagosomes in 20 μL of 1× PBS and mount on a microscope slide using a cover slip and mounting

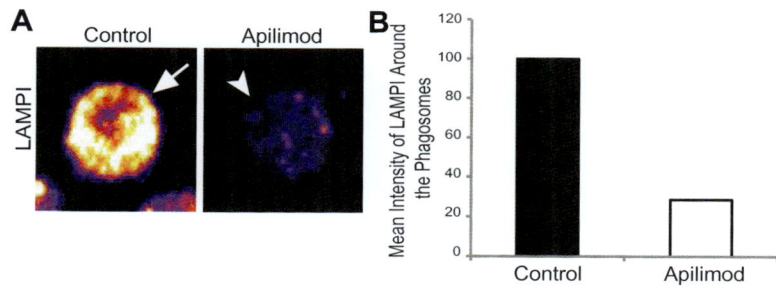

Fig. 2 Immunofluorescence of isolated phagosomes. (**a**) Matured phagosomes isolated from control and PIKfyve-inhibited (Apilimod) RAW macrophages and stained with LAMPI antibodies. *Arrow* points to the LAMPI-positive phagosome and *arrow* head points to the LAMPI-negative phagosome. (**b**) Quantification of the intensity of LAMPI in isolated phagosomes in control and Apilimod-treated RAW macrophages; mean intensity of LMAPI was measured in each phagosome

media and use a microscope to visualize the isolated phagosomes, or resuspend them in 500 μL of 1× PBS for flow cytometry (*see* **Note 27**).

3.2.4 Microscopy and Analysis of Isolated Phagosomes

1 Acquire z-stack images of 10–15 fields of view per condition using a confocal microscope (*see* **Note 28**).

2 Open the images using Image J and make a single image out of the z-stack (image J: Stacks → Z-project → assign the star and stop slice → in the projection type select sum slices and press OK; Fig. 2a).

3 For semi-quantitative analysis, use the pseudo-color processing (*see* **Note 15**) to help determine the intensity of LAMP1 around the phagosome (Image J: Image → look-up tables → fire).

4 For quantitative analysis, select a phagosome (ROI) and measure the mean intensity of LAMPI (Fig. 2b).

3.2.5 Flow cytometry and Analysis of Isolated Phagosomes

1 Resuspend the phagosome prep by agitating and then inject into the flow cytometer.

2 Run the flow cytometer to measure the total intensity of LAMPI per phagosome (*see* **Note 29**).

3 While running the flow cytometer (*see* **Note 30**), select the population of the isolated phagosomes by drawing a gate and measure the intensity of the fluorescent probe of the selected region (*see* **Note 31**).

4 Using the available flow cytometry software, obtain either a histogram and/or density plot, followed by statistical values such as mean and median intensity of the fluorescent probe (*see* **Note 32**).

4   Notes

1. Other antibodies to LAMPI exist. In addition, antibodies to other markers (e.g. EEA1, LAMP2, CD63) could be employed. Lastly, as long as the antibodies are raised in different species, one could employ double immunofluorescence staining.

2. Homogenization buffer has to be prepared fresh. However, stock solution of Tris can be made and stored at 4 °C for a few weeks, which can be used to make the homogenization buffer. It is very important to adjust the pH of Tris to 7.4 with HCl.

3. Other cells such as primary macrophages, neutrophils, and dendritic cells can also be used as professional phagocytes to perform this experiment.

4. For a six-well plate add 2 mL of cell suspension into each well containing a glass coverslip.

5. If cells are to be transfected the next day, then the confluency of the cells should be adjusted to 30–40% while seeding onto a glass cover slip.

6. This is for a 12-well plate, if using 6-well simply double the volume of all the reagents.

7. After opsonization of the beads, it is very important to wash the beads with 1× PBS, ideally three times to remove the unbound/excess IgG. The excess IgG will occupy the Fcγ receptors and may reduce the number of internalized beads.

8. Polymer beads can be substituted with other particles such as IgG-opsonized sheep red blood cells or bacteria.

9. Phagocytosis can be synchronized by placing the plate containing cells and beads on ice for 10 min or centrifuge the plate at $400 \times g$ for 5 min at room temperature. Wash cells three times with 1× PBS and add complete DMEM and incubate at 37 °C and 5% CO_2.

10. Chase time is highly depended on the purpose of the experiment. For example, if one is interested to look at the initial events of phagosome maturation and stain for early phagosomal markers such as EEA1, then the chase time can be as short as 5 min.

11. Triton is a typical permeabilization reagent that is mostly used in the lab. However, permeabilization depends on antibodies. For this specific LAMPI antibody, it is recommended that 100% ice-cold methanol be used to permeabilize cells. Therefore, it is important to look at antibody specifications before proceeding with this proposed method.

12. It is very important to use the right dilution of the antibodies. Thus, check the data sheet provided by the antibody supplier or optimize by trying different dilution factors.

13. Secondary antibodies need to be matched to the species in which the primary antibody was raised. For double labeling experiments, the secondary antibodies must be cross-absorbed against IgGs from the other species to prevent cross-reactivity with other secondary antibodies. Always check the data sheet provided by the supplier to ensure cross-absorption was performed.

14. Acquiring single images in a single XY plane if often sufficient for quantifying phagosome maturation (rings of LAMP1 around beads, for example) as long as only particles in focus with that plane are counted. Alternatively, one can obtain and collapse a full z-stack set to quantify all phagosomes within a field-of-view, but this is not always necessary.

15. Epi-fluorescent microscope reflect light, so using polymer beads will results in a high background noise and will make it difficult to determine whether a phagosome is positive for LAMPI or not. Therefore, confocal microscopy is ideal when using polymer beads. Epifluorescence microscopy can be used when using red blood cells or bacteria for phagocytosis though.

16. Using pseudo-coloring is a semi-quantitative method to determine whether a phagosome is matured or not. Data can be quantitatively analyzed by selecting a region of interest around each phagosome and measure the mean intensity of the fluorescent probe, in this case LAMPI, and subtract the background (Fig. 1b).

17. White-yellow color indicates high intensity of LAMP1 with grayscale intensities of 255-180 and is designated as LAMP1-positive, orange to red indicates the partial intensity of LAMP1 (grayscale intensities of 180-80) and purple to blue indicates the absence of LAMP1 (grayscale intensities of 80-1) around the phagosomes, and can be designated as LAMP1-negative (Fig. 1a, c).

18. The centrifuge should be balanced. So after pipetting 1 mL of sucrose into each tube, use a balance to measure the weight of each tube and adjust by adding or removing sucrose solution to ensure the tubes are within 0.5 g from each other.

19. Pulse time can be different based on the purpose of the experiment. For synchronization, place the flask on ice for 10 min (bead binding time), then wash off the unbound beads three times with cold PBS and add prewarmed DMEM medium into each flask and incubate for a specific chase time. Alternatively, spin beads onto cells if a plate-rotor adaptor is available.

20. Chase time is dependent on the purpose of the experiment. If looking at late maturation, then a one hour chase is enough for the phagosomes to mature. However, if one is interested to look at the early stages of maturation, then cells can be incubated for a shorter (5–10 min) period of time.

21. Alternative to syringing technique, French press can also be used to disrupt the cells. But the orifice has to be adjusted to avoid breaking phagosomes.

22. All the solutions need to be ice-cold, and the tubes containing the cells need to be placed on ice throughout the syringing process.

23. It is very important not to disturb the sucrose gradient. Therefore, add the sample dropwise onto the gradient.

24. Immunostaining of isolated phagosomes require multiple steps of washing and centrifugation. It is very important to pay attention when removing the supernatant and not to disturb the pellet after each centrifugation. If the pellet is disrupted, sample needs to be centrifuged again.

25. Since this is a long experiment, this is a perfect step in which one can pause and leave the beads on glycine at 4 °C until the next day to perform immunofluorescent.

26. There is no need to add 1 mL of primary antibody dilution into each tube—200 µL is sufficient. After adding the antibody solution, resuspend the pellet to ensure homogeneous staining of beads.

27. The volume of PBS being added to the phagosomes for flow cytometry is highly dependent on the amount of phagosomes obtained at the end of the experiment (200–500 µL of PBS is ideal).

28. The thickness of each slice is dependent on the size of the beads being used in the experiment. Since beads are very small (2–4 µm), we typically use 0.3 µm thickness for each slice.

29. Before running the sample, it is very important to vortex the tube containing isolated phagosomes and PBS for 5 s to obtain a homogenous solution.

30. Run your sample at a low speed; however, if your solution containing the isolated phagosomes is dilute and the counter-window shows a count of less than 100 phagosomes/s then switch to medium speed or high speed.

31. While running the sample, one might see two populations parted from each other. It is very important to select the right population (singlet phagosomes). Sometimes phagosomes can be associated and show up as a doublet (might look bigger in size and appears brighter than single phagosomes). It is easy to distinguish phagosome aggregates from singlets, since they constitute a very small population and appear bigger in the forward scatter axis.

32. There are several flow cytometry analysis software available. Older flow cytometers may not provide median intensity of the fluorescent probe. However, software such as *FCS Express* can use the raw data and provide the median values.

Acknowledgements

R.M.D. is supported through a Doctoral Scholarship from the Canadian Institutes of Health Research. R.J.B. is supported through a Canada Research Chair and Early Researcher Awards, and by grants from the Canadian Institutes of Health Research and the Natural Sciences and Engineering Research Council of Canada.

References

1. Aderem A (2003) Phagocytosis and the inflammatory response. J Infect Dis 187(0022–1899):S340–S345, Print

2. García-García EI, and Rosales C. (2002) Signal transduction during Fc receptor-mediated phagocytosis. J Leukoc Biol 72(6):1092–1108

3. Bongrand P (1994) Relationship between phagosome lysosome fusion, and mechanism. J Leukoc Biol 55:729–734

4. Flannagan RS, Jaumouillé V, Grinstein S (2012) The cell biology of phagocytosis. Annu Rev Pathol 7:61–98

5. Vieira OV, Botelho RJ, Rameh L, Brachmann SM, Matsuo T, Davidson HW, Schreiber A, Backer JM, Cantley LC, Grinstein S (2001) Distinct roles of class I and class III phosphatidylinositol 3-kinases in phagosome formation and maturation. J Cell Biol 155(1):19–25

6. Kim GHE, Dayam RM, Prashar A, Terebiznik M, Botelho RJ (2014) PIKfyve inhibition interferes with phagosome and endosome maturation in macrophages. Traffic 3:1–21

7. Huotari J, Helenius A (2011) Endosome maturation. EMBO J 30(17):3481–3500

8. Vergne I, Fratti RA, Hill PJ, Chua J, Belisle J, Deretic V (2004) Mycobacterium tuberculosis phagosome maturation arrest: mycobacterial phosphatidylinositol analog phosphatidylinositol mannoside stimulates early endosomal fusion. Mol Biol Cell 15:751–760

9. Aoki V, Sousa JX, Fukumori LMI, Périgo AM, Freitas EL, Oliveira ZNP (2010) Direct and indirect immunofluorescence. An Bras Dermatol 85(4):490–500

10. St Croix CM, Shand SH, Watkins SC (2005) Confocal microscopy: comparisons, applications, and problems. Biotechniques 39(6):S2–S5

11. Brown M, Wittwer C (2000) Flow cytometry: principles and clinical applications in hematology. Clin Chem 46(8 Pt 2):1221–1229

Chapter 9

Using Fluorescence Resonance Energy Transfer-Based Biosensors to Probe Rho GTPase Activation During Phagocytosis

Veronika Miskolci, Louis Hodgson, and Dianne Cox

Abstract

The p21-family members of Rho GTPases are important for the control of actin cytoskeleton dynamics, and are critical regulators of phagocytosis. The three-dimensional structure of phagosomes and the highly compartmentalized nature of the signaling mechanisms during phagocytosis require high-resolution imaging using ratiometric biosensors to decipher Rho GTPase activities regulating phagosome formation and function. Here we describe methods for the expression and ratiometric imaging of FRET-based Rho GTPase biosensors in macrophages during phagocytosis. As an example, we show Cdc42 activity at the phagosome over Z-serial planes. In addition, we demonstrate the usage of a new, fast, and user-friendly deconvolution package that delivers significant improvements in the attainable details of Rho GTPase activity in phagosome structures.

Key words Macrophages, Phagosome, Ratiometric imaging, FRET, Biosensors, Z-stack, Deconvolution

1 Introduction

Phagocytosis is a critical function of our immune system performed by phagocytes to eliminate foreign invaders such as bacteria and fungi. Phagocytosis targets particles larger than 0.5 μm and is triggered by direct ligand–receptor contacts between the particle and the phagocyte [1]. While phagocytosis can be triggered via several different receptors, one of the best-studied phagocytic receptors is the Fc gamma receptor (FcγR), which recognizes particles opsonized by IgG antibodies. FcγR-triggered phagocytosis involves a series of complex changes in cell morphology with an absolute dependence on actin reorganization [1].

Master regulators of actin dynamics are the members of the p21 Rho family of small GTPases, belonging to the Ras superfamily of small GTPases [2, 3]. These GTP-binding signaling molecules alternate between GDP-bound inactive and GTP-bound active

Roberto Botelho (ed.), *Phagocytosis and Phagosomes: Methods and Protocols*, Methods in Molecular Biology, vol. 1519,
DOI 10.1007/978-1-4939-6581-6_9, © Springer Science+Business Media New York 2017

state, in which they bind and activate wide array of downstream effector molecules to elicit a cellular response [4, 5]. The Rho GTPase family has 20 known members, and the canonical RhoA, Rac1, and Cdc42 that were first identified, are the best studied in the family [6]. Both Rac1 and Cdc42 have been shown to be required for FcγR-mediated phagocytosis [7, 8].

The traditional workhorse technique to study GTPase activation during cellular responses has been an affinity-based precipitation assay [9]. While this assay is highly useful, it provides only ensemble averages from whole cells that lack discrete resolution in space and time, which is not ideal for studying highly compartmentalized three-dimensional cellular structures, such as phagosomes. Forster resonance energy transfer (FRET)-based biosensors have become powerful tools to decipher spatial and temporal activation dynamics of Rho GTPases at high resolution on a single-cell basis, allowing researchers to gain further insights into their functional roles [10]. A previous study analyzing Rac1 and Cdc42 activity in macrophage phagosomes used bi-molecular versions of FRET biosensors, where the FRET donor and acceptor halves are on separate molecules [11]. This approach, while useful, involves cumbersome data analysis due to the non-equimolar distribution of the two separate FRET donor/acceptor components. We have overcome this issue by the development of fully genetically encoded, single-chain, FRET-based Cdc42 and Rac1 biosensors, applicable for fixed- and live-cell imaging [12, 13]. Importantly, our design maintains the C-terminal polybasic region of the Rho GTPases and allows for correct native intracellular localization and interaction with upstream regulators, including guanosine nucleotide dissociation inhibitor (GDI).

Here we detail approaches for expression of FRET-based biosensors in hematopoietic cells, using a murine monocyte/macrophage RAW 264.7 cell line as a model system and discuss important considerations relevant for successful expression of full-length biosensor that forms the critical basis for proper biosensor readout and data interpretation. In this chapter we describe methods for the implementation of biosensors to study phagocytosis and provide a specific example using the Cdc42 biosensor that shows Cdc42 activity at the phagosome in macrophages [12].

2 Materials

1. RAW/LR5 cells derived from RAW 264.7 [8].
2. GP2-293 packaging cell line (Clontech).
3. RAW/LR5 growth medium: RPMI 1640 with L-glutamine supplemented with 10% newborn calf serum (NBCS), 1% penicillin/streptomycin (Pen/Strep).

4. RAW/LR5 induction medium: RPMI 1640 with L-glutamine supplemented with 10% fetal bovine serum (FBS), 1% Pen/Strep.

5. RAW/LR5 infection medium: RPMI 1640 with L-glutamine supplemented with 5% NBCS, 1% Pen/Strep.

6. GP2-293 growth medium: DMEM supplemented with 10% FBS, 1% Pen/Strep, 1% Glutamax.

7. 0.05% Trypsin/EDTA.

8. Buffer with divalent (BWD): 125 mM NaCl, 5 mM KCl, 1 mM KH_2PO_4, 5 mM glucose, 10 mM $NaHCO_3$, 1 mM $MgCl_2$, 1 mM $CaCl_2$, and 20 mM HEPES.

9. Phosphate buffered saline (PBS) without calcium and magnesium.

10. 10 mM EDTA in PBS: dilute 0.5 M EDTA (pH 8) in PBS.

11. 3.7% formaldehyde in BWD.

12. 50% glycerol in PBS.

13. Tris-buffered saline (TBS): 137 mM NaCl, 24.7 mM Tris-base, pH 7.4.

14. OptiMEM.

15. FuGENE HD transfection reagent (Promega).

16. Lipofectamine 2000 transfection reagent (Invitrogen).

17. Retro-X virus concentrator (Clontech).

18. 0.01% poly-L-lysine (Sigma); dilute 1:10 in PBS for working solution.

19. 1.5 mL microcentrifuge tubes.

20. 15 and 50 mL polypropylene tubes.

21. 6 and 10 cm cell culture dishes.

22. 0.45 μm Surfactant-free Cellulose Acetate SFCA-membrane 28 mm syringe.

23. 12-well and 24-well cell culture plates.

24. 12 mm round glass coverslips.

25. $3'' \times 1'' \times 1$ mm microscope slides, plain.

26. 8 mg/mL Polybrene aqueous stock solution.

27. 1 mg/mL Doxycycline (Dox) aqueous stock solution.

28. 100 mg/mL G418 (neomycin) aqueous stock solution.

29. 100 mg/mL Zeocin aqueous stock solution.

30. Sheep blood alsevers.

31. Rabbit IgG anti-sheep red blood cell.

32. 18 Gauge needle with syringe.

33. DNA constructs: VSV-g, gag/pol, Rev, Tet, Cdc42, or other biosensors [12].

3   Methods

3.1   Expression of Rho GTPase Biosensors in a Macrophage Cell Line

3.1.1   Transient Expression of Rho GTPase Biosensors in RAW/LR5 Cells

As an example, we will use our protocol for the Cdc42 biosensor [12].

1. (Day 1) Plate RAW/LR5 cells in a 12-well plate the day before transfection so that they will be approximately 60–80% confluent for transfection the next day.

2. (Day 2) Let FuGENE HD (*see* **Note 1**) and OptiMEM come to room temperature (RT), ~10 min.

3. Prepare FuGENE HD transfection mix (DNA:FuGENE HD ratio 1:3): in 100 μL OptiMEM add 1 μg Cdc42 biosensor expression plasmid (*see* **Note 2**), vortex for 10 s; add 3 μL FuGENE HD and pipet to mix (do not vortex at this point). Incubate for 15 min at RT. The transfection mix can be scaled based on surface area if smaller or larger cell numbers are needed.

4. During incubation of the transfection mix, rinse cells once with PBS and add 500 μL of complete medium for transfection.

5. Add transfection mix dropwise. Swirl gently to mix.

6. Incubate 2–3 h (*see* **Note 3**) at 37 °C and 5% CO_2.

7. Transfer medium containing the transfection mix to a 15 mL tube to collect cells that may have detached during the incubation. Then lift adherent cells by adding 10 mM EDTA/PBS to the well and incubate for ~5 min at 37 °C and 5% CO_2. Tap gently to lift cells and transfer to the 15 mL tube. Rinse the well once with complete medium to collect all cells and add to the 15 mL tube. Spin cells at $300 \times g$ for 3 min.

8. Aspirate the fluid and resuspend the cell pellet in 2 mL complete medium.

9. Set up sterile 12 mm round coverslips in 24-well plate (4 coverslips / 12-well transfection) (*see* **Note 4**). Add 500 μL of cell suspension per coverslip.

10. Let cells recover overnight at 37 °C and 5% CO_2.

11. (Day 3) Next day, assess transfection efficiency and perform phagocytosis assays.

3.1.2   Generation of Stable RAW/LR5 with Inducible Expression of Rho GTPase Biosensors

There are several important considerations for successful creation of an inducible system for the expression of single-chain Rho GTPase biosensors in general, as well as specifically in hematopoietic cells (*see* **Notes 5** and **6**). Generating stable cell lines with inducible expression of a biosensor requires the creation of a double-stable, RetroX-Tet-OFF Advanced/gene of interest (GOI) cell line, involving two consecutive retroviral transductions of the target cell line (*see* **Notes 7** and **8**). Below we outline the protocols we use routinely in our laboratory:

1. (Day1) Start virus production by setting up 4×10 cm dish per GOI (*see* **Note 9**) to plate GP2-293 packaging cells (*see* **Note 10**).

2. Coat 10 cm dish with 0.001% poly-L-Lysine solution diluted in 1× PBS for at least 10 min at RT.

3. Plate 6×10^6 GP2-293 cells per coated 10 cm dish and incubate overnight at 37 °C and 5% CO_2.

4. (Day 2) Transfect GP2-293 cells using Lipofectamine 2000; per 10 cm dish mix (*see* **Notes 11** and **12**): in *tube 1* add 4 μg pVSV-g + 4 μg gag/pol + 16 μg Cdc42 biosensor [12], or other carrier DNA in 500 μL OptiMEM (for lentivirus: 4 μg pVSV-g + 2 μg gag/pol + 2 μg Rev + 2 μg Tet + 14 μg carrier DNA). In *tube 2* add 60 μL of Lipofectamine 2000 in 500 μL optiMEM.

5. Vortex tubes 1 and 2 for 10 s, spin and incubate 5 min at RT.

6. Combine tubes 1 and 2, vortex for 10 s, spin and incubate 20 min at RT (*see* **Note 13**).

7. During incubation of transfection mix, wash the cells with 1× PBS and add 4 mL DMEM containing 10% FBS with *no antibiotics*.

8. Add 1 mL transfection mix to cells dropwise. Gently swirl to mix.

9. Incubate overnight at 37 °C and 5% CO_2.

11. (Day 3) Supplement dishes with 3 mL serum-free DMEM to adjust serum concentration to 5%.

12. Transfer dishes to 32 °C since retroviruses are more stable at 32 °C (for lentivirus keep at 37 °C).

13. Allow cells to produce virus for 48 h.

14. (Day 5) Start harvesting virus by pooling supernatants from the 4 dishes into a 50 mL tube; swirl plate a little before pipetting the supernatant.

15. Spin 3 min at $300 \times g$.

16. Filter supernatant using 0.45 μm SFCA-membrane 28 mm syringe filter. Filter slowly, dropwise along the sides of a 50 mL tube.

17. Estimate the final volume of the filtrate and divide the volume by 3; that is the volume of Retro-X virus concentrator solution to add to the filtered viral titer.

18. Add virus concentrator straight into filtrate, pipet up and down to clean the pipet of all the virus concentrator solution. Invert gently a few times to get a homogenous solution.

19. Let it sit overnight at 4 °C.

20. Meanwhile, plate RAW/LR5 cells for infection in 6- or 10 cm dishes at approximately 20% confluency for next day. Low confluency is recommended as infection might occur over a 3-day period, depending on GOI to be expressed.

21. (Day 6) Collect virus by centrifuging tubes (from **step 19**) at $1500 \times g$ for 45 min at 4 °C.

22. Resuspend the virus pellet very gently by pipetting up and down in 200 μL of RPMI without serum (or the base medium in which the target cells are normally grown) per 10 cm dish (i.e., if 4×10 cm dishes are pooled, add 800 μL). At first the pellet will come apart in pieces, but will become a clear solution while pipetting up and down.

23. Aliquot at 100-120 μL.

24. Store aliquots at -80 °C. If transducing cells on the same day, keep aliquots at 4 °C until use. Freeze-thawing one time reduces the viral infectivity by at least 50%.

25. To transduce cells (*see* **Note 14**), start by rinsing cells once with PBS.

26. For transduction in 10 cm dish, add 5 mL RAW/LR5 infection medium, 5 μL of 8 mg/mL polybrene (final concentration of 8 μg/mL), and 1 aliquot of virus (for 6 cm dishes: 3 mL medium, 3 μL of polybrene stock and 1 aliquot of virus). The use of polybrene increases infectivity by 1000 times.

27. Incubate cells at 32 °C and 5% CO_2 (for lentivirus keep at 37 °C). Virus will have infectivity for 6–8 h.

28. At the end of the day, aspirate medium and repeat **step 26** for overnight incubation (*see* **Note 15**). Repeat morning and evening dosing of virus for *up to* 3 days to maximize infection efficiency (for lentivirus, do one dose per day for 2 consecutive days maximum). For generating the stable Tet-OFF tetracycline Trans-Activator (tTA)-expressing cell line, infect for 3 days (6 doses total) to achieve optimal integration of tTA for most efficient inducible system. Check infection efficiency for biosensor expression by fluorescence microscopy. If expression is very good (meaning most cells express and they are bright enough), then there is no need to infect on the 3rd day. Usually we start seeing expression after 3-4 doses (2 days of infection). If expression is low even after a 3-day infection, then it is an indication of a problem (*see* **Note 16**).

29. Let cells recover from infection for one day in complete medium. To repress biosensor expression, supplement medium with 2 μg/mL Dox. It is not necessary to supplement with Dox during generation of stable tTA-expressing cell line.

30. Start selection. For stable tTA integration select with G418 starting at 500 μg/mL and increase by doubling the concen-

tration, up to 2 mg/mL. For stable biosensor integration select with Zeocin starting at 250 μg/mL (in the presence of Dox) and increase by doubling the concentration, up to 1 mg/mL as cells become tolerant of the lower concentration. Selection may take up to 3 weeks. During selection cells will look stressed, however their morphology and growth will return to normal by the end of selection process.

31. For normal culture of double-stable RetroX Tet-OFF Advanced/ biosensor RAW/LR5 cells, maintain cells in 2 μg/mL Dox, 1 mg/mL G418 and 600 μg/mL Zeocin (*see* **Note 17**).

3.1.3 Induction of Biosensor Expression (See Note 18)

1. Wash cells once with PBS.

2. Aspirate and add 1 mL trypsin. Spread evenly and wait ~3 min. Tap side of the dish to dislodge cells. Check that the cells are detached and flowing smoothly by light microscopy or by holding the plate up against the light (*see* **Note 19**).

3. Add 4 mL RAW/LR5 induction medium. Pipette up and down and transfer to a 15 mL tube (*see* **Note 20**).

4. Rinse the plate with another 5 mL of induction medium and pool together in the 15 mL tube.

5. Spin 300 × *g* for 3 min.

6. Rinse with 7–10 mL induction medium and spin as above.

7. Aspirate and resuspend cell pellet in induction medium.

8. Plate cells at 1:10 dilution in 10 cm dish (this is based on the 50% confluency, *see* **Note 17**). For most efficient induction it is best to pass the cells at low confluency. One 10 cm dish will yield enough induced cells for a typical imaging session for ratiometric imaging. For western blot analysis of biosensor expression set up 4 dishes to yield enough cells for the lysate preparation.

9. Next day, repeat the trypsinization as above (**steps 1–7**). For ratiometric imaging the following day, pass at 1:3 or 1:4 onto 12 or 25 round mm coverslip. For preparing lysates for western blot analysis of biosensor expression levels over 72 h period, pass at 1:3 in 6 × 10 cm dishes for the 24 h time point, 1:3 in 3 × 10 cm dishes for 48 h time point, and 1:3 in 1 × 10 cm dishes for the 72 h time point after the **2nd** trypsinization (*see* **Note 21**).

3.2 Phagocytosis Assay

While there are several ways to assay phagocytosis, here we detail synchronized phagocytosis as performed in Hanna et al. [12].

3.2.1 Opsonization of Red Blood Cells (RBCs)

This procedure typically yields approximately $1\text{-}2 \times 10^8$ IgG-coated RBCs in 1 mL.

1. Using sterile technique, draw some blood from sheep blood alsevers vial using an 18 gauge needle with syringe. Transfer 250 μL sheep blood into 1.5 mL microcentrifuge tube.

2. Pellet RBCs by spinning for 10 s at maximum speed using a bench-top mini microcentrifuge.

3. Wash RBCs pellet twice with 1 mL BWD. Spin as above.

4. Resuspend RBCs in 1 mL BWD.

5. In parallel, add 5 µL rabbit IgG anti-sheep RBC (*see* **Note 22**) to 9 mL BWD in 15 mL polypropylene tube and mix by inverting few times.

6. Add RBCs to the antibody solution and mix immediately by inverting gently few times.

7. Incubate for 20 min at 37 °C water bath.

8. Incubate for 20 min on ice.

9. Pellet IgG-coated RBCs by centrifuging at $1250 \times g$ for 5 min.

10. Aspirate supernatant. Resuspend RBCs in 1 mL BWD and transfer to 1.5 mL microcentrifuge tube. Spin as in **step 2**.

11. Wash IgG-coated RBCs twice with 1 mL BWD. Spin as in **step 2**.

12. Resuspend in 1 mL BWD.

13. Store at 4 °C. Use within 7 days.

3.2.2 Synchronized Phagocytosis Assay

1. Cells expressing the biosensor either by transient expression or following stable induction are plated on 12 mm round coverslips in wells of a 24-well plate and allowed to attach and spread overnight at 37 °C.

2. Remove coverslips from 24-well plate and chill on ice (*see* **Note 23**) by replacing medium with 35–50 µL ice-cold BWD and let sit for 5–10 min on ice.

3. Add 25 µL of opsonized RBCs into 500 µL of ice-cold BWD and gently tap to mix.

4. Replace plain BWD with 35-50 µL of ice-cold BWD containing opsonized RBCs and incubate coverslip for 15 min on ice to allow RBCs to bind to the cells.

5. During **step 4** prewarm 500 µL BWD in 24-well plate in 37 °C water bath.

6. Rinse cells with ice-cold BWD on ice three times to wash away unbound RBCs.

7. Transfer coverslips to wells of 24-well plate containing prewarmed BWD in water bath for 1 min, or the desired time (1 min for early events and 5 min for maximal cup formation), before fixing with 3.7% formaldehyde in BWD.

8. Fix for at least 10 min. At this point, coverslips can be stored in BWD overnight at 4 °C or directly stained for imaging.

9. Permeabilize with 0.2% Triton X100 in BWD for 10 min.

10. Stain RBCs and actin by adding TBS containing Alexa Fluor 568 anti-rabbit IgG antibody (1:400) and Alexa Fluor 680-phalloidin (1:20) and incubate for 20-30 min at RT (*see* **Note 24**).

11. Rinse three times with TBS.

12. Mount on glass slide in 50 % glycerol in PBS.

3.3 Fixed Cell Imaging and Data Analysis

Activation of Rho GTPases at individual phagocytic events, using the Cdc42 biosensor in this example, is measured by observing the ratio of FRET emission to the donor mCerulean emission. We briefly outline the steps below:

1. Using a 60× magnification objective lens (60× DIC N/A 1.45), acquire mCerulean (CFP), FRET, and mVenus (YFP) emission images upon excitation by mCerulean, mCerulean, and mVenus excitation wavelengths, respectively (*see* **Note 25**). As shown in our previous publication [12], Fig. 1a shows a representative ratiometric image of localized Cdc42 activity with an F-actin-rich phagocytic cup at the F-actin focal plane (*see* **Notes 26** and **27**).

2. The optimal focal plane for the phagocytic cup can be set by observation of co-localization of F-actin staining (phalloidin) with a bound RBC. Alternatively, confocal imaging could be used to reduce the out of focus light when obtaining a Z-stack (*see* **Note 28**).

3. Obtain the shading correction image set (*see* **Note 29**), and the camera noise correction image set (*see* **Note 30**) as previously described [14, 15], taking care to maintain the exposure and imaging conditions identical to the actual data acquisition. If imaging a Z-stack, obtain appropriate shading correction image sets at the correct Z-distance positions.

4. Process the background subtraction (*see* **Note 31**), threshold-masking (*see* **Note 32**), and ratio calculations (*see* **Note 33**) at each Z-position as previously described [14, 15].

5. Apply a linear pseudocolor lookup table to the ratio images and adjust the image scaling appropriately to visualize the Cdc42 activation patterns at phagocytic cups. Make sure to apply identical scaling limits to all planes if imaging a Z-series.

6. Alternatively, process the raw Z-series data set using a deconvolution algorithm to enhance the signal to noise ratio and then process the Z-series as in **steps 3–5** above. This deconvolution enhances the information obtained in the ratio images. Compare Fig. 1 and the same images that have been subsequently deconvolved and shown in Fig. 2 (*see* **Notes 34** and **35**).

4 Notes

1. Macrophage cell lines are challenging to transfect in general. We have had the most success with FuGENE HD, giving a transfection efficiency of ~10 %. However, the transfection effi-

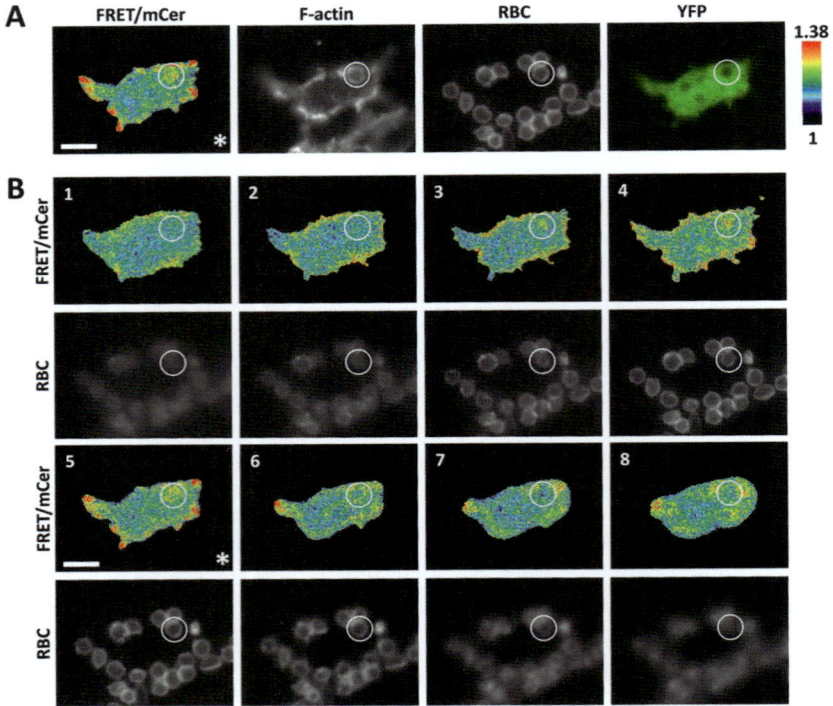

Fig. 1 Cdc42 activity over a stack of Z-serial planes containing a developing phagosome. RAW/LR5 cells transiently expressing the Cdc42 biosensor were imaged (**a**) at optimal F-actin focal plane and (**b**) in Z-series at 1 μm-steps where the focal plane from A was set as the center (*). Planes 4-1 progress down towards the base of the phagocytic cup and below, while planes 6–8 move upwards from the F-actin plane. FRET/mCerulean ratio image, phalloidin staining of F-actin, red blood cell (RBC) staining and YFP for biosensor localization are shown; representative image set of *n* = 6 cells. Scale bar = 10 μm. Modified from Hanna et al. [12]

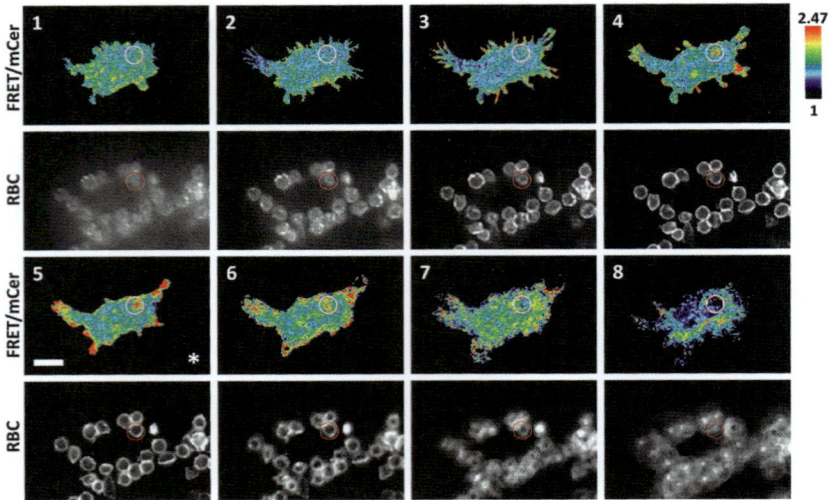

Fig. 2 Cdc42 activity over serial planes of the phagosome post image deconvolution. An identical set of raw input images presented in Fig. 1 was processed for deconvolution using the Microvolution GPU-deconvolution package [28]. Image processing and ratiometric calculations were performed as in Fig. 1

ciency also varies greatly depending on the GOI being expressed. The DNA amount and DNA:FuGENE HD ratio have been carefully optimized for RAW/LR5 cells, thus the protocols described may be suboptimal for other cell types.

2. The Cdc42 biosensor cassette was subcloned into the pTriEX-4 vector backbone (Novagen). The pTriEX-4 vector can be used for transient overexpression of the biosensor in mammalian cells but it does not contain any mammalian antibiotic resistance gene to select for stable expression.

3. Overnight incubation of the cells increases transfection efficiency, while cell health is adversely affected and preactivation can also be a confounding issue. For a simple protein expression analysis an overnight incubation is tolerable. However, for downstream cellular assays we routinely opt for shorter transfection times, in the order of 2–3 h. For example, if one is interested in studying podosome function, short incubation times with the transfection mixture is critical. We found that podosomes are no longer observable if cells are incubated longer than 2–3 h with the transfection mixture.

4. Keep stock of 12 mm round coverslips in 100% ethanol and flame them to sterilize.

5. Using viral transduction for the generation of stable cell lines of GOIs is problematic for constructs that contain tandem, repeated sequences (such as our FRET biosensors) since they are susceptible to homologous recombination from the intrinsic properties of retrovirus [16, 17]. This may result in internal deletions of the repeated sequences within the GOI. Single-chain biosensors incorporate a FRET pair of two fluorescent proteins that are highly homologous. In addition, our designs for Rac1 and Cdc42 biosensors include two tandem repeats of the p21 binding domain of PAK1 [12, 13]. The presence of two sets of repeated sequences places the biosensors at extremely high risk for homologous recombination. We recently reported a generalizable solution, in which "synonymous modification" of the DNA sequence encoding the biosensor offered a simple means to overcome this significant problem [18]. We now routinely apply synonymous modification to all of our biosensor systems.

6. Another important factor in stable biosensor expression is proper promoter usage for ectopic gene expression. Different promoters have variable resistance against potential promoter silencing by host cells [19]. In our second-generation Tet-OFF system we express tTA under the human elongation factor 1α (EF1α) promoter, as the traditionally used CMV promoter is subject to silencing in hematopoietic cells [20, 21].

7. In our laboratory we use the second-generation Tet-OFF inducible system from Clontech for stable expression of all of our biosensors. Generating stable cell lines with inducible expression of biosensor requires the creation of a double-stable, RetroX-Tet-OFF Advanced/GOI cell line, involving two consecutive infections of the target cell line. First infection generates a cell line with stable, constitutive expression of tTA. The stable tTA cell line then serves as the subsequent base cell line for a second round of infection for the stable incorporation of any GOI for inducible expression. Clontech now has available a more advanced third-generation Tet-ON 3G inducible system [22, 23], however we find that this system is not particularly suitable for FRET biosensor expression. The Tet-ON system requires the presence of Dox in medium for the induction of the GOI. Importantly, Dox possesses intrinsic fluorogenic properties [24]. Macrophages, as phagocytes, internalize and accumulate considerable amount of Dox, creating measurable background auto-fluorescence that overlaps with the spectral properties of the fluorescent protein FRET pair (i.e. mCerulean and mVenus pair) used in our Rho GTPase biosensors, which impacts the ratiometric calculations. Consequently, it is best to use a Tet-OFF system for biosensor expression that requires removal of Dox for induction of the GOI.

8. This procedure involves production of retrovirus that is considered a biosafety level 2 organism by the National Institute of Health and Center for Disease Control. This requires the observance of Biosafety Level 2 practices. The viral packaging system used consists of retroviral components that are on separate plasmids, and the resulting retroviruses produced are replication-incompetent, adhering to the Biosafety Level 2 requirements.

9. The number of plates can be scaled as needed. If using HEK293 for virus production, we recommend supplementing transfection mixture either with pCL-Eco for ecotropic or pCL-Ampho for amphotropic packaging system, depending on the target cell. Ecotropic infects rodents (but not hamster) only, while amphotropic infects rodents and human (but not hamster). Some cells have a block on retroviral infection (i.e. neurons) so those must be infected with lentivirus. VSV-g -pseudo-typed virus will infect everything including hamster cells.

10. It is important to achieve a single cell suspension (no clumps) of GP2-293 cells both for passaging during normal cell culture and plating for virus production. In addition, GP2-293 cells can only be kept in culture for a limited number of passages after thawing, approximately 3–4 weeks since virus production decreases as they age. One indication for discontinuing their culture is that they proliferate at higher rate. While the manufacturer does recommend plating them on gelatin-coated dishes for culture purposes we find that this is not necessary.

11. In our hands Lipofectamine 2000 is most optimal for highest expression by transient transfection of GP2-293 cells. In addition, we find Lipofectamine 2000 better suited for homogenous transfection of multiple-DNA plasmids in a single transfection mixture.

12. This protocol is for the production and transduction of retroviruses. Modifications relevant for lentivirus will be noted in the protocol. Scale amount of transfection mixture as needed and prepare a master mixture.

13. Transfection mixture is viable for 4–6 h.

14. If frozen stable tTA expressing cells are to be infected with a GOI, thaw out the cells at least a week ahead of the planned infection and keep them in G418 at 2 mg/mL. During selection for GOI, do not use G418 as all three antibiotics at these high concentrations are too toxic for the cells to tolerate.

15. It is not a good idea to add a new aliquot of virus to the existing virus-containing medium because too much virus could be toxic. Therefore it is best to aspirate the medium prior to applying an additional dose of virus.

16. There are several possible reasons for poor biosensor expression. It may be that the GP2-293 cells are too old or not handled properly during normal culture. GP2-293 cells should not be allowed to overgrow and it is important to obtain single cell suspension both during normal passage and plating for virus preparation. It may be possible that the viral aliquot is too old since frozen virus loses infectivity over time. Also, it is possible that the plasmid preparation may contain contaminants that interfere with downstream applications. We have found that it is critical to use purified water free of any bioorganic contaminants during any cloning or plasmid preparation procedures; use either commercially available molecular biology grade water or in-house purified water (such as using Millipore system) that has been certified to be free of bioorganic contaminants.

17. The stable inducible RAW/LR5 cells grow at somewhat slower rate than the parental counterpart. This is due to the presence of three antibiotics in the medium, and not the inducible system itself. Once the antibiotics are removed for induction, the growth rate returns to that of the parental line. Furthermore, it is not advisable to overgrow the inducible cell line; we routinely maintain them in culture at 50% confluency.

18. When a stable inducible RAW/LR5 cell line is first created and then induced for biosensor expression, the cells will induce at variable expression levels, ranging from no expression to dim and moderately high expression. It is possible to sort cells by FACS to obtain a near 100%-expressing population. Optimal expression level can be determined empirically based on the desired signal-

to-noise level during FRET imaging, avoiding dominant negative effects from overexpression of the biosensor.

19. In our experience we find that detaching with RAW/LR5 cells with 10 mM EDTA/PBS, as used during normal culture, prevents robust induction and the cells require trypsinization instead. In addition, a complete removal of Dox from the medium is essential for a successful induction. Residual Dox could remain internally within cells or as cell-membrane associated fraction. We find that, in the case of RAW/LR5 cells, a complete removal of Dox requires two rounds of trypsinization over a 48 h period in combination with very sparse plating (<20% confluency).

20. It is imperative that the serum used for induction has been checked for residual tetracycline. If there are trace amounts of tetracycline in the serum, it will likely prevent induction. We find that NBCS is suboptimal for induction, as it appears to contain trace amounts of tetracycline depending on the particular lot. We routinely use FBS that is tested to be tetracycline-free in the induction medium.

21. For some imaging applications, it may be that a dilution of 1:3 or 1:4 results in coverslips that are too sparse. However, for robust biosensor induction to occur the cells need to be passaged at low confluency post 2nd trypsinization. If a higher confluency is needed the cells should be trypsinized for a third time and then plated at desired confluency for next-day imaging.

22. It is necessary to titrate every new batch of rabbit anti-sheep RBC IgG as too little antibody will decrease the efficiency of phagocytosis. However, excess antibody will cause aggregation of the RBCs during preparation. For each new batch of antibody make small batches of IgG-coated RBCs varying the concentration of the antibody in a range from 1 to 10 μL with 5 μL being the anticipated optimal concentration. Examine the IgG-coated RBC for aggregation with light microscopy.

23. It is best not to place the 12 mm round coverslips directly on ice. We normally use the lid of a cell culture plate (rectangular) lined with a piece of parafilm to provide a nonslip surface.

24. While other fluorochromes can be utilized, these wavelengths were chosen to be compatible for imaging with CFP-YFP FRET pair of the Cdc42 biosensor. In addition, optimal filter settings are required to minimize spectral bleedthroughs.

25. For fixed cell imaging of macrophages we have used a single CoolsnapHQ2 camera (Roper Photometrics) attached on the bottom 100%-throughput port of the microscope. In this configuration, excitation and emission filterwheels allow for the switching of appropriate filter sets to acquire mCerulean,

FRET, and mVenus emissions. The optical specifications for this configuration are detailed in Spiering et al. [25]. Transilluminated DIC images can also be obtained.

26. As phagosomes are three-dimensional structures we also imaged the same phagosome over serial planes in Z-direction, where the F-actin focal plane from Fig. 1a was set as the center position of the Z-stack (Fig. 1b) [12]. For the collection of Z-stack serial planes the optimal focal plane is set as the center of the stack we have found that eight planes at 1 μm steps was sufficient.

27. It is important to control for the relative expression levels when imaging fixed macrophages transiently overexpressing the biosensor. It is necessary to image only those cells that have a fluorescent intensity sufficient to fill approximately 80% of the dynamic range of a detector to maximize the signal to noise ratio of the data being captured [14, 15]. For our optical setup, this requirement translates to using excitation intensities of 0.4–1.0 mW/cm^2 at the specimen plane with a camera exposure time in the range of 500–1000 ms.

28. Because of the three-dimensional structure of phagosomes, epifluorescence wide-field imaging, as in Fig. 1, cannot clearly resolve the overall structure of phagocytic cups due to out of focus light being included in the imaging focal plane. This problem can be exacerbated further due to the mechanics of ratiometric analysis where FRET and donor emissions could scatter differentially because of lateral chromatic aberration effects resulting in unequal levels of out of focus light being present within a ratiometric set of images at each Z-section. To address this, confocal imaging is often the first choice in attempting to resolve and to reconstruct these three-dimensional cellular structures by directly removing the out of focus light. However, the available lasers used for excitation of FRET and donor mCerulean are often suboptimal (i.e., 405 nm and/or 458 nm) in a typical laser-based confocal system. These two laser lines miss the optimal excitation peak of mCerulean excitation spectra and result in either: (a) requiring overexpression of the biosensor to compensate for weak excitation; or (b) reduced excitation of FRET and mCerulean that produces a suboptimal signal to noise ratio in the final ratiometric data. Therefore, we recommend sourcing a 445 nm laser that will maximally excite FRET and mCerulean during imaging to allow for better final signal to noise levels in the ratiometric data set.

29. In a wide-field microscopy, evenness of field illumination within the imaging field of view is not always guaranteed. This is especially the case if an arc-lamp coupled to a condenser sys-

tem is used for illumination (typically, the center of the field of view will be brighter than the edges). Even in cases where a light source enters the system via a fiber or a liquid light guide (i.e., lasers, monochromators, light engines etc.) coupling to a beam expander-collimator assembly, the evenness of the field illumination should also be checked and corrected, as small imperfections in alignment will create measurable effects in the final ratio data if uncorrected. To perform this correction, images of empty fields of view are collected at the same exposure conditions as the experimental image acquisitions, and then used to divide out any unevenness in the field brightness (typically termed "flatfield correction" or "shading correction") [14, 15].

30. For a typical cooled CCD camera (sCMOS camera will be similar, albeit typically with less associated noise levels in general) used in image acquisitions, the camera "noise" is present within each frame of image set. This is a combination of the image "shot noise" associated with the actual foreground data within the image (differences in the arrival times of photons at every element contributes to this noise, thus longer exposure time averages out this effect), the number of times the data in the detector elements are "read" for digitization ("read noise"; this depends on the architecture of the data reading buffer system of the detector chip as well as the build and shielding quality of the digitization circuitry), and the time duration of exposure per frame which contributes to the production of thermal noise ("dark current noise"; thus, cooling the sensor chip is effective at reducing this noise as it is temperature dependent). The combined noise image acquired in absence of all light hitting the detector chip but using an identical exposure time as the experimental image acquisition condition produces the camera noise image containing appropriate noise levels for the particular imaging condition used for the actual experiment. The noise image is subtracted from the raw data images and the shading correction images [14, 15].

31. The average background intensity value is subtracted from the shading corrected FRET, mCerulean and mVenus images. The average background value is determined by drawing a small region of interest away from any foreground signal and averaging the gray values within such a region of interest. Because the shading correction and noise subtraction were applied to the data image sets, the background intensity levels should be more or less uniform everywhere within the image [14, 15].

32. Even after the background subtraction, regions outside of the cell are not going to be uniformly at zero pixel intensity but rather at a stochastic distribution of low pixel intensity values. If such image sets were divided against each other for ratiomet-

ric calculation, the stochastic nature of the intensity distribution will produce speckles in regions outside of the cell edge, making data interpretation difficult. To avoid this effect, a histogram of pixel intensity values is manually thresholded such that the "background" regions outside of the cell edges can be set to zero pixel intensity. This is used to produce a binary cell mask, in which regions outside of the cell is now uniformly set to zero and the region inside of the cell is set uniformly to one. This binary cell mask is then multiplied into the shade-corrected, background-subtracted data sets prior to the final ratio calculations [14, 15].

33. For the ratio calculations, the masked FRET image is divided by the masked donor mCerulean image. In this step, it is critical to ascertain that pixel-to-pixel match is achieved between the two images to be divided. It is not correct to assume that both images are properly aligned with each other just because a single camera is used to acquire all of the data sets. Wavelength-dependent shifts could be present in the optical setup, depending on the thickness differences and the imperfections in the mounting angles of the bandpass filters within the focused section of the microscope, field lateral chromatic dispersion effects of the objective lens, and dichoric mirror mounting imperfections, etc., which could impact pixel-to-pixel alignment of the two images to be divided. This can be a priori calibrated and corrected using multispectral beads and nonlinear coordinate transformation approaches, and using a cross-correlation-based approach to optimize the X-Y translational alignment within 1/20th subpixel accuracy [14, 26, 27].

34. Another approach to resolve the three-dimensionality of phagosomes, which we show an example here (Fig. 2), is to use a deconvolution technique on the wide-field epifluorescence Z-series data. Deconvolution is a mathematical operation in which the out of focus light at any Z-position is removed through inverting the effect of the point-spread function of the objective lens on the specimen. When a point-source specimen is imaged through an objective lens, the point-spread function of the lens will describe how such a point-source will spread and blur in x, y, and z directions. If this function is known, then it is possible to iteratively and quantitatively solve for the inverse function to reconstruct the original point-source specimen image, effectively removing the out of focus light. The main advantage of the deconvolution technique is that if it is used on a wide-field epifluorescence data set there will be no loss of light, unlike confocal imaging in which a physical pin-hole is used to remove out of focus light. This yields better signal to noise and reduces the level of biosensor expression needed to achieve workable fluorescence intensity levels.

35. The major drawback of using the deconvolution approach is the high computational load. Often, iterative calculations of a data set would take minutes to hours in order to obtain the final result, so this approach has not been amenable to live-cell imaging and analysis on-the-fly. This problem could be addressed by parallel computing in which calculations are parsed over many computers and central processing units (CPUs), but such arrangements are cost prohibitive for most ordinary researchers. This requirement has now changed through the utilization of graphical processing units (GPUs), resident on computer video cards. These newer GPU cards, which are quite affordable at only a few hundred dollars, contain thousands of parallel computing processing units to accelerate the video graphics capability of a computer. A new deconvolution software package from Microvolution LLC [28] is now taking advantage of GPUs to accelerate the deconvolution calculations by several orders of magnitudes. This package was used to deconvolve the same Z-series used in Fig. 1 to show the difference in the resulting ratiometric data (Fig. 2). The deconvolution processing using the GPU-based system from Microvolution took approximately 5 s each, and ratio calculations were performed on the deconvolved images. Immediately, we observed an improvement in the signal to noise ratio, resulting from the removal of out of focus light. This improvement was effectively in the order of a fourfold increase in the final ratio dynamic range (original normalized ratios ranged from 1.0 to 1.38; post-deconvolution range: 1.0–2.47). In addition, we now observe clear filopodial structures (Fig. 2, panels 2 and 3) in which Cdc42 is activated as expected [6, 29]. Furthermore, we observe an apparent dorsal cup closure event with associated Cdc42 activity (Fig. 2, panels 4–6), which is less clear from the ratio images prior to deconvolution (Fig. 1). We also observe additional activation "hot spots" in regions where RBCs are contacting the cell edges. These results indicate the ability of the deconvolution technique to improve the signal to noise ratio, contributing to a more precise interpretation of biosensor readouts in three dimensions. Furthermore, the speed improvement in the processing by several orders of magnitudes offered by the Microvolution system now enables an in-line processing of the Z-series, on-the-fly, as the image acquisitions are performed on the microscope. This allows integration of deconvolution processing directly in the actual image data acquisition work flow for the first time.

Acknowledgements

This work was supported by National Institutes of Health grants T32GM007491 to VM, GM071828 to DC, and GM093121 to LH.

References

1. Rougerie P, Miskolci V, Cox D (2013) Generation of membrane structures during phagocytosis and chemotaxis of macrophages: role and regulation of the actin cytoskeleton. Immunol Rev 256:222–239

2. Hall A (1998) Rho GTPases and the actin cytoskeleton. Science 279:509–514

3. Wennerberg K, Rossman KL, Der CJ (2005) The Ras superfamily at a glance. J Cell Sci 118:843–846

4. Jaffe AB, Hall A (2005) Rho GTPases: biochemistry and biology. Annu Rev Cell Dev Biol 21:247–269

5. Bishop AL, Hall A (2000) Rho GTPases and their effector proteins. Biochem J 348(Pt 2):241–255

6. Heasman SJ, Ridley AJ (2008) Mammalian Rho GTPases: new insights into their functions from in vivo studies. Nat Rev Mol Cell Biol 9:690–701

7. Caron E, Hall A (1998) Identification of two distinct mechanisms of phagocytosis controlled by different Rho GTPases. Science 282:1717–1721

8. Cox D, Chang P, Zhang Q, Reddy PG, Bokoch GM, Greenberg S (1997) Requirements for both Rac1 and Cdc42 in membrane ruffling and phagocytosis in leukocytes. J Exp Med 186:1487–1494

9. Jennings RT, Knaus UG (2014) Rho family and Rap GTPase activation assays. Methods Mol Biol 1124:79–88

10. Pertz O (2010) Spatio-temporal Rho GTPase signaling—where are we now? J Cell Sci 123:1841–1850

11. Hoppe AD, Swanson JA (2004) Cdc42, Rac1, and Rac2 display distinct patterns of activation during phagocytosis. Mol Biol Cell 15:3509–3519

12. Hanna S, Miskolci V, Cox D, Hodgson L (2014) A new genetically encoded single-chain biosensor for Cdc42 based on FRET, useful for live-cell imaging. PLoS One 9, e96469

13. Moshfegh Y, Bravo-Cordero JJ, Miskolci V, Condeelis J, Hodgson L (2014) A Trio-Rac1-Pak1 signalling axis drives invadopodia disassembly. Nat Cell Biol 16:574–586

14. Spiering D, Bravo-Cordero JJ, Moshfegh Y, Miskolci V, Hodgson L (2013) Quantitative ratiometric imaging of FRET-biosensors in living cells. Methods Cell Biol 114:593–609

15. Hodgson L, Shen F, Hahn K (2010) Biosensors for characterizing the dynamics of rho family GTPases in living cells. Curr Protoc Cell Biol. Chapter 14, Unit 14 11 11–26

16. An W, Telesnitsky A (2002) Effects of varying sequence similarity on the frequency of repeat deletion during reverse transcription of a human immunodeficiency virus type 1 vector. J Virol 76:7897–7902

17. Delviks KA, Pathak VK (1999) Effect of distance between homologous sequences and 3' homology on the frequency of retroviral reverse transcriptase template switching. J Virol 73:7923–7932

18. Wu B, Miskolci V, Sato H, Tutucci E, Kenworthy CA, Donnelly SK et al (2015) Synonymous modification results in high-fidelity gene expression of repetitive protein and nucleotide sequences. Genes Dev 29:876–886

19. Qin JY, Zhang L, Clift KL, Hulur I, Xiang AP, Ren BZ et al (2010) Systematic comparison of constitutive promoters and the doxycycline-inducible promoter. PLoS One 5, e10611

20. Ramezani A, Hawley TS, Hawley RG (2000) Lentiviral vectors for enhanced gene expression in human hematopoietic cells. Mol Ther 2:458–469

21. Salmon P, Kindler V, Ducrey O, Chapuis B, Zubler RH, Trono D (2000) High-level transgene expression in human hematopoietic progenitors and differentiated blood lineages after transduction with improved lentiviral vectors. Blood 96:3392–3398

22. Loew R, Heinz N, Hampf M, Bujard H, Gossen M (2010) Improved Tet-responsive promoters with minimized background expression. BMC Biotechnol 10:81

23. Zhou X, Vink M, Klaver B, Berkhout B, Das AT (2006) Optimization of the Tet-On system for regulated gene expression through viral evolution. Gene Ther 13:1382–1390

24. Khader H, Solodushko V, Al-Mehdi AB, Audia J, Fouty B (2013) Overlap of doxycycline fluorescence with that of the redox-sensitive intracellular reporter roGFP. J Fluoresc 24:305–311

25. Spiering D, Hodgson L (2012) Multiplex imaging of Rho family GTPase activities in living cells. Methods Mol Biol 827:215–234

26. Shen F, Hodgson L, Rabinovich A, Pertz O, Hahn K, Price JH (2006) Functional proteometrics for cell migration. Cytometry A 69:563–572

27. Danuser G (1999) Photogrammetric calibration of a stereo light microscope. J Microsc 193:62–83

28. Bruce MA, Butte MJ (2013) Real-time GPU-based 3D deconvolution. Opt Express 21:4766–4773

29. Ridley AJ (2011) Life at the leading edge. Cell 145:1012–1022

Chapter 10

Analysis of LC3-Associated Phagocytosis and Antigen Presentation

Laure-Anne Ligeon, Susana Romao, and Christian Münz

Abstract

The noncanonical macroautophagy pathway, LC3-associated phagocytosis (LAP) has recently emerged as an important catabolic process involved during exogenous antigen processing. It has been described that in human macrophages and dendritic cells the direct recruitment of LC3 to the phagosomal membrane is associated with its maturation impairment, allowing the stabilization of the cargo to prolong antigen presentation on major histocompatibility complex (MHC) class II molecules.

In this chapter, we describe methods to monitor, manipulate, and understand the role of LAP during MHC class II presentation. We show how to enhance LAP formation resulting in antigen presentation by using zymosan or beads coated with *Candida albicans* extract. Then, we describe how to determine the localization of Rab7 or Lamp2 on LC3-phagosomes by confocal microscopy, a useful technique to follow phagosome maturation. Finally, we propose an assay to understand how MHC class II antigen presentation can be modulated by the LAP pathway.

Key words Autophagy, Confocal microscopy, LC3-associated phagosome, MHC class II, MHC class II antigen presentation, Phagosome maturation, Phagosome assay

1 Introduction

T cells are a key component of the adaptive immune response, because of their ability to recognize antigen fragments presented to them by major histocompatibility complex (MHC) molecules. CD8[+] T lymphocytes recognize proteins, degraded by the proteasome, as peptides loaded onto MHC class I molecules, while products of lysosomal degradation are bound to MHC class II molecules and recognized by CD4[+] helper T cells. MHC class II molecules mostly present antigens derived from exogenous proteins acquired by phagocytosis or endocytosis by antigen presenting cells (APC). These endocytosed antigens are then delivered to lysosomes for processing. However, there are also different intracellular transport pathways, by which antigens can be degraded and delivered to MHC class II loading compartments, one of them being macroautophagy.

Roberto Botelho (ed.), *Phagocytosis and Phagosomes: Methods and Protocols*, Methods in Molecular Biology, vol. 1519, DOI 10.1007/978-1-4939-6581-6_10, © Springer Science+Business Media New York 2017

During macroautophagy more than 30 autophagy related (*atg*) gene products are involved in generating a double-membrane surrounded vesicle, called autophagosome. The microtubule-associated protein light chain 3 (LC3, mammalian homologue of Atg8) exists in two forms, a cytosolic form (LC3-I) and a lipidated form (LC3-II). LC3-II is conjugated to the forming autophagosome membrane and thought to facilitate membrane fusion events. In a final step, the autophagosome fuses with lysosomes for degradation of its cargo and the inner autophagosome membrane [1]. Cytoplasmic protein aggregates and damaged organelles are the main targets of macroautophagy. In fact, Atgs have been identified as regulators of endogenous [2, 3] and exogenous antigen processing [4] for MHC class II presentation. For instance, macroautophagy was described to play a role in delivering Epstein Barr virus antigen for MHC class II presentation to CD4+ T cells [2]. More recently, we have shown that in addition to the function of classical macroautophagy in MHC class II antigen processing, the LC3-associated phagocytosis (LAP) pathway also contributes to exogenous antigen processing [4]. Briefly, LAP discovered in 2007 by Sanjuan and collaborators, is characterized by the direct recruitment of lipidated LC3 (LC3-II) to the phagosomal membrane [5]. In addition, cargo that for example engages the C-type lectin receptor Dectin-1 and/or the toll like receptor TLR2 is able to activate the LAP pathway, resulting in the direct LC3 recruitment to the phagosomal membrane, for durable antigen presentation by MHC class II molecules [4, 6]. In fact, we described that the involvement of the LAP pathway in human monocyte-derived macrophages and dendritic cells (DCs) allowed for the stabilization of TLR2-containing phagosomes for prolonged MHC class II antigen presentation [4]. It has been previously shown that potent antigen presenting cells (APCs), such as DCs are able to preserve antigen for continuous MHC class II loading [7, 8], while macrophages are known to rapidly degrade endocytosed material (Table 1). The discovery that the subset of LC3-positive phagosomes surrounding TLR2-binding cargo allowed antigen preservation for

Table 1
Choice of cell types: macrophages versus dendritic cells

Macrophages	Dendritic cells
– Professional Antigen presenting cells – Degrade endocytic material rapidly – After 18 h the *C. albicans* T-cells clone are still activated [4]	– Potent professional antigen presenting cells – Limit the degradation in their endocytic compartment [8] – Maintain and sustain the antigen presentation longer [7] – After 36 h the *C. albicans* T-cells clone are still activated [4]

longer MHC class II loading in macrophages supports the notion that macrophages can also be potent APCs [4, 7]. Our TLR and Dectin-1 receptor targeting methods take advantage of this knowledge and tools being available to study macroautophagy in this process. In fact, the main protein involved in the LAP pathway is LC3. The molecular mechanism involved during LC3 coupling to the autophagosome membrane is quite well understood and many fluorescent constructs are now available for this protein. Furthermore, one of the Atg proteins implicated in LC3 coupling to the autophagosome is the Atg5 protein. Atg5 can be targeted by specific shRNAs to inhibit LAP and allows us to identify the role of LAP during MHC class II antigen presentation (Table 2).

The respective assays are discussed in this chapter. First, we propose different stimuli that engage Dectin-1 or TLR2 and that can be used to enhance LC3-associated phagosome formation. In fact, zymosan particles of the yeast cell wall were traditionally used to promote LAP formation but in our previous study we also used latex beads coated with Pam3CSK4 or *C. albicans* extract to stimulate LC3 recruitment to the phagosome (Fig. 1). Secondly, we discuss how protein markers can be used to monitor the maturation of the subset of LC3-associated membranes that surrounds TLR2-bindings cargo (Table 3). And finally, we describe how the involvement of the LAP pathway for prolonged antigen presentation by MHC class II molecules can be studied by using MHC class II antigen presentation assays with *C. albicans* specific CD4+ T-cell clones and by measuring the cytokines IL-17 and IFN-γ, which are released by these T cells. These assays can be a valuable tool to elucidate the role of LAP during adaptive immune responses.

2 Materials

2.1 Cell Cultures

All the experiments are performed with human primary cells. For the choice of cell type *see* Table 1.

2.1.1 Human Macrophages

1. Peripheral blood mononuclear cells (PBMCs) are isolated from leukocyte concentrates by density-gradient centrifugation on Ficoll-Paque PREMIUM (GE Healthcare).

2. CD14+ positive cells are isolated by positive magnetic cells separation using an autoMACS cell separator (Miltenyi Biotec).

3. CD14+ monocytes are differentiated into monocyte-derived macrophages with human granulocyte-macrophage colony stimulating factor (GM-CSF) and matured as described in Subheading 3.1.1. Differentiated macrophages are used between day 6 and 8 of culture.

4. Macrophage cell culture medium: Dulbecco's modified Eagle's medium (DMEM,) supplemented with 10% heat-inactivated

Table 2
Advantages and disadvantages to study the expression of proteins using overexpression, endogenous expression, or after RNA silencing

	Overexpression of Atg proteins	Endogenous expression of Atg	Silencing of Atg5
Advantage	– Directly fluorescent tagged proteins – Direct and live analysis of the protein localization in response to the stimuli – Analysis of protein-protein interaction – Investigate the protein roles if no antibody is available	– Observe the physiological phenotype – Physiological localization of the proteins – Observe the interaction with other proteins	– Investigate the role of the autophagy machinery – Investigate the importance of the proteins in the investigated biological process – Can be coupled with the overexpression of other proteins
Disadvantage	– The overexpression of the protein can cause artifacts – Wrong localization of the proteins due to overexpression – Toxicity of overexpression for the cells	– Sometimes no antibodies are available for the investigated protein or the choice of the antibody source species is limited – No live cell imaging – Phenotype is too rare to be easily observed	– Lack of the proteins can be toxic for the cells – Difficult to transduce cells – Efficiency of shRNA can be low
Which kind of application	Immunofluorescence staining Immunoprecipitation Western blot Live imaging	Immunofluorescence staining Immunoprecipitation Western blot	Immunofluorescence staining Immunoprecipitation Western blot Live cell imaging
Notes	*For a complete and systematic analysis of LAP formation, it is important to combine all these techniques to observe the expression, the localization and the role of LC3 during LAP formation. In our case, we only present the silencing of the atg5 gene, but for a complete study of the autophagy involvement during MHC class II antigen presentation other Atg proteins should be silenced in addition*		

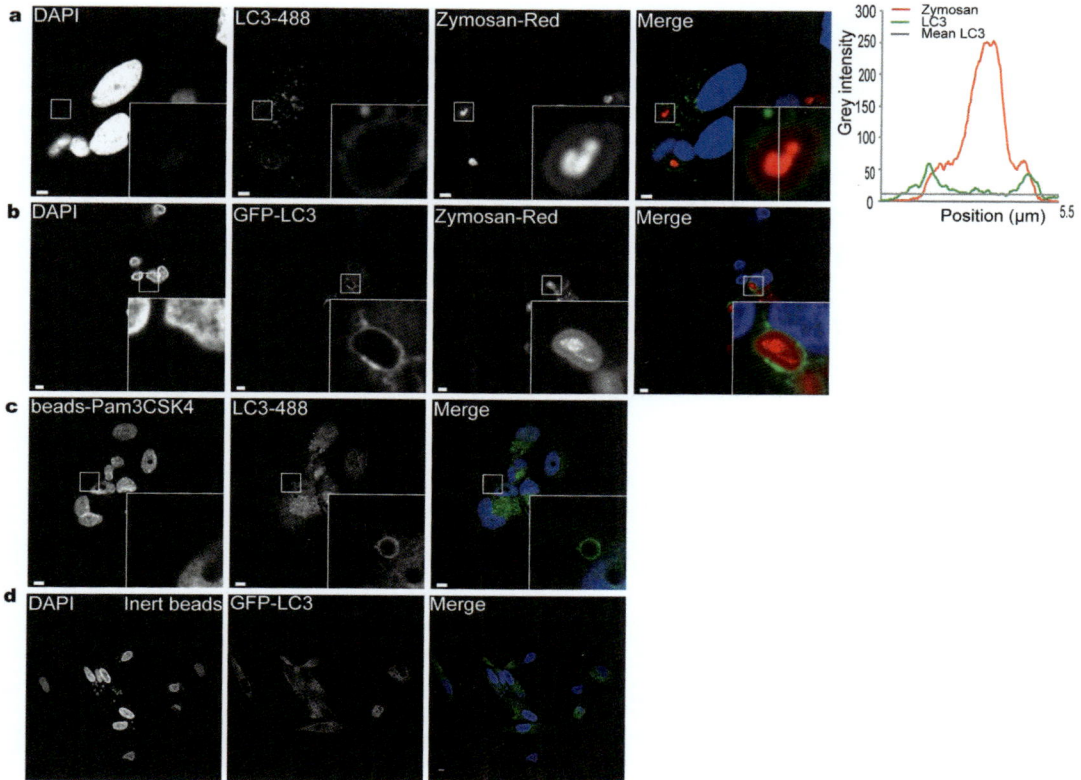

Fig. 1 Various stimuli can be used to enhance LC3-positive phagosome formation in human macrophages. (**a**) Human macrophages stimulated with zymosan coupled with Texas Red fluorophores and then processed for endogenous LC3 staining. The recruitment of LC3 was analyzed by confocal microscopy. The DAPI staining delineates the nucleus, the green channel (488 or GFP) shows an LC3 ring, the red channel is used to detect zymosan. Finally, the merged image, corresponding to the compilation of the three images, demonstrates the recruitment of LC3 to the zymosan containing phagosome. Fluorescence intensity was recorded for *red* (zymosan) and *green* (LC3) along the trajectory through the LAP phagosome, as depicted in the magnified image of the merged picture. (**b**) Human macrophages transduced with a lentiviral construct encoding GFP-LC3 and then stimulated with zymosan coupled with Texas Red fluorophores. (**c**) Human macrophages stimulated with Pam3CSK4 coating beads and immunostaining for endogenous LC3. (**d**) Human macrophages stimulated with blue non-coated beads and immunostaining for endogenous LC3. In this case, the LC3 rings around the phagocytosed beads were not observed

fetal bovine serum (FBS), 2 mM glutamine, 110 μg/ml sodium pyruvate, and penicillin/streptomycin. 1000 U/ml recombinant human GM-CSF was added at days 0, 3, and 5.

2.1.2 Human Dendritic cells

1. PBMCs are isolated as described above in Subheading 2.1.1, and then CD14+ monocytes are differentiated to monocyte-derived dendritic cells (DCs) with 500 U/ml recombinant human IL-4 and 500 U/ml recombinant human GM-CSF added on days 0 and 3 and cultured as described in Subheading 3.1.2.

Table 3
Different tools to follow LC3-phagosome maturation

	GFP-Rab7	Lamp2	LysoTracker
Role of proteins or probes	– Late endosome marker	– Lysosomal associated membrane protein – Lysosome marker	– Fluorescent probes to visualize acidic organelles
Immature LC3-phagosomes	– Neutral pH – Is not associated with late endosome – No GFP-Rab7 is observed on phagosomes	– Neutral pH – Has not fused yet with the lysosome – No Lamp2 is observed on phagosomes	– Neutral pH – Not staining by LysoTracker
Mature LC3-Phagosome	– Acidic pH – Fused with the late endosome – Phagosome membrane displays GFP-Rab7	– Acidic pH – Fused with the lysosome – Phagosome displays Lamp2	– Acidic pH because phagosome has fused with the lysosome – Staining by LysoTracker Red – Red phagolysosomes are observed
Advantages	– Assesses phagosome maturation	– Demonstrates fusion of phagosomes with lysosomes	– Specifically detects acidification of phagosomes
Disadvantages	– Rab7 is not the only marker that is required for late phagosome fusion with lysosomes	– Phagosome maturation can occur without the Lamp2 acquisition	– LysoTracker is a pH sensor, a phagosome can mature and just block is acidification by inactivation of the vesicular ATPase – Not specific to the LC3 phagosome but detects all acidic organelles
Application	Immunofluorescence staining Live cell imaging	Immunofluorescence staining	Live cells imaging
Notes	*To have a comprehensive view of the LC3-phagosome maturation, it is important to combine all these tools*		

2. DCs culture medium: RPMI 1640 medium supplemented with 2% heat-inactivated pooled human serum (Invitrogen) and 2 μg/mL gentamicin.

2.1.3 Candida albicans Specific CD4+ T-Cell Clones

1. PBMCs are isolated from donors with known HLA type, for example HLA-DRB1*0401 (*see* **Note 1**) by density-gradient centrifugation on Ficoll-Paque PREMIUM. T-cells producing IL-17 are specifically selected by positive magnetic selection with an IL-17 cytokine capture assay after overnight restimulation with *C. albicans* extract. From the positively selected T cell population CD4+ T-cells clones are generated as described in Subheading 3.7.1.

2. Medium: RPMI 1640 supplemented with 5% human serum and 25 μg/ml gentamycin.

3. Positive control stimulus: Heat killed *C. albicans* particles (30 min-45 min at 100 °C) used at three particles per macrophage or soluble *C. albicans* extract used at final concentration of 20 μg/ml.

4. MACS buffer: PBS supplemented with 1% human serum and 2 mM EDTA.

5. IL-17 catch reagent and IL-17 detection antibody coupled with PE fluorophore (Miltenyi kit) (*see* **Note 2**).

6. Selected T-cells secreting IL-17 are cultured with a mixture of irradiated PBMCs and B-LCLs cells, as described in Subheading 3.7.1.

7. Irradiated PBMCs and EBV transformed lymphoblastoid cell lines (B-LCLs) were cultured in R8 medium: RPMI 1640 supplement with 8% pooled human serum (PHS) and 25 μg/ml gentamicin.

8. Coculture medium: RPMI 1640 supplemented with 8% PHS, 25 μg/ml gentamicin, 450 U/ml IL-2, and 3 μg/ml PHA.

9. After 14 days of culture, the growing T-cell clones are screened for their ability to produce IL-17 and IFN-γ in response to *C. albicans* extract using a human IL-17A ELISA kit (ELISA MAX Deluxe, BioLegend, used according to manufacturer's recommendation) or an IFN-γ ELISA kit from (Mabtech) as described in Subheading 3.7.1.

2.1.4 Cytokines and Buffers

1. Human granulocyte-macrophage colony-stimulating factor (GM-CSF), stock solution 20 ng/μl; stored at –20 °C.

2. Recombinant human interleukin-4 (IL-4) (Peprotech), stock solution 20 ng/μl; stored at –20 °C.

3. Recombinant human interleukin-2 (IL-2) (Peprotech), 5×10^6 U/ml; stored at –20 °C.

4. Phytohemagglutinin (PHA, Sigma), stock solution 1 mg/ml in PBS; stored at –20 °C.

5. Phorbol-12-myristate 13-acetate (PMA), stock solution 1 mg/ml in DMSO; stored at –20 °C.

6. Ready-to-use ionomycin 1 mg/ml (Sigma); stored at –20 °C.

2.2 Beads

1. Zymosan (Invivogen), stock solution 1 mg/ml in sterile water; store at 4 °C for 1 month and used to feed cells to the final concentration of 100 μg/ml.

2. Texas Red zymosan (Molecular Probes), stock solution 10 mg/ml in 2 mM solution of PBS-sodium azide; stocked at 4 °C and used to feed cells to the final concentration of 100 μg/ml.

3. Latex beads, carboxylate-modified polystyrene (Bangs Laboratories), stored at room temperature (RT) in the dark.

4. Pam3CSK4, a synthetic triacylated lipopeptide (Invivogen), stock solution 1 mg/ml in endotoxin-free water and stored at –20 °C. Reconstituted Pam3CSK4 is coupled to beads as described in Subheading 3.4.2 and used at 1:1 macrophage–bead ratio.

5. *Candida albicans* extract is reconstituted to a concentration of 0.4 mg/ml and 125 μl aliquots are kept at 4 °C. Reconstituted *C. albicans* extract is coupled with beads as described in Subheading 3.4.3 and used at 20 μg/ml.

6. Latex beads, amine-modified polystyrene, fluorescent blue (L0280, Sigma-Aldrich) are stocked at RT in the dark and used at a 1:1 macrophages/ beads ratio, as described in Subheading 3.4.4.

2.3 Probes

1. Antibody anti-LC3 from Medical and Biological Laboratories used at 1:200 dilution for immunofluorescence.

2. Antibody anti-Lamp2 clone H4B4 (SouthernBiotech) at 1:100 dilution for immunofluorescence.

3. Mouse monoclonal anti-HLA-DR molecule, clone L243 (BioLegend), at 1:100 dilution for immunofluorescence.

4. Alexa Fluor® 488 or 555 conjugated goat anti-rabbit (Invitrogen) to be used at 1:500 dilution (*see* **Note 3**).

5. Alexa Fluor® 488 or 555 conjugated goat anti-mouse (Invitrogen) to be used at 1:500 dilution (*see* **Note 3**).

6. 1 mM LysoTracker® Red DND-99 (L7528, Invitrogen).

2.4 Lentiviral Constructs of Fluorescent-Tagged Proteins and shRNA Silencing

1. Lentiviral constructs:

   (a) Lentiviral construct of fluorescence GFP-Atg8 is obtained as previously described in Schmid et al. [3].

   (b) Lentiviral construct of fluorescence mRFP-Atg8 and GFP-Rab7 are obtained as described in Romao et al. [4].

(c) The pLKO.1-puromycin containing shRNA *Atg5* is a gift from J. Tschopp (University of Lausanne, Lausanne, Switzerland).

2. Lentiviral particles are generated as described in Schmid et al. [3] and kept at −80 °C (*see* **Note 4**). Lentiviral particles are used for transduction of primary human cells as described in Subheading 3.2.

2.5 Immunofluorescence and Confocal Microscopy

1. 8-Well chamber slides containing 2.5×10^5 cells per well, in a final volume of 200 µl.

2. 4% parafomaldehyde solution in PBS (PFA). Prepare a fresh solution at 3% in PBS for each experiment.

3. Permeabilization solution: 0.01% Triton X-100 in PBS.

4. Image-iT® FX signal enhancer (Invitrogen).

5. Blocking buffer: PBS supplemented with 10% NGS (normal goat serum) and 1% bovine serum albumin (BSA).

6. Wash buffer: PBS with 0.01% saponin.

7. Primary antibodies are diluted at 1:100 or 1:200 in PBS with 0.01% saponin, 10% NGS, and 1% BSA.

8. Secondary antibodies are diluted at 1:500 in PBS with 0.01% saponin.

9. 4,6-diamidino-2-phenylindole (DAPI) nucleic acid stain, stock solution 5 mg/ml in ddH$_2$O; stored at −20 °C. Work solution: 1:5000 dilution in PBS.

10. Mounting medium: Prolong Gold antifade reagent (Invitrogen-Molecular Probes).

2.6 MHC Class II Presentation Assay Using C. albicans-Specific T-Cells Clones

1. *C. albicans specific* T-cells clones are cocultured with human macrophages or DCs previously feed with soluble or *C. albicans* extract-coated beads as described in Subheading 3.7.2.

2. The detection of IL-17 and IFN-γ in the supernatant of cocultures using ELISA assays is the direct read out of the ability of the human macrophages or DCs to present *C. albicans* antigen as described in Subheading 3.7.2.

3 Methods

3.1 Cells Cultures

The following two first steps were common for both monocyte-derived macrophages and monocyte-derived DCs (For choice of cell type *see* Table 1).

1. To isolate PBMCs from leukocyte concentrates, 35 ml of blood are gently overlaid on 15 ml of Ficoll-Paque PREMIUM and centrifuged for 25 min at $800 \times g$.

2. CD14$^+$ monocytes are isolated from the PBMCs using positive magnetic cell selection technology (AutoMacs, Miltenyi Biotec).

3.1.1 Monocyte-Derived Macrophages

1. The CD14⁺ monocytes are plated in the adequate experimental plate. For instance, for immunofluorescence or live imaging experiments, cells are plated in 8-chamber slides at a density of 2.5×10^5 cells per well, in a final volume of 200 μl and cultured for at least 6 days in macrophage cell medium (DMEM, 10% FBS, 2 mM glutamine, 110 μg/ml sodium pyruvate, and penicillin/streptomycin).

2. GM-CSF is added to the culture medium on day 0 at the concentration of 1000 U/ml.

3. On day 3 and 5, the culture medium is replaced by fresh medium supplemented with GM-CSF (1000 U/ml).

4. After 6 days of culture, monocyte-derived macrophages are ready to be used.

3.1.2 Monocyte-Derived DCs

1. The CD14⁺ monocytes are plated in the adequate plate and cultured for 5 days in dendritic cell medium (RPMI 1640, 2% heat-inactivated pooled human serum, gentamycin)

2. Recombinant human IL-4 (500 U/ml) and recombinant human GM-CSF (500 U/ml) are added to the culture medium on days 0 and 3.

3.2 Transduction of Primary Human Cells with Lentiviruses

The human DCs are transduced with lentiviral construct of fluorescence-tagged proteins or shRNA following the same protocol applied to transduce human macrophages, described bellow.

3.2.1 Lentiviral Construct of Fluorescent-Tagged Proteins

1. CD14⁺ monocytes are plated in 24-well plate at the density of 5×10^5 cells per well, in 1 ml of culture medium. The monocyte-derived macrophages are used between day 5 and 7.

2. Primary human macrophages are washed with complete medium and incubated with SIV virus like particles at 37 °C for 3 h, to optimize the entry of lentiviruses permitting efficient delivery of the target construct.

3. Macrophages are infected with the correspondent lentiviruses at a multiplicity of infection (MOI) of 5, together with 6 μg/ml of polybrene and then the plates are centrifuged at $1900 \times g$ for 45 min at 37 °C.

4. Macrophages can be double transduced with two different lentiviral constructs (*see* **Note 5**).

5. Twenty-four hours post-transduction the medium is replaced, and then the cells are incubated for additional 24 h or longer times at 37 °C, 5% CO_2 and are ready to be used.

3.2.2 Lentiviral Construct of shRNAs

1. The two first steps are the same as previously described in Subheading 3.2.1.

2. The primary cells are infected only with the correspondent lentiviral shRNA construct at a MOI of 5. Then the plates are centrifuged at $1900 \times g$ for 45 min at 37 °C.

3. Twenty-four hours post-transduction the medium is replaced, then the cells are incubated at 37 °C, 5% CO_2 for a total of either 3 to 5 days.

3.3 Track Autophagy role in LAP by Silencing Atg5

1. CD14$^+$ monocytes are plated in 24-well plate at the density of 5×10^5 cells per well, in 1 ml of culture medium.

2. After 5 days of culture, human macrophages are transduced with lentiviral construct of shRNA *Atg5* following the protocols described in Subheading 3.2.2.

3. After 4 days of culture the macrophages or DCs are ready to be stimulated with zymosan, other stimuli or used in the experiment described below (*see* **Note 6** and Table 2).

3.4 LC3-Associated Phagosome (LAP) Formation

3.4.1 Zymosan Stimulation

1. LC3-asssociated phagosome formation can be enhanced using zymosan particles (*see* **Note 7**). Depending on the experimental setting, it is possible to use uncoupled zymosan or zymosan coupled with the Texas Red dye. Using Texas Red zymosan allows the direct observation of the phagocytosed cargo.

2. Human primary cells are cultured for at least 6 days before stimulation with zymosan at a final concentration of 100 μg/ml, and then incubated at 37 °C, 5% CO_2 for 1 h up to 12 h or longer.

3. Afterwards, cells can be observed in live cell imaging or fixed to be processed for immunofluorescence staining.

3.4.2 Coating of Beads with Pam3CSK4 for Phagocytosis

1. Modifying beads for phagocytosis allows enhancement of LAP formation. For this purpose, we use Pam3CSK4, a TLR2 ligand (*see* **Note 8**). Polystyrene beads are coated with 50 μg of Pam3CSK4 per 10^8 beads. Conjugated beads are prepared fresh for each individual experiment.

2. Count 3×10^8 beads and put it in 1 ml of PBS pH 7.4.

3. Centrifuge at $600 \times g$ for 5 min and carefully discard the supernatant and resuspend the beads in 1 ml of PBS.

4. Repeat **step 3**, twice. Resuspend the beads in 850 μl of PBS.

5. Add 850 μl of beads to 150 μl of Pam3SKC4 at a concentration of 1 mg/ml. For an efficient coupling of TLR stimuli to beads it is important to keep this order (*see* **Note 9**).

6. The solution of Pam3CSK4 and beads is incubated at room temperature under rotation for 1–2 h and then rotated at 4 °C overnight.

7. The following day, the beads are centrifuged at $600 \times g$ for 5 min, the supernatant carefully aspirated and beads resuspended in 1 ml of PBS. This step is repeated five times to wash the excess of ligand thoroughly.

8. Finally, the beads are counted and used at the ratio range between 1:1 and 0.5:1 (bead–macrophage).

3.4.3 Coating of Beads with C. albicans Extract for Phagocytosis

1. LAP formation can be enhanced by using other stimuli such as *C. albicans* extract (*see* **Note 10**).

2. Prepare freshly 50 µg of *C. albicans* extract per 10^8 beads.

3. Count 1×10^8 beads and put them in 1 ml of PBS, pH 7.4.

4. Spin the beads down at $600 \times g$ for 5 min and carefully discard the supernatant.

5. Resuspend the beads in 1 ml of PBS and repeat **steps 3** and **4** twice.

6. After the last centrifugation the beads are resuspended in 375 µl of PBS.

7. Add the amount of beads to 125 µl of *C. albicans* extract (concentration: 0.4 mg/ml). For an efficient coupling keep this order (*see* **Note 9**).

8. The mix is incubated at RT for 1–2 h under rotation and then rotated at 4 °C overnight.

9. The beads are centrifuged at $600 \times g$ for 5 min, the supernatant carefully aspirated and beads resuspended in 1 ml of PBS. Repeat this **step 5** times to wash the excess of ligand thoroughly.

10. Finally, the beads are counted and used at the ratio 1:1 (bead–macrophage).

3.4.4 Inert Beads

1. To investigate the role of autophagy proteins in the maturation of LAP phagosomes, it is important to use a negative control. In the literature, latex beads are used as negative control [9]. In fact, inert beads are not able to stimulate the LAP pathway, since LC3 recruitment to phagosomes containing such beads has not been observed.

2. Count 3×10^8 beads and put them in 1 ml of PBS-0.5 % BSA, then incubate 1–2 h at RT under rotation.

3. Centrifuge at $600 \times g$ for 5 min, carefully remove the supernatant and resuspend the beads in 1 ml of PBS. Repeat this step three times.

4. Count beads and use them at 1:1 beads per macrophage ratio.

5. As previously described, the human macrophages are fed with inert beads. The cells are incubated at 37 °C, 5 % CO_2 for the adequate time periods during the experiment.

6. Cells can be directly fixed and processed for immunofluorescence.

3.5   Confocal Microscopy to Analyze LC3-Phagosomes

1. Cells can be plated in 8-chamber slides or on sterile round 1.5 mm microscopy coverslips, which are placed into 24-well plate wells (*see* **Note 11**). Process additional wells/slides that will be used for control stainings (*see* **Note 12**).

2. After CD14[+] positive monocyte selection the cells are directly plated in the appropriate wells (8-chamber slide or 24-well plate with cover slide).

3. There are two possibilities to observe LC3's recruitment to phagosomal membranes: (a) investigating endogenous LC3, or (b) overexpressing fluorescent protein constructs fused to LC3 (*see* Table 2).

3.5.1   Recruitment of Endogenous LC3 to LAP

1. After 6 days, human primary macrophages are treated with the stimuli of interest depending on the experimental set up (in our case with zymosan, Pam3CSK4 coupled beads or inert beads for 8 h). Treat two wells for each condition and leave at least two wells of cells untreated as controls (*see* **Note 12**).

2. Cells are then gently washed twice with PBS (8-chamber = 150 μl/ well, P24 plates = 0.5 ml/well) and fixed in 3 % PFA for 20 min at 4 °C (8-chamber = 75 μl/well, P24 plates = 250 μl/well) (*see* **Note 13**).

3. Fixed cells are washed twice with PBS (same volume used in **step 2**) and afterwards permeabilized with 0.01 % Triton X-100 for 1 min at RT and then washed with PBS.

4. Apply a sufficient volume of Image-iT® FX signal enhancer (Invitrogen) to cover each coverslip or well and incubate for 30 min at room temperature (*see* **Note 14**). Cells are washed with PBS supplemented with 1 % fish skin gelatin and 0.02 % of saponin.

5. Add the blocking buffer (8-chamber = 75 μl/well, P24 plates = 250 μl/well) and incubate for 30 min to 1 h at room temperature.

6. The primary antibodies are diluted in PBS with 0.01 % saponin, 10 % NGS, and 1 % BSA (*see* **Note 15**).

7. Antibody solution is added directly to the cells for the 8-chamber or put in contact with the coverslips (*see* **Note 16**) for 1 h at RT or for longer periods up to overnight in humid environment at 4 °C. Co-staining of LC3 with other proteins is possible *see* **Note 17**.

8. The cells are washed three times with PBS supplemented with 1 % fish skin gelatin and 0.02 % of saponin and incubated for 5 min each time.

9. The secondary antibodies are diluted in blocking buffer (1:500) and added to cells as previously described in **step 7**. After 1 h, the secondary antibody is removed and then the

DAPI solution (1:5000) is added for 5 min at room temperature to stain nucleic acids.

10. The cells are washed three times with PBS and then processed for mounting:

(a) Coverslips are mounted on microscopy specific slides by inverting them onto a drop of Prolong gold antifade reagent, up to three coverslips per slide. Each coverslip is carefully and gently pressed down and left to dry at room temperature in the dark (*see* **Note 18**).

(b) For the 8-chamber slides, first remove the plastic wells and rubber sealing attached to the slide. Secondly remove all PBS traces by aspiration and all the plastic well remains until obtaining a clear slide. Next, put some drops of mounting medium on the slide-chamber footprints and then mount a coverslip of 1 mm thickness, carefully press down and leave it to dry at room temperature in the dark.

3.5.2 Overexpression of Fluorescent Protein Tagged LC3

1. At day 5 of culture human primary macrophages are transduced with lentiviral constructs of fluorescent protein tagged LC3 as described in Subheading 3.2.

2. After 48 h the transduced macrophages are stimulated with the appropriate stimuli (zymosan or coated beads).

3. **Steps 2** at **5** are the same as described above (Subheading 3.5.1).

4. The cells can be directly stained with the DAPI solution or developed with the normal staining protocol (described in Subheading 3.5.1) to stain for a second marker.

5. Slides are mounted following the same steps as described in Subheading 3.5.1.

3.5.3 Immune Fluorescence Analysis

1. Cells are analyzed with an inverted confocal laser-scanning microscope (SP5, Leica), using a 63×, 1.4 NA oil immersion lens. The excitation is performed at 405 nm to elicit the DAPI fluorescence (blue emission). To elicit the GFP or Alexa 488 nm fluorescence (green emission), an excitation at 488 nm is used. Finally, excitation at 543 elicits the mRFP or Alexa 555 nm fluorescence (red emission). Images are acquired with Leica software (Leica), then analyzed and assembled using Image J software.

2. To quantify the fluorescence of LC3 and of different markers around the phagosome, intensity values are recorded by the Image J software for the corresponding region of interest (for example a part of the phagosomal membrane). The data are represented in profile plots of the fluorescent intensity across trajectories through the LAP phagosomes.

3. Several experiments are shown in Fig. 1: An experiment with human macrophages stimulated with Texas Red zymosan and stained for endogenous LC3 is shown in Fig. 1a. Figure 1b shows human macrophages transduced with GFP-LC3, then stimulated with zymosan particles. Figure 1c shows human macrophages fed with Pam3CSK4 coated beads and then stained for endogenous LC3. Finally, Fig. 1d shows the negative control, when human macrophages overexpressing GFP-LC3 are exposed to inert beads. The formation of LC3-phagosomes by the direct recruitment of LC3 to the zymosan-containing or Pam3CSK4-containing phagosome is shown in Fig. 1a–c. The LC3 ring around the phagosome, characteristic of the LAP phagosome is not observed when the human macrophages are fed with inert beads (Fig. 1d).

3.6 LAP Phagosome Maturation

LC3 phagosome maturation can be observed using different strategies: (1) observe the presence of the late endosome marker Rab7 or acquisition of the lysosome marker Lamp2 or (2) characterize the acidification of the phagosome with LysoTracker (*see* Tables 3).

3.6.1 Costaining of LC3 with Rab7 or Lamp-2

1. Phagosome maturation can be assessed by colocalization of LC3 with other proteins previously described to play a role in the late stages of phagosome maturation, namely fusion with lysosomes, such as Rab7 and Lamp 2 (Table 3).

2. Rab7 protein is a late endosome marker (Table 3).

   (a) At day 5 of culture, human macrophages are double transduced with lentiviral construct for GFP-Rab7 and mRFP-LC3 (*see* Subheading 3.2).

   (b) Two days post transduction the cells are stimulated with uncoupled zymosan for 8 h.

   (c) Cells are fixed with 3% PFA for 15 min at 4 °C and permeabilized for 1 min with 0,01% Triton X-100 before adding DAPI (1: 5000).

   (d) Observe localization of GFP-Rab7 and mRFP-LC3 by confocal microscopy. Presence of both indicates that the LC3-phagosome has matured and fused with late endosomes and lysosomes. If only LC3 is observed, the maturation might be slowed down or blocked.

3. Lamp2 protein is a lysosomal marker (Table 3):

   (a) At day 5 of culture human macrophages are transduced with lentiviral construct mRFP-LC3, (*see* Subheading 3.2).

   (b) Forty-eight hours post transfection the cells are stimulated with uncoupled zymosan for 8 h.

   (c) Fix the cells for 15 min with 3% PFA.

(d) Then process for Lamp2 immunofluorescence staining and used the antibody Alexa 488 as secondary antibody (*see* Subheading 3.5.1 and **Notes 3** and **12**).

(e) Afterwards, the slides are investigated by confocal microscopy. The absence of Lamp2 at the LC3-phagosome can indicate a delay of maturation for LC3 associated phagosomes.

3.6.2 LysoTracker and Live Cell Imaging

1. CD14⁺ monocytes are plated in 8-well chamber slides with 2.5×10^5 macrophages per well in a final volume of 200 µl. Macrophages are derived as described in Subheading 3.1.

2. At day 5 of culture, the cells are transduced with the lentiviral GFP-LC3 construct (*see* Subheading 3.2) and incubated for 48 h at 37 °C and 5 % CO_2.

3. Dilute the 1 mM LysoTracker solution at the final concentration of 50 nM in pre-warmed DMEM phenol-red free (*see* Subheading 2.3, **Note 3**).

4. Wash the transduced macrophages and add the Red LysoTracker-containing medium.

5. Cells are incubated for 30 min to 2 h at 37 °C.

6. Replace the medium with fresh medium without LysoTracker.

7. Add 100 µg of zymosan or other stimuli.

8. Go directly to the microscope and observe the cells through a 63× 1.4 NA oil immersion lens with an inverted confocal microscope at 37 °C and 5 % CO_2 atmosphere.

9. Live cell imaging analysis is performed using the Imaris software.

10. Phagosomes displaying the LysoTracker-Red dye have already fused with lysosomes and acquired an acidic pH, whereas LysoTracker negative LC3-phagosomes have not yet fused with lysosome (Table 3).

3.7 MHC Class II Antigen Presentation Assay

3.7.1 Generation of C. albicans Specific T-Cell Clones

Step I: Selection of IL-17 secreting T cells

1. Collect 50 ml of blood from donor with known HLA type, for example HLA-DRB1*0401 positive (*see* **Note 1**).

2. PBMCs are isolated by density-gradient centrifugation on Ficoll-Paque PREMIUM.

3. Isolated PBMCs are centrifuged at $200 \times g$ for 10 min, and then carefully remove the supernatant containing platelets.

4. Cell pellet is washed with the cell medium (RPMI, 5 % human serum and 25 µg/ml gentamycin) and centrifuged at $300 \times g$ for 5 min.

5. Aspirate the supernatant and resuspend cells in medium at a final cell concentration of 1×10^7 cells/ml.

6. Plate 1 ml of cell suspension per well in 24-well plates (approximately 4 wells).

7. Cells are stimulated with 3×10^7 heat killed *C. albicans* particles (3 heat killed particles per cell) and plates are incubated at 37 °C and 5% CO_2 for 16 h. Always have negative controls, including a well without antigen stimulation.

8. Carefully collect the cells by using a cell scraper or by pipetting up and down.

9. Spin cells down at $500 \times g$ at 4 °C for 5 min, and then resuspended 1×10^7 cells in 80 μl of cold MACS buffer (PBS supplemented with 1% human serum and 2 mM EDTA).

10. Add 20 μl of IL-17 catch reagent per 1×10^7 cells (*see* **Note 2**).

11. Mix the cell suspension well and incubate for 5 min on ice.

12. Dilute cells with warm medium:
    <1% frequency of specific cells: dilution of 10^6/ml, add 10 ml of medium to 10^7 cells.
    >1% frequency of specific cells: dilution of 10^5/ml, add 100 ml of medium to 10^7 cells.

13. Cells are incubated for 45 min at 37 °C under slow continuous rotation.

14. Put the tube on ice and fill up with cold MACS buffer, followed by centrifugation at $400 \times g$ for 5 min at 4 °C.

15. Cells are resuspended in 80 μl of MACS buffer per 10^7 cells.

16. Add 20 μl of IL-17 detection antibody coupled to PE per 10^7 cells and incubate 10 min on ice (*see* **Note 2**).

17. Add 10 ml of cold MACS buffer and centrifugation at $400 \times g$ at 4 °C for 10 min.

18. The cell pellet is then resuspended in 80 μl of cold MACS buffer per 10^7 cells.

19. Add 20 μl of anti-PE beads per 10^7 cells and mix well before incubation for 15 min at 4 °C (not on ice).

20. Cells are washed by adding 10 ml of cold MACS buffer per 10^7 cells and then spin down at $400 \times g$ at 4 °C for 10 min.

21. Cells are resuspended in 500 μl of cold MACS buffer (for more than 5×10^7 cells, resuspended the cells at a dilution of 1×10^8/ml).

22. Proceed with magnetic separation in the autoMACS separator. Choose positive selection "posseld" program.

23. Take the positive fraction, centrifuge it at $400 \times g$ at 4 °C for 10 min. The negative fraction can be frozen down.

24. Carefully remove the supernatant and resuspended cells in 100 μl of PBS and then count cells. These cells are enriched for T cells producing IL-17 in response to heat killed *C. albicans*

stimulation and they will be used for the generation of *C. albicans* specific T-cell clones.

Step II: Generation of T-cell clones

1. Collect PBMCs from 2 to 3 donors.

2. Irradiate 1.5×10^8 PBMCs at 3000 rad in R8 medium (RPMI 1640, 8% PHS, 25 µg/ml gentamicin) and distribute into 96-well U bottom plates at a final concentration of 10^5 cells per wells (*see* **Note 19**).

3. Irradiate 3×10^7 B-LCLs at 20,000 rad in R8 and distribute into 96-well U bottom plates to a final cell concentration of 2×10^4 cells per well (*see* **Note 19**).

4. Prepare a mixture of irradiated cells containing 2×10^6 PBMCs per ml and 4×10^5 B-LCLs per ml, in a final volume of 50 ml of R8 medium.

5. Take 8 plates of 96-well U bottom plates and add 50 µl of cell suspension into each well.

6. Use the isolated IL-17 producing T cells, sorted previously (*see* Subheading 3.7.1, **step I**), to perform limiting dilutions following this scheme:
   Use the 8 plates prepared in **step 5** (*see* **Note 20**).

   (a) One 96-well plate containing 10 T-cells per well:
      - Prepare 12 ml of solution (called solution A) with 200 T-cells per ml of R8 medium. Add 50 µl of the solution A per well (*see* **Note 20**).

   (b) Two 96-well plates containing 1 T-cell per well:
      - Prepare 12 ml of solution (called solution B) with 20 T-cells per ml of R8 medium. Add 1.2 ml of the solution A in 10.8 ml of R8 medium. Add 50 µl of the solution B per well.

   (c) Five 96-well plates containing 0.3 T-cells per well:
      - Prepare 50 ml of solution (called solution C) with 6 T-cells per ml. Add 1.2 ml of the solution A in 38.8 ml of R8 medium. Add 50 µl of the solution C per well (*see* **Note 20**).

7. Add 50 µl of coculture medium (RPMI 1640 supplemented with 8% PHS, 25 µg/ml gentamicin, 450 U/ml of IL-2, and 3 µg/ml of PHA) to every well and incubated the cells for 5 days at 37 °C and 5% CO_2 (*see* **Note 21**).

8. At day 5 of culture, centrifuge the plates at $400 \times g$ at 4 °C for 5 min and then remove 70 µl of medium.

9. Add 75 µl of fresh coculture medium without PHA and put the plates back at 37 °C and 5% CO_2.

10. Check every well daily, and if necessary split them into new wells/plates in RPMI 1640 with PHS, 25 µg/ml gentamicin and 450 U/ml of IL-2.

11. Finally, after 14 days the T-cell clones are tested for IL-17 production upon restimulation (as described below) and selected for preservation by freezing down aliquots.

Step III: Test T-cell clones for IL-17 production by ELISA

After 14 days of culture approximately 20 clones are tested for IL-17 production by ELISA assays using the Human IL-17A ELISA Kit, according to manufacturer's instructions.

1. Each clone will be tested in four conditions:
   (a) T-cells + HLA-matched CD14$^+$ cells.
   (b) T-cells + HLA-matched CD14$^+$ cells + PMA/ionomycin.
   (c) T-cells + HLA-matched CD14$^+$ cells + C. *albicans* extract.
   (d) T-cells + HLA-matched CD14$^+$ cells + PMA/ionomycin + C. *albicans* extract.

2. In 96-well flat bottom plates 1×10^4 of CD14$^+$ cells are plated per well in a final volume of 200 µl of R8 medium.

3. Add 20 µg/ml of C. *albicans* extract according to the scheme described in 1 and incubate for 4 h.

4. Cells are centrifuged at $400 \times g$ at 4 °C for 5 min and 130 µl of medium R8 (RPMI 1640 with 8% PHS and 25 µg/ml gentamicin) is added.

5. Add 1×10^5 T-cell clones to each well to an effector to target ratio of 1:2 and leave the coculture for 29 h.

6. The day before the ELISA, add PMA (prepare 1 µl in 999 µl medium, apply 5 µl per well) and ionomycin (prepare 10 µl in 990 µl medium, apply 6.7 µl per well) to certain wells (according to the scheme) and incubate overnight.

7. At the day of the ELISA assay, spin down cells at $500 \times g$ at 4 °C for 5 min and collect supernatants.

8. To the pre-coated ELISA plates, add 80 µl of supernatant per well (do duplicates for each condition) and then follow the ELISA kit's instructions.

9. T-cell clones that produce IL-17 in response to C. *albicans* extract are kept and expanded.

Step IV: Expansion of T-cell clones

1. T-cell clones with the desired specificity are expanded.

2. Irradiate 1.5×10^8 PBMCs at 3000 rad (final: 10^5 cells /well) and 3×10^7 B-LCLs at 20,000 rad (final: 2×10^4 cells/well) in R8 medium (RPMI 1640, 8% PHS, 25 µg/ml gentamicin).

3. Prepare a mixture of irradiated PBMCs (at 2×10^6/ml) and B-LCLs (at 4.10^5/ml) and put 50 µl of this cell suspension into each well of 96-well U bottom plates.

4. Add to each well 50 µl of coculture medium containing 450 U/ml IL-2 and 3 µg/ml PHA.

5. Add between 3000 and 5000 clonal T cells per well in 50 µl of medium and incubated at 37 °C with 5% CO_2.

6. At day 5 of culture, monitor clones and added 75 µl of fresh coculture medium without PHA.

7. Depending on their growth split daily using RPMI 1640 with PHS, 25 µg/ml gentamicin and 450 U/ml IL-2.

8. After 14 days freeze some of the clones or use them for MHC class II antigen presentation experiments.

3.7.2 MHC Class II Antigen Presentation Assays

1. Isolate CD14+ monocytes from healthy donors, HLA matched to the T-cell clone donor, and then differentiate them into macrophages or DCs.

2. Monocyte-derived macrophages or DCs are fed with *C. albicans* extract-coated beads (prepared as described in subheading 3.4.3) at a ratio of 1:1 and incubate for 4 h.

3. The cells are washed and fresh RPMI 1640 + 5% FBS and gentamycin is added.

4. Specific T-cell clones are added at an effector to target ratio of 1:1, and cocultured for 0 h, 3 h, or 18 h with macrophages and for 0 h, 18 h, or 38 h with DCs.

5. Eighteen hours post coculture, the supernatants are harvested and spin down at 500 g and 4 °C for 5 min to be tested for IL-17 and IFN-γ production by ELISA assays.

6. The supernatants are collected, diluted and plated directly on previously coated ELISA plates.

7. Perform the IL-17 or IFN-γ specific ELISA assays according to the manufacturer's instructions. Recombinant human IFN-γ and IL-17 are used as standards.

8. Detection of IL-17 and IFN-γ in the supernatant of the coculture indicates that the antigen presenting cells are able to present *C. albicans* antigen on MHC class II molecules. To investigate the involvement of the LAP pathway in MHC class II antigen presentation, it is important to RNA silence Atg proteins in the macrophages or DCs and follow the impact of this treatment on IL-17 and IFN-γ secretion by the specific T-cell clones.

4   Notes

1. To produce *C. albicans* specific T-cell clones, the PBMCs are isolated from healthy lab donors previously tested to be able to respond to *C. albicans* extract. Further, characterization of the T-cell clones revealed that the restriction element for their recognition of *C. albicans* antigen was the HLA-DRA1*0101/ HLA-DRB1*0401 molecule. However, as long as the MHC class II molecules are known and can be matched to the target cells, the identity of the matched MHC class II molecules is less important.

2. The human interleukin 17 is a member of the IL-17 family of cytokines and it is produced by CD4+ helper T cells in response to extracellular bacterial or fungal (*C. albicans)* infections. The IL-17 specific catch reagent is attached to the surface of all leukocytes. Then the stimulation with *C. albicans* will enhance IL-17 secretion, which will bind to the IL-17 catch reagent allowing for the positive selection of IL-17 producing cells. To sort this cell population, they are labeled with a second antibody, the IL-17 detection antibody conjugated to PE. Living IL-17 coated cells can be sort by flow cytometry or with the autoMACS separator after addition of anti-PE antibodies coupled to paramagnetic beads.

3. For the choice of the fluorophore associated with the secondary antibodies, first, consider which panel and filter set are available on your confocal microscope. Second, in case of co-localization studies, you should select two fluorophores that have nonoverlapping emission spectra.

4. Lentivirus production is performed following protocols of Schmid et al. [3]. Briefly, 293T cells are cotransfected with lentiviral constructs and two helper plasmids (pCMVdeltaR8.91 and pMDG) by calcium phosphate transfection. After 2 days, culture supernatant containing recombinant viral particles are collected, filtered, aliquoted, and frozen at −80 °C.

5. Human Macrophages can be double transduced with two different lentiviral constructs following the same protocol used to a single transduction. For example, macrophages are simultaneously transduced with GFP-Rab7 and mRFP-LC3 constructs for co-localization analysis.

6. In all the experiment described above, it is important to include cells in which macroautophagy has been compromised by RNA silencing. This will highlight the role of the autophagy machinery in LAP formation, maturation and MHC class II antigen presentation. The difference of phenotype between wild type and Atg silenced cells is key to understand the role of this pathway (*see* Table 2).

7. Zymosan is a component of the cell wall of the yeast *Saccharomyces cerevisiae* and known to activate macrophages via the TLR2 or Dectin1 receptors. It has been shown by different groups and ours, that the phagocytosis of zymosan by murine and human macrophages induces the direct recruitment of the autophagic protein LC3 to the phagosomal membrane. This new pathway is called LC3-associated phagocytosis [4, 5, 9].

8. Pam3CSK4 is a synthetic triacylated lipopeptide, that activates macrophages *via* the TLR2 receptor. In macrophages, the literature demonstrated that Pam3CSK4 coated beads induce the direct recruitment of LC3 proteins to the phagosomal membrane, containing these beads [4, 5].

9. The order of adding the bead to the stimulus is crucial for coating beads. In fact, by adding the microsphere to the ligand the efficiency is maximized and even distribution of adsorption is more likely.

10. Soluble *Candida albicans* or *Candida albicans* extract coated beads are also known to enhance the LAP pathway. In the study by Romao and colleagues, it is shown by electron microscopy that *C. albicans* is contained inside a single membrane surrounded phagosome displaying LC3 [4].

11. The 1.5 mm coverslips are sterilized with 70 % Ethanol for 20 min and then washed twice with sterile PBS before plating the cells on them. Any traces of ethanol should be removed by completely aspirating ethanol and washing solutions with a vacuum suction flask.

12. For correct interpretation of the result, the following controls should be included:

    (a) One slide should be incubated without primary or secondary antibodies but only with blocking solutions.

    (b) One slide should be incubated with only the secondary antibody (no primary antibody).

    (c) Non-transfected cells or/and untreated cells should be used as controls.

    (d) In case of multiple stainings, include slides with single staining and check the bleed through fluorescence in adjacent channels that will be used for other marker detections. For instance, Alexa 488 will be analyzed with the red channel and Alex-555 with the green channel. No significant signal should be detected in these channels for these fluorochromes.

13. After fixation with PFA, cells can be handled on a laboratory bench. Vacuum suction flask can be used to change solution but change tip between each condition and solution (antibodies, blocking solution etc.).

14. Image-iT® FX signal enhancer is applied directly on coverslips containing fixed and permeabilized cells prior to staining with fluorescent probes. Use Image-iT® FX to decrease the background from the fluorophore conjugated goat anti-rabbit IgG or from GFP.

15. Normal goat serum is used as blocking solution because the used secondary antibodies are derived from goats. If a secondary antibody is used that was generated in another species, it is recommended to use normal serum from that species as blocking reagent. Another solution is to use 5% BSA as blocking solution.

16. To save antibodies, a drop of 30 μl of antibody solution is put on Parafilm, take out the coverslips from the well of P24-well plate, invert and put it on the drop (the cells are in direct contact with the antibody). For washing, place the coverslips in the well (invert it again).

17. Co-staining of LC3 and other proteins is possible by using the anti-LC3 rabbit antibody and a mouse antibody specific for the second proteins of interest. For instance, use an anti-Lamp2 mouse antibody to investigate phagosome maturation or use an anti-HLA-DR mouse antibody to observe the role of LAP in delivering antigen to MHC class II loading compartments.

18. Prolong gold antifade mounting medium (Invitrogen) should dry at room temperature, overnight in the dark. During this time, the mounting medium will polymerize and its refractive index increases. With the Prolong Gold medium it is not necessary to seal the cover slides with the nail polish, but this is recommended for other, water-based mounting media.

19. Irradiated PBMCs and B-LCLs are added to the T-cell clone as feeder cells, which allow the expansion of the T-cell clones. The irradiated cells are alive but are not able to multiply. The irradiation time is determined by the strength of the radiosource. The longer cells get irradiated the more rad they accumulate.

20. It is expected to have T-cell clones from the plates prepared in (c) and sometimes from the plates prepared in (b) but never from the first plate prepared. The limiting dilution condition in which no more than one-third of the wells grow can be considered as yielding T-cell clones.

    Plate A is used to maintain polyclonal T cell lines of the sorted cells. All wells should grow, but this plate can be further analyzed in case cloning efficiency is lower than expected due to counting or treatment mistakes.

    In plate C some wells contain only the irradiated feeder cells (PBMCs and B-LCLs cells) and every third should contain irradiate feeder cells plus a single T cell.

21. Phytohemagglutinin stimulates T cell proliferation by glycosylated surface receptor cross-linking, including the T cell receptor.

Acknowledgments

Research in our laboratory is supported by grants from Cancer Research Switzerland (KFS-3234-08-2013), Worldwide Cancer Research (14-1033), KFSP^{MS} and KFSP^{HHLD} of the University of Zurich, the Sobek Foundation, the Swiss Vaccine Research Institute and the Swiss National Science Foundation (310030_162560 and CRSII3_160708).

References

1. Mizushima N, Yoshimori T, Ohsumi Y (2011) The role of Atg proteins in autophagosome formation. Annu Rev Cell Dev Biol 27:107–132

2. Paludan C, Schmid D, Landthaler M et al (2005) Endogenous MHC class II processing of a viral nuclear antigen after autophagy. Science 307:593–596

3. Schmid D, Pypaert M, Münz C (2007) Antigen-loading compartments for major histocompatibility complex class II molecules continuously receive input from autophagosomes. Immunity 26:79–92

4. Romao S, Gasser N, Becker AC et al (2013) Autophagy proteins stabilize pathogen-containing phagosomes for prolonged MHC IIantigen processing. J Cell Biol 203:757–766

5. Sanjuan MA, Dillon CP, Tait SWG et al (2007) Toll-like receptor signalling in macrophages links the autophagy pathway to phagocytosis. Nature 450:1253–1257

6. Ma J, Becker C, Lowell CA et al (2012) Dectin-1-triggered recruitment of light chain 3 protein to phagosomes facilitates major histocompatibility complex class II presentation of fungal-derived antigens. J Biol Chem 287:34149–34156

7. Delamarre L, Couture R, Mellman I et al (2006) Enhancing immunogenicity by limiting susceptibility to lysosomal proteolysis. J Exp Med 203:2049–2055

8. Savina A, Jancic C, Hugues S et al (2006) NOX2 controls phagosomal pH to regulate antigen processing during crosspresentation by dendritic cells. Cell 126:205–218

9. Martinez J, Malireddi RKS, Lu Q et al (2015) Molecular characterization of LC3-associated phagocytosis reveals distinct roles for Rubicon, NOX2 and autophagy proteins. Nat Cell Biol 17:893–906

Chapter 11

Quantitative Spatiotemporal Analysis of Phagosome Maturation in Live Cells

Laura Schnettger and Maximiliano G. Gutierrez

Abstract

Phagocytosis and phagosome maturation are central to the development of the innate and adaptive immune response. Phagosome maturation is a continuous and dynamic process that occurs rapidly. In this chapter, we describe fluorescence-based live cell imaging methods for the quantitative and temporal analysis of phagosome maturation of latex beads and *M. tuberculosis* as two phagocytic targets. We also describe two simple protocols for monitoring phagosome maturation: the use of the acidotropic probe LysoTracker and analyzing the recruitment of EGFP-tagged host proteins by phagosomes.

Key words Phagosome, Macrophage, Live cell imaging, Lysosome, Mycobacterium

1 Introduction

Phagocytosis is the process by which professional and non-professional phagocytes ingest particles whose size normally exceeds one micrometer [1]. The resulting intracellular organelles, termed phagosomes, progress through biochemical changes that modify the composition of both their limiting membrane and contents, by a sequence that resembles the progression of the endocytic pathway [2]. This process is referred to as phagosome maturation, and bestows the phagosome with degradative properties, which are central to the microbicidal function of phagosomes and the first line of defense against infection in multicellular organisms [3, 4]. During evolution, many intracellular pathogens developed strategies to survive within host cells and manipulate phagosome maturation. A good example is the intracellular pathogen *Mycobacterium tuberculosis* (Mtb) that manipulates the cellular trafficking machinery to survive within macrophages [5].

Phagosome maturation is a highly complex and dynamic pathway, being the result of multiple rounds of vesicular fusion but also phagosomal membrane fission (e.g. recycling) and other types of transient interactions such as "kiss and run" fusion or tubular

Roberto Botelho (ed.), *Phagocytosis and Phagosomes: Methods and Protocols*, Methods in Molecular Biology, vol. 1519, DOI 10.1007/978-1-4939-6581-6_11, © Springer Science+Business Media New York 2017

contacts [2, 6]. Thus, in addition to complete fusion with other organelles, transient and rapid contacts are also responsible for many aspects of the maturation process [6, 7].

As well as being dynamic, maturation of a phagosome to a phagolysosome can take different amounts of time depending on its contents and the ligand that triggers internalization. Monitoring the process by specific late endocytic markers for example shows that formation of a phagolysosome can take between 15 and 60 min [8, 9]. Therefore, studying the association of specific endocytic markers with phagosomes by valuable methods such as indirect immunofluorescence or immunogold labeling of ultrathin thawed cryosections in fixed cells has technical limitations. For example, rapid events (during the first two to three minutes of phagosome formation) or transient events such as "kiss and run" fusion or tubular structures cannot be analyzed in real time using fixed samples.

When compared to techniques with low temporal resolution such as immunolabeling of fixed cells, western blot of isolated phagosomes, proteomics, or flow cytometry [10–12], live cell imaging provides a complementary tool to investigate phagosome maturation progression continuously over long periods of time, providing important information about bacterial replication and host cell activation in real time [13]. Compared to "high content" techniques such as flow cytometry of phagosomes (phagoFACS), live cell imaging is time consuming. However, when combined with automated or semi-automated quantitative analysis, live cell imaging is very powerful and provides critical, and sometimes unique, information about changes in composition of phagosomes in real time at the single cell and individual phagosome level [14, 15].

Here, we describe two simple methods to investigate the dynamic association of late endocytic markers with phagosomes using live cell imaging, as a measurement of phagosome maturation. We also explain the process required for semi-automated quantitative image analysis in a typical experiment.

2   Materials

2.1   Cells, Buffers, and Solutions

2.1.1   Cells

1. RAW264.7 mouse macrophages.
2. Bone marrow from C57BL/6 mice.
3. *Mycobacterium tuberculosis* H37Rv expressing EGFP (*see* **Note 1**).

2.1.2   Culture Media

1. *RAW264.7 complete medium*: Dulbecco's Modified Eagles Medium (DMEM), supplemented with 10% heat-inactivated fetal calf serum (FCS). Store at 4 °C.

2. *Bone marrow-derived macrophages (BMM) complete medium:* Roswell Park Memorial Institute (RPMI) 1640 Medium, supplemented with GlutaMAX, 10% FCS. During differentiation steps medium is further supplemented with 20% L929 fibroblast culture supernatant. Store at 4 °C (*see* **Note 2**).

3. *7H9 medium for mycobacteria:* Middlebrook 7H9 broth base (Fluka Analytical, Sigma-Aldrich, Germany) in distilled water, supplemented with BBL Middlebrook ADC Enrichment (BD Diagnostics, USA), 0.05% (v/v) tween 80 and 0.2% glycerol. Store at 4 °C.

2.1.3 Reagents for the Preparation of Phagocytic Targets

1. 2.5% (w/v) 3 μm polystyrene beads (Krisker Biotech, Germany).

2. MES buffer: 1 M, pH 6.7 (Sigma-Aldrich, Germany).

3. EDAC (1-ethyl-3-[3-dimethylaminopropyl] carbodiimide hydrochloride): 10 mg/ml in water.

4. Stop buffer: 1% Triton X-100 in 10 mM Tris–HCl, pH 9.4.

5. Mouse IgG or fluorophore-conjugated mouse IgG.

6. 2.5–3.5 mm sterile glass balls (WVR International, USA).

2.1.4 Labeling of Intracellular Organelles

1. LysoTracker solution: Aliquot 1 M LysoTracker® Red DND-99 in DMSO (Invitrogen, USA), store at –20 °C and protect from light.

2. JetPEI®-Macrophage (Polyplus-transfection, France).

3. 150 mM sterile NaCl.

4. DNA/plasmids of fluorescent genes.

2.1.5 Preparation of BMM

1. Phosphate buffered saline pH 7.4 (PBS).

2. 70% ethanol.

2.2 Equipment

2.2.1 Preparation of BMM

1. Two pairs of autoclaved scissors and forceps.

2. Petri dishes.

3. Syringe and 25G 5/8 (0.5×16 mm) needle (Terumo, Germany).

4. 35 mm glass bottom dishes with 12 mm or 22 mm glass aperture, Thickness #1.5 (WillCo Wells, Germany).

2.2.2 Microscope and Environmental Chamber

1. Leica TCS SP5 AOBS laser scanning confocal microscope equipped with HyD detectors (Leica Microsystems, Germany).

2. HC PL APO CS2 63.0x/1.40 OIL objective.

3. Argon (488 nm), DPSS (561 nm), HeNe (633 nm) lasers.

4. Environmental control chamber (37 °C, 5% CO_2, 20–30% humidity; *see* **Note 3**).

5. Type 37 immersion oil (Cargille Laboratories, USA).

6. Leica Applications Suite (LAS) software for image acquisition.

7. Fiji (Fiji is just ImageJ), available for download at http://fiji.sc (NIH, USA).

3   Methods

3.1   Preparation of Cells

In order to carry out phagocytosis experiments, it is possible to use either mouse macrophage cell lines such as RAW264.7 macrophages or primary mouse bone marrow derived macrophages (BMM). The use of mouse primary cells is considered to be more physiologically relevant regarding macrophage function. In cases where a particular knockout mouse is available, they can be used to analyze the role of specific host factors in phagosome biology. Primary cells also avoid the problem of bias of clonal selection. However, BMM are more challenging to use for experiments requiring expression of fluorescently tagged markers since transfection efficiency is consistently very low. The use of lentiviral-based expression vectors significantly improves transfection yields, but sometimes with noticeable macrophage activation [16]. Nucleoporation is a good alternative, although macrophage functionality can be compromised. In cases where fundamental questions regarding intracellular trafficking are addressed, RAW264.7 macrophages represent a suitable model to study the localization of specific markers and their kinetics of association to phagosomes. In fact, many important discoveries in phagosome maturation studies in RAW264.7 macrophages were reproduced in BMM [17].

3.1.1   Preparation of Bone-Marrow Derived Macrophages

1. Perform culling of the mice by cervical dislocation (*see* **Note 4**).

2. After culling, use a set of sterile/autoclaved scissors and forceps to dissect and remove the femur and tibia bones from the mouse hind legs. The hind legs are longer and will give a higher BMM yield.

3. Remove the mouse tissue from the bones and collect them in PBS.

4. To reduce risk of contamination, perform all following steps in a class II biosafety cabinet.

5. In a sterile petri dish, rinse the bones with 70% ethanol and then transfer to a new petri dish and repeat washing once more with 70% ethanol and once with PBS. Then transfer the bones to a new petri dish containing culture medium.

6. Cut off the ends of the bones to make the bone marrow accessible (*see* **Note 5**).

Flush the medullary cavity with ice-cold culture media using a syringe attached to a 25G needle. If the bones are flushed properly they will appear translucent/white.

7. Centrifuge the cells at $350 \times g$ for 10 min at 4 °C to obtain a pellet. The pellet is then resuspended in BMM complete medium supplemented with L929 fibroblast supernatant and plated in sterile microbiology uncoated 9 cm petri dishes. Plate approximately 4×10^6 cells in 10 ml BMM culture media per dish. Incubate the cells at 37 °C in 5% CO_2 atmosphere.

8. Every 48 h aspirate 70–80% of the medium from the petri dishes and replace with fresh BMM complete medium supplemented with L929 fibroblast supernatant for at least 5 days to differentiate into macrophages.

9. After differentiation, cells should be transferred into glass bottom dishes at the concentration required for the experiment.

10. Wash BMM twice with PBS to remove non-adherent cells. Add 5 ml of ice-cold PBS to each dish and leave them on ice at 4 °C until all cells are completely detached. Cells should be detached after 10–15 min.

11. Carefully collect the cells by washing the dishes and centrifuge at $350 \times g$ for 10 min. Resuspend the pellet in BMM complete medium. Count and dilute BMM to the required density and transfer the cells into the recess of the glass bottom live cell dishes. For live cell imaging, 4×10^6 cells per 22 mm aperture dish in 1 ml medium are transferred and subsequently incubated at 37 °C in 5% CO_2 atmosphere (*see* **Note 6**).

3.1.2 Preparation of RAW264.7 Macrophages

1. Culture RAW264.7 macrophages in RAW264.7 complete medium. Cells are cultured in T-75 or T-25 vented cell culture flasks.

2. Every 2–3 days wash macrophages twice with PBS before scraping them into medium.

3. Resuspend cells in medium to avoid clumping and passage 1:10 into a new flask. Add medium to the flask to complete the total volume.

4. When plating the RAW264.7 macrophages for an experiment, scrape, count and dilute cells in RAW264.7 complete medium and then transfer the cells into the recess of the glass bottom live cell dishes. For live cell imaging, plate 3×10^6 cells per 22 mm aperture dish in 1 ml medium and subsequently incubate at 37 °C in 5% CO_2 atmosphere.

3.2 Preparation of Phagocytic Targets

There are several particles that can be used to study phagosome maturation by live cell imaging. Here we will be describing two different methods. Firstly, the use of inert particles as a model of phagocytosis and secondly, mycobacteria (*Mycobacterium tuberculosis* H37Rv) as an intracellular pathogen that modulates normal phagosome maturation.

3.2.1 Polystyrene Beads

Polystyrene beads are inert particles that can be used to study phagosome maturation. Fcγ-receptor mediated internalization is one the best characterized internalization mechanisms involved in phagocytosis and further phagosome maturation [2]. Opsonization of the polystyrene beads by coating them with the appropriate IgG is required to ensure that they are recognized and internalized by mouse macrophages. Furthermore they can be coated with fluorescent dyes to visualize them.

1. Mix 400 μl polystyrene beads (2.5% (w/v), 3 μm) with 100 μl MES buffer, pH 6.7 and 50 μg of mouse IgG.

2. Mix the solution of beads for 15 min in a rotating wheel at room temperature.

3. Add 14 μl of EDAC (10 mg/ml in sterile water) to the beads and mix the whole solution in a rotating wheel for 1 h at room temperature. Then add another 14 μl of EDAC and mix for a further 1 h.

4. The coupling reaction is stopped by washing with stop buffer three times. During these washing steps the bead solution is centrifuged and the supernatant is removed before the beads are resuspendend in 1 ml Stop buffer.

5. The IgG coated beads are stored in PBS at 4 °C. PBS is added to the beads to give a final concentration of 1% (w/v).

3.2.2 Preparation of Mycobacteria for Infection

Mycobacterium tuberculosis H37Rv (Mtb) expressing EGFP was used in the experiments shown in Figs. 1 and 2 to analyze the association of LysoTracker or various phagosomal markers to bacteria.

1. Grow mycobacteria in rolling 50 ml falcon tubes to mid-exponential phase (OD_{600} 0.5–1) in 7H9 medium at 37 °C (*see* **Note 7**).

2. Pellet bacteria in a 50 ml falcon tube by centrifuging at $2000 \times g$ for 5 min at room temperature.

3. Wash the pellet once with PBS and once in complete medium.

4. Add 2.5–3.5 mm sterile glass balls to mycobacterial pellets. The amount of glass balls added depends on the size of the pellet; the volume of both pellet and balls should be similar.

5. Break bacterial clumps by vigorously shaking the bacterial pellet together with the glass balls for 1 min (*see* **Note 8**).

6. Resuspend bacteria in complete BMM or RAW264.7 cell culture medium.

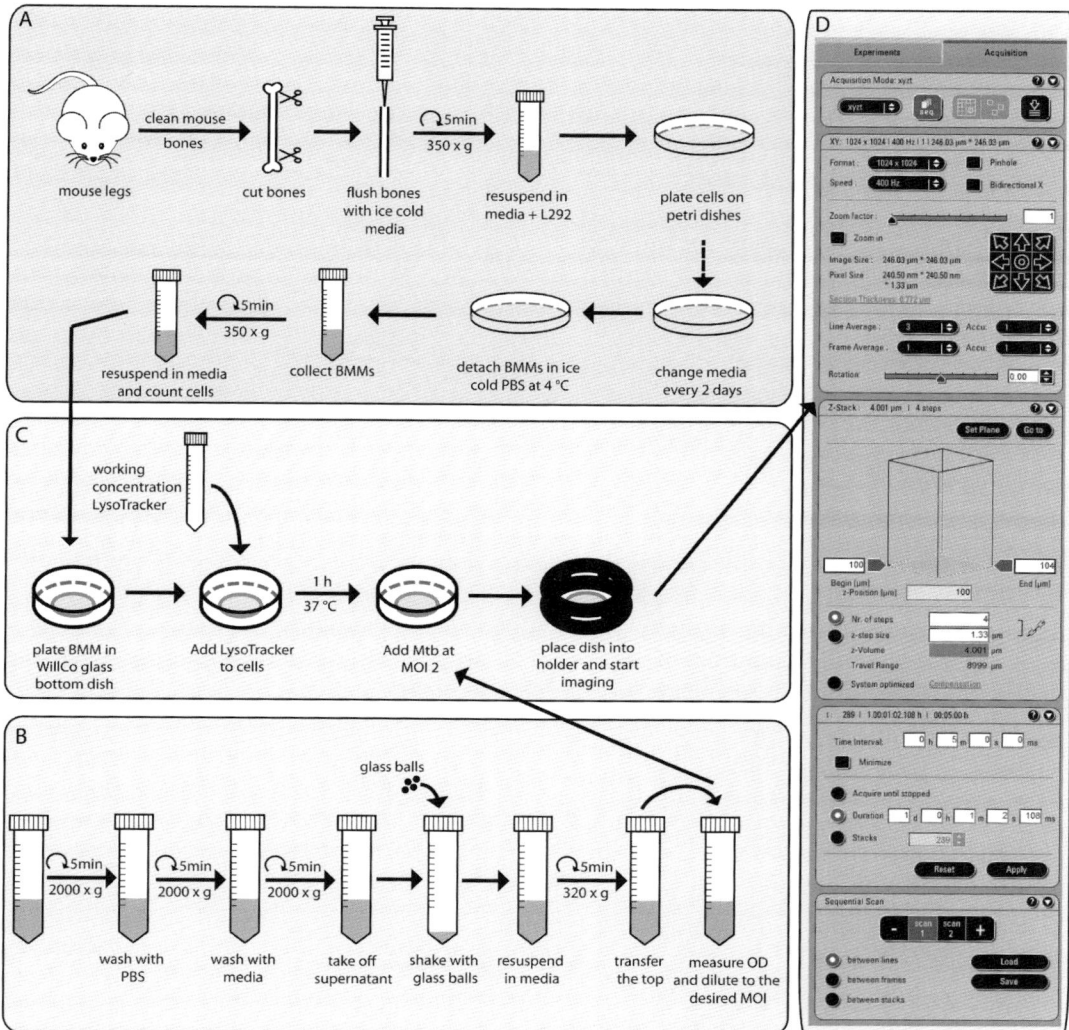

Fig. 1 Overview of the experimental setting in live cell imaging of phagosome maturation. (**a**) Preparation of bone marrow derived macrophages from mouse legs. (**b**) Overview of the sequential steps involved in the preparation of mycobacteria, here shown for Mtb, for the infection of macrophages. (**c**) Overview of experimental set up. BMM prepared in (**a**) are plated in a glass bottom live cell dish. On the day of the experiment LysoTracker is added to the cells for 1 h prior to infection with Mtb prepared in (**b**). The dish is placed in a holder and transferred to the microscope for imaging. (**d**) Image settings used for acquisition of time-lapse movies using a Leica TCS SP5 confocal microscope

7. Spin down any remaining clumps by centrifugation at $320 \times g$ for 5 min.

8. The upper part of the supernatant after centrifugation now contains a suspension of single bacteria (rather than clumps of bacteria).

9. Measure the OD_{600} (blank using cell culture medium) and calculate the volume needed to add to the macrophages assuming that an OD_{600} of 1 represents 1×10^8 bacteria. For live cell imaging in BMM with LysoTracker we added Mtb at a multiplicity of infection (MOI) of 2 in 1 ml BMM medium (*see* **Note 9**).

3.3 Labeling of Intracellular Organelles

There are multiple methods to investigate live dynamics of phagosome maturation; here we will be focusing on two methods: one using a fluorescent probe that labels acidic organelles and another using expression of fluorescently tagged proteins in macrophages. Additionally, transfer of preloaded fluorescent dextran into phagosomes can also be used which has previously been described in detail [6, 14].

3.3.1 LysoTracker Delivery into Phagosomes

LysoTracker is a fluorescent dye linked to a weak base and is used to label and track acidic compartments in live cells [13, 15]. The weak base causes it to be only partially protonated at neutral pH, allowing LysoTracker to freely cross cell membranes in live cells. The probe is then trapped in acidic compartments following protonation of the base. There are different commercially available LysoTracker probes: LysoTracker Red, Green, and Blue. In our experience the most reliable probe is the LysoTracker Red, since it gives a bright fluorescent signal, without fast bleaching or high cytotoxicity. LysoTracker Blue might be required in cases were red and green channels are already used. However, LysoTracker Blue has the disadvantage of bleaching faster and being more cytotoxic and therefore only short time periods can be imaged using this dye. It is very important to consider that LysoTracker association with beads or bacterial phagosomes could be both due to acidification of this compartment and therefore direct LysoTracker accumulation or due to delivery by fusion with LysoTracker positive compartments. In the second case, the dye is trapped in an acidic compartment by protonation of its base and then gets delivered into the phagosome [15].

1. LysoTracker solution is diluted to 50 nM in complete medium (1:20,000 dilution) (*see* **Note 10**).

2. LysoTracker solution is added to macrophages 1 h before addition of beads/mycobacteria in 1 ml media.

3. LysoTracker containing medium is replaced and cells are washed three times with PBS before beads or mycobacteria are added as described in Subheading 3.2.

3.3.2 Cell Transfection

To monitor association of cellular markers to phagosomes, RAW264.7 macrophages are transfected with plasmids expressing fluorescently tagged proteins. Here we describe one protocol to efficiently transfect RAW264.7 macrophages using the transfection reagent JetPEI (*see* **Note 11**).

1. RAW264.7 macrophages are plated on glass bottom live cell dishes in RAW264.7 complete medium and incubated at 37 °C, 5 % CO_2 for at least 10 h before transfection.

2. For each transfected dish, 1–2 μg plasmid DNA (vectors expressing fluorescently tagged proteins) and 3 μl JetPEI reagent is each added to a total volume of 50 μl of 150 mM sterile NaCl. The solutions are vortexed followed by a short centrifugation.

3. Add the solution containing the JetPEI reagent to the DNA. The transfection mixture is then vortexed followed by spinning.

4. Leave the transfection mixture at room temperature for 15–20 min to allow the transfection complexes to from.

5. Add the mixture dropwise onto the cells and incubate for approximately 16 h at 37 °C, 5 % CO_2.

6. Add fresh complete medium and let the cells rest for a few hours (*see* **Note 12**).

3.4 Live Cell Imaging

There are various suitable microscopes available to perform live cell imaging, including wide-field fluorescent microscopes, spinning-disk and confocal laser scanning microscopes. There are both advantages and disadvantages between the different systems and important parameters to consider are stability, photocytotoxicity, photobleaching, and resolution [14]. For the live cell imaging studies described here, we used a confocal laser scanning microscope since it provides high resolution with relatively low phototoxicity. In combination with a microscope environmental chamber, it provides a very stable and reliable system. An overview of the experimental design is shown in Fig. 1.

1. Transfer cells into glass bottom live cell dishes as described in Subheadings 3.1.1 and 3.1.2.

2. Prepare a single cell mycobacterial solution and add bacteria at a MOI of 2 in 1 ml total volume to the BMM for live cell imaging as described in Subheading 3.2.2 (*see* **Note 13**).

3. For live cell imaging an environmental control chamber providing 37 °C, 5 % CO_2 and 20–30 % humidity in a closed chamber is needed.

4. Adjust the imaging settings according to the experiment and start the image acquisition (*see* **Notes 14** and **15**).

5. Acquire images from the bright-field channel to check both for cell morphology and in the case of beads for their localization (*see* **Note 16**).

6. Always save the data in the native file-format of the imaging system used.

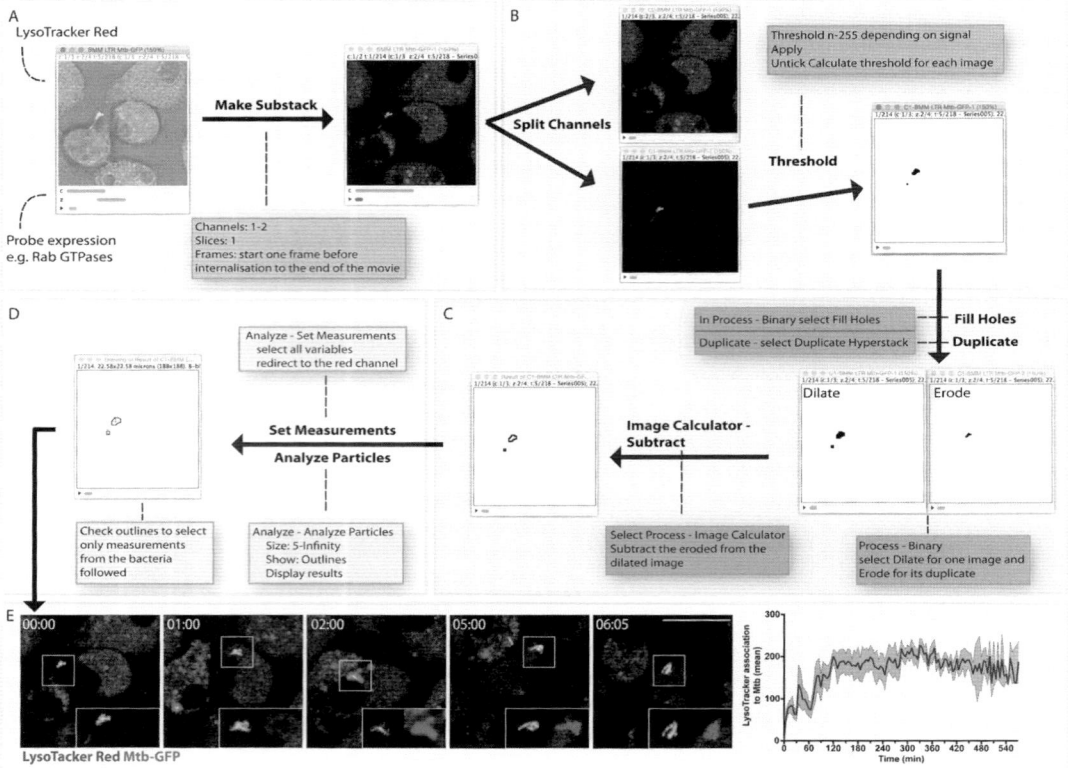

Fig. 2 Diagram of the image analysis flow and representative data. Overview of the different steps during the image analysis of association of LysoTracker to Mtb. Detailed description can be found in Subheading 3.5.2. (**a**) Image preparation for analysis. The image is opened in Fiji and one focal plane is selected for the following analysis. In the images LysoTracker Red DND-99 is shown in red and EGFP-Mtb in green. (**b**) Selection of the bacteria. The channels with the different colors are split and bacteria are thresholded. (**c**) Image calculations. The thresholded image is dilated and eroded and then the two images are subtracted from each other to result in the selection of only the regions surrounding the bacteria. (**d**) Analysis. The measurements are set and the association of LysoTracker to Mtb is analyzed. (**e**) Association of LysoTracker with EGFP-Mtb H37RV in BMM. Quantification shows mean ± SEM from 10 cells in two biological experiments

3.5   Image Analysis

The image analysis can be performed using various software packages; here we will describe the use of the freely available java-based Fiji platform. Fiji is an ImageJ distribution containing additional preloaded plugins. It is available for download at http://fiji.sc. Macros and plugins used are mentioned in notes. An overview of the imaging analysis flow and a representative experiment is shown in Fig. 2.

3.5.1   Analysis of the Association of Various Markers to IgG-Coated Beads

We have previously described in detail the image analysis of the association of various markers to bead-containing phagosomes [14]. Therefore, here we only briefly outline the steps required for this analysis and focus in the next section on bacteria-containing phagosome analysis that is more challenging. In general, many steps in the analysis of association of different markers to beads or mycobacteria are very similar [14].

1. Open the image saved in the native file-format in Fiji using the "LOCI"—"Bio-Formats Importer" and split the different channels into separate windows. The details of these steps can be found in Subheading 3.5.2

2. Set the measurement parameters ("Analyze"—"Set Measurements…") and redirect to the channel containing the signal from the probe or marker.

3. Select the channel with the beads—bright field or fluorescence depending on experimental set up—and select a circular region of interest (ROI) on the internalized bead using the "oval selection tool".

4. Start the analysis one frame before complete internalization, similar to analysis with mycobacteria. Starting with that frame select "Analyze"—"Measure." A table with the results for this measurement should appear.

5. Repeat the process for the following image time frames. As the beads move inside the cell the ROI might need to be adjusted for localization exactly on the bead.

6. Plot the measurements obtained in the results table against the time after internalization (*see* **Note 17**).

3.5.2 Analysis of the Association of Various Markers to Mycobacteria

1. Open the file to be analyzed in Fiji by either using the plugin "LOCI"—"Bio-Formats Importer" or by dragging and dropping the file into Fiji which should automatically use this plugin (*see* **Note 18**).

2. Choose the following options for opening the file: "view as hyperstack" and in the case of files larger than the available amount of random access memory (RAM) for the computer, select "virtual" (*see* **Note 19**).

3. Create a duplicate image containing only the one focal plane with the bacterium. If the bacterium analyzed changes focal plane throughout the time-lapse movie split the movie into several shorter files using the "Make Substacks…" command ("Image—Stack—Tools—Make Substack"). Select only the channels, focal plane and time frames containing the bacterium and marker to be analyzed. Start the time-lapse movie one or two frames before the bacterium is completely internalized (*see* **Note 20**).

4. To facilitate further analysis the short time lapse-movies can be made into a single file again using the "Concatenate…" command ("Image—Stacks—Tools—Concatenate"). Select the files in the order they appear in the original time-lapse movie.

5. Split the different channels ("Image—Color—Split Channels").

6. Select the channel containing the bacteria. Adjust the threshold to select only the bacterial region ("Image—Adjust—Threshold").

Select a thresholding setting that will result in thresholding of bacteria over the complete time-lapse movie. After selecting "ok" make sure that in the window that opens "Calculate threshold for each image" is NOT ticked (*see* **Note 21**).

7. Starting from the thresholded image, select "Fill Holes" ("Process—Binary—Fill Holes") and process all images.

8. Duplicate the thresholded image selected ("Image—Duplicate"). Tick the box next to "Duplicate Hyperstack".

9. Select the duplicated image and reduce the size of the thresholded area by 1 pixel. ("Process—Binary—Erode").

10. Select the original thresholded image and increase the size of the thresholded area by 1 pixel ("Process—Binary—Dilate").

11. Subtract the duplicated eroded image from the original dilated one ("Process—Image Calculator..."). In the window that opens select "Create new window" and then "Process all images."

12. Select the newly generated image, which should now show a thresholded area around the location of the bacteria. The size of the ring can be adjusted by multiple rounds of dilation or erosion before subtracting the two thresholded images from each other.

13. Go to "Analyze"—"Set Measurements..." and select area, mean, gray value, standard deviation, integrated density, display label and any other parameter that you wish to measure.

14. In the same window click "redirect to" and select the channel containing the signal from the probe or marker. In Fig. 2 we redirected to the channel containing the signal from LysoTracker.

15. Go to "Analyze"—"Measure particles...". Select the size of the particles to be analyzed: "Size (pixel∧2): 5—Infinity," "Circularity: 0.00–1.00," "Show: Outlines," "Display results". The lower size can be adjusted depending on the size of the particle/bacteria measured. Selecting a value between 5 and 10 in most cases is suitable to stop measurement of thresholding artifacts.

16. A results table will appear with the values for the signal intensity in the channel with the marker in each thresholded area in every single image. In the case of bacteria the area of the thresholded regions can change significantly depending on bacterial size and orientation. To compare between different results use a measurement that is independent of area, e.g., mean intensity (*see* **Note 22**).

17. Check in the Results window showing the outlines that only one thresholded area corresponding to the location of the analyzed bacterium is measured. In some cases cells will internalize more

than one bacterium, which can result in the measurement of the signal from 2 particles. In this case only select the measurements obtained from the same single bacterium.

18. Plot the measurement against time. In Fig. 2 we plotted the mean intensity against time after internalization.

4 Notes

1. *Mycobacterium tuberculosis* H37 is one of the most commonly used laboratory strains. The H37 strain is further distinguished as "virulent" (Rv) or "avirulent" (Ra).

2. Macrophage colony-stimulating factor (M-CSF) is secreted by L929 cells and is used in the form of L929-conditioned medium. Supernatants of L929 cells provide a cheap source of M-CSF but contain many other growth factors. A more expensive alternative is to use purified growth factors such as granulocyte-macrophage colony-stimulating factor (GM-CSF) or (M-CSF).

3. Temperature stability is critical for long-term imaging since very subtle variations in temperature during image acquisition will cause focus drift. Alternatively, many companies offer autofocus correction. Different types of environmental boxes are available in the market; we use a customized box from EMBL technology transfer (Heidelberg, Germany).

4. Animal care and all procedures described in this chapter follow the institutional and national guidelines.

5. Use a different set of sterile/autoclaved scissors and forceps then during the removal of the bones from mice to avoid contamination. Macrophages are sensitive to activation by lipopolysaccharides and glassware should be avoided.

6. Cells can be counted using various methods. A counting chamber or automated cell counter from for example Bio-Rad both use image-based analysis to determine the cell number. To further assess cell viability, trypan blue can be added to the cell solution in a 1:1 ratio before counting, cells stained blue should be excluded from the viable cell count.

7. *M. tuberculosis* is a biosafety level 3 (BSL3) pathogen and should be handled in a BSL3 laboratory obeying all appropriate regulations.

8. *M. tuberculosis* is notorious for clumping, it is particularly important to have a single cell suspension for live cell imaging in order to image single events.

9. *M. tuberculosis* should be added at a low MOI where possible. Depending on the cell type used for infection the MOI

might have to be adjusted. If there are too many extracellular bacteria, reduce the MOI and/or wash cells before the image acquisition.

10. Very high concentrations of LysoTracker can result in labeling of the bacteria.

11. In addition to JetPEI other transfection reagents can be used. Lipofectamine® reagents (Thermo Fisher Scientific Inc, UK) are another good example of transfection reagents that can be used to efficiently transfect macrophages.

12. When expression induces toxicity, reduce the time of transfection. Avoid the use of high concentrations of DNA to minimize cytosolic background fluorescence.

13. In the case of infection with Mtb, add Parafilm around the dish and place it in a live cell dish-clamping holder.

14. In the experiment shown in Fig. 2 we used the following settings which can also be seen in the panel on the right side of the figure: Acquisition mode: xyzt, resolution 1024×1024 pixels, 400 Hz scanning speed, 3x line averaging, sequential mode acquisition, z-stack with 4 steps, interval between image acquisitions 5 min.

15. The imaging settings should be adjusted according to the experiment. If the acquisition of a very fast event is sought, it might be necessary to adjust both the resolution/line averaging and number of z-planes to decrease the interval between image acquisitions. For longer time-lapse movies with bacteria it is best to acquire images as a z-stack choosing multiple focal planes as both cells and bacteria can move during the image acquisition and can otherwise not be followed over time.

16. In the case of transfected cells bright-field channels can also be used for looking at the localization of surrounding non-transfected cells.

17. In contrast to bacteria, the size and shape of inert particles will not change during the imaging process. Therefore, instead of plotting the "mean" value obtained, it is also possible to plot the intensities independent of the area ("IntDen" column in the results table).

18. The file should have been saved in the native file-format of the imaging system used as mentioned in Subheading 3.4. In the case of Leica confocal microscope it will be a .lif file.

19. Choosing to view the file as hyperstack will make it easier to follow the time-lapse movie. Furthermore, this step is essential for splitting the image for analysis of data from one focal plane. Viewing the file as "virtual" reduces the capacity needed to open and then view the file and therefore helps speeds up the analysis process for large files.

20. It is essential to analyze the association of markers to bacteria from only one focal plane to avoid the signal from another location in the cell from resulting in a false positive bacterial association signal.

21. During the thresholding step select a lower limit of a value n depending on the signal intensity in the images and the maximal value 255 as the upper limit.

22. The mean intensity refers to the total signal intensity measured for each particle divided by the area of the particle.

Acknowledgments

We thank the Host–Pathogen Interactions in Tuberculosis Laboratory for useful discussions and comments on the manuscript. This work was supported by the Francis Crick Institute which receives its core funding from Cancer Research UK (FC001092), the UK Medical Research Council (FC001092), the Wellcome Trust (FC001092) and by the UK Medical Research Council (MC_UP_12012/11) to MGG.

References

1. Fairn GD, Grinstein S (2012) How nascent phagosomes mature to become phagolysosomes. Trends Immunol 33:397–405

2. Flannagan RS, Jaumouille V, Grinstein S (2012) The cell biology of phagocytosis. Annu Rev Pathol 7:61–98

3. Haas A (2007) The phagosome: compartment with a license to kill. Traffic 8:311–330

4. Jutras I, Desjardins M (2005) Phagocytosis: at the crossroads of innate and adaptive immunity. Annu Rev Cell Dev Biol 21:511–527

5. Flannagan RS, Cosio G, Grinstein S (2009) Antimicrobial mechanisms of phagocytes and bacterial evasion strategies. Nat Rev Microbiol 7:355–366

6. Kasmapour B, Gronow A, Bleck CK, Hong W, Gutierrez MG (2012) Size-dependent mechanism of cargo sorting during lysosome-phagosome fusion is controlled by Rab34. Proc Natl Acad Sci U S A 109:20485–20490

7. Stuart LM, Ezekowitz RA (2005) Phagocytosis: elegant complexity. Immunity 22:539–550

8. Hoffmann E, Marion S, Mishra BB, John M, Kratzke R, Ahmad SF, Holzer D, Anand PK, Weiss DG, Griffiths G et al (2010) Initial receptor-ligand interactions modulate gene expression and phagosomal properties during both early and late stages of phagocytosis. Eur J Cell Biol 89:693–704

9. Lu N, Zhou Z (2012) Membrane trafficking and phagosome maturation during the clearance of apoptotic cells. Int Rev Cell Mol Biol 293:269–309

10. Colas C, Menezes S, Gutierrez-Martinez E, Pean CB, Dionne MS, Guermonprez P (2014) An improved flow cytometry assay to monitor phagosome acidification. J Immunol Methods 412:1–13

11. Rogers LD, Foster LJ (2007) The dynamic phagosomal proteome and the contribution of the endoplasmic reticulum. Proc Natl Acad Sci U S A 104:18520–18525

12. Trost M, English L, Lemieux S, Courcelles M, Desjardins M, Thibault P (2009) The phagosomal proteome in interferon-gamma-activated macrophages. Immunity 30:143–154

13. de Souza Carvalho C, Kasmapour B, Gronow A, Rohde M, Rabinovitch M, Gutierrez MG (2011) Internalization, phagolysosomal biogenesis and killing of mycobacteria in enucleated epithelial cells. Cell Microbiol 13:1234–1249

14. Bronietzki M, Kasmapour B, Gutierrez MG (2014) Study of phagolysosome biogenesis in live macrophages. J Vis Exp 10:85

15. Pei G, Repnik U, Griffiths G, Gutierrez MG (2014) Identification of an immune-regulated

phagosomal Rab cascade in macrophages. J Cell Sci 127:2071–2082

16. Zhang X, Edwards JP, Mosser DM (2009) The expression of exogenous genes in macrophages: obstacles and opportunities. Methods Mol Biol 531:123–143

17. Gutierrez MG, Master SS, Singh SB, Taylor GA, Colombo MI, Deretic V (2004) Autophagy is a defense mechanism inhibiting BCG and Mycobacterium tuberculosis survival in infected macrophages. Cell 119: 753–766

Chapter 12

Measuring Phagosomal pH by Fluorescence Microscopy

Johnathan Canton and Sergio Grinstein

Abstract

Dual wavelength ratiometric imaging has become a powerful tool for the study of pH in intracellular compartments. It allows for the dynamic imaging of live cells while accounting for changes in the focal plane, differential loading of the fluorescent probe, and photobleaching caused by repeated image acquisitions. Ratiometric microscopic imaging has the added advantage over whole population methods of being able to resolve individual cells and even individual organelles. In this chapter we provide a detailed discussion of the basic principles of ratiometric imaging and its application to the measurement of phagosomal pH, including probe selection, the necessary instrumentation, and calibration methods.

Key words Phagosome, Dual wavelength ratio imaging, Fluorescence microscopy, pH, Fluorescein, Oregon Green, SNARF1

1 Introduction

Phagocytosis denotes a complex biological process in which relatively large particles (>0.5 μm) can be bound and engulfed by a eukaryotic cell. The process itself is evolutionarily ancient and represents the primary means by which single-celled phagotrophic eukaryotes, such as those belonging to the *Dictyostelium* genus, acquire nutrients [1]. In metazoans, phagocytosis is central to the maintenance of homeostasis, serving to scavenge and clear dead cells and particulate debris. In our bodies, for example, approximately 200–300 billion cells undergo apoptosis or some other form of cell death each day [2]. It is through phagocytosis (referred to as efferocytosis when targeting dead cells) that these cells are cleared and their components recycled [2]. Other homeostatic by-products cleared by phagocytosis include fibrillar β-amyloid in the brain, the shed outer segments of photoreceptor cells in the eye, and the nuclei extruded by erythroblasts at sites of erythroid maturation [3–5]. Phagocytosis also plays a central role in both the execution of innate immune defense and the initiation of the adaptive phase. Invading pathogens can be recognized by specialized

Roberto Botelho (ed.), *Phagocytosis and Phagosomes: Methods and Protocols*, Methods in Molecular Biology, vol. 1519,
DOI 10.1007/978-1-4939-6581-6_12, © Springer Science+Business Media New York 2017

receptors on the surface of phagocytic cells; these specific receptors not only immobilize the microorganism but also transduce signals that result in its engulfment and elimination of the associated threat. Enclosure of the pathogen in a membrane-bound intracellular compartment, termed a phagosome, serves two functions: (i) it allows for the compartmentalized killing of the invading organism, thereby limiting the release of inflammatory agents and, (ii) it targets antigenic material to a compartment where it can be loaded onto major histocompatibility molecules for presentation to cells of the adaptive immune system [6].

All of the abovementioned functions of phagocytosis rely on one salient feature—the acquisition of an acidic phagosomal lumen ($\leq$pH 5) (Fig. 3b). Various degradative enzymes are sorted into phagosomes as they mature; however, some are zymogens and are only processed into their mature, active forms at low pH. Such is the case for most of the cathepsin family [7]. Other phagosomal enzymes have acidic pH optima, and therefore only function effectively in an acidic environment [6]. Only fully acidified phagosomes possess the potent hydrolytic capacity required for the degradation of internalized particles. Moreover, resultant degradation products such as nucleosides, iron and amino acids are translocated into the cytosol by proton-coupled transporters [8, 9]. It is not surprising then that dissipating phagosomal pH with ionophores or weak bases, such as chloroquine, has been used historically to inhibit functions associated with phagosomal processing [10–12].

It is through the acquisition of the vacuolar-type ATPase (V-ATPase) that luminal acidification is achieved. Shortly after sealing, a series of fusion and fission interactions with organelles of the endocytic pathway result in the accumulation of V-ATPases on the phagosomal membrane [13]. The V-ATPase is a multimeric pump that transports protons from the cytosol to the lumen of the phagosome using energy derived from the hydrolysis of ATP [14]. Since the movement of protons across the phagosomal membrane is an electrogenic process, continued activity of the V-ATPase, without some form of charge compensation, would result in the generation of a self-inhibitory electrical potential [15–20]. Although there is a dearth of information on how the necessary charge compensation is achieved on phagosomes, it can be inferred from work done in lysosomes that either the inward movement of anions (such as Cl^-) or the outward movement of cations (such as Na^+ and K^+) can provide a counter-ion current, thereby limiting the buildup of an inhibitory potential [20].

Attempts to measure phagosomal pH have been made from the very dawn of macrophage biology. Indeed, shortly after discovering phagocytes, Élie Metchnikoff himself made the first approximations of phagosomal pH. By allowing phagocytic protists to engulf bits of litmus paper, he observed a color change in the sealed phagosome indicative of an acidic lumen [21]. Later, investigators began to adsorb indicator dyes such as neutral red, bromophenol blue, and bromocresol green to microbes that could be ingested by mammalian phagocytes [22–25]. The resultant color changes of the indicator dyes gave rough estimates of changes in pH over time. These measurements, however, were purely qualitative and not amenable to precise calibrations. Around this time Ohkuma and Poole pioneered a more accurate, sensitive, and quantitative method to estimate the pH of a different intracellular organelle, the lysosome [26]. Their technique involved targeting a pH-sensitive fluorophore, in this case fluorescein, to the lysosomal compartment. Fluorescein has a peak fluorescence emission at 520 nm when excited at 490 nm (Fig. 1a). That emission is exquisitely sensitive to changes in pH (Fig. 1a). However, other factors, such as photobleaching of

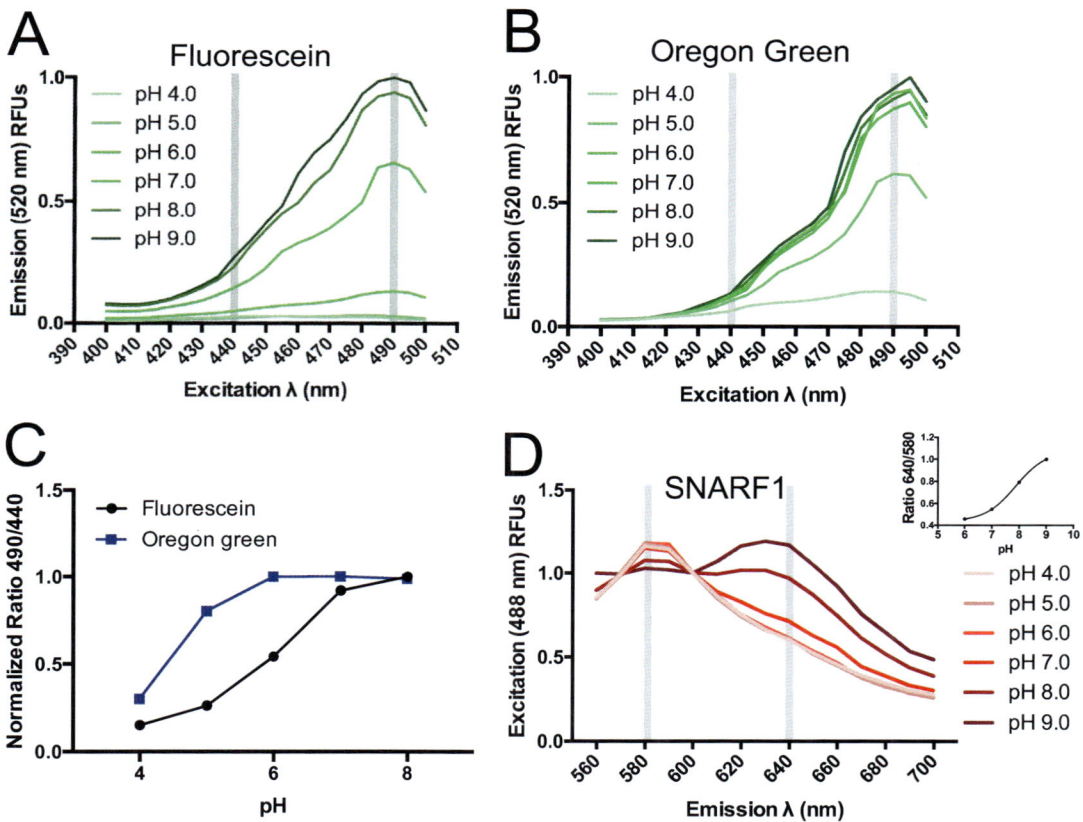

Fig. 1 Fluorescent pH sensors are most sensitive near their pK_a. (**a**) Fluorescein (10 µg/mL) was dissolved in calibration buffers at the indicated pH values. An excitation wavelength spectral sweep was performed from 400 to 500 nm at 10 nm intervals, recording emission at 520 nm using the SpectraMax Gemini EM fluorescence plate reader. (**b**) Oregon Green (10 µg/mL) was dissolved in calibration buffers at the indicated pH. An excitation wavelength spectral sweep was performed from 400 to 500 nm at 10 nm intervals, recording emission at 520 nm, as above. (**c**) Background-subtracted fluorescence values at 490 and 440 nm excitation from (**a**) and (**b**) were used to generate a ratio 490/440 vs. pH curve. (**d**) SNARF1 (5 µg/mL) was dissolved in calibration buffers at the indicated pH. An emission wavelength spectral sweep between 560 and 700 nm was performed with excitation at 488 nm, as above. Background-subtracted fluorescence values at the 640 and 580 nm emission wavelengths were used to generate a ratio 640/580 vs. pH curve (inset)

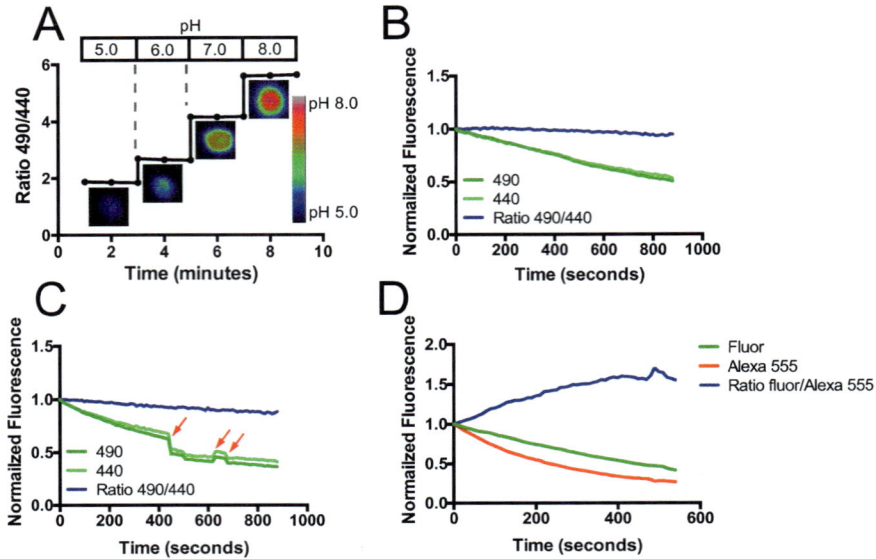

Fig. 2 Basic principles of dual-wavelength ratio imaging. (**a**) Fluorescein-labeled zymosan particles were attached to an 18 mm coverslip and imaged in a series of calibration buffers at the indicated pH values. 490/440 ratios were plotted against time. Images of individual fluorescein-labeled zymosan particles acquired in each calibration buffer with both 490 and 440 nm excitation were used to generate ratio images using the RatioPlus macro in ImageJ. The resulting pseudo-colored ratio images are depicted (insets). (**b**) Fluorescein-labeled zymosan particles attached to a coverslip were imaged for 15 min, alternating between 490 and 440 nm excitation wavelengths every minute. The background-subtracted and normalized fluorescence for the 490 and 440 nm wavelengths alone as well as the ratio 490/440 are shown. The traces illustrate the effect of progressive photobleaching on the fluorescence recorded at the individual wavelengths and on the ratio 490/440 (**c**) Fluorescein-labeled zymosan particles were imaged as in (**b**), but the plane of focus was changed during the acquisition at the three points indicated by *red arrows* to illustrate the effect of changes in the focal plane on the fluorescence recorded at the individual wavelengths and on the ratio 490/440. (**d**) Zymosan particles co-labeled with fluorescein and Alexa 555, a red dye used as reference, were imaged as in (**b**). The differential photobleaching over time of fluorescein and Alexa 555 is shown and the resulting effect on the ratio FITC/Alexa 555 is shown. Note that application of the ratio does not adequately compensate for the bleaching induced loss of fluorescence, which occurs at different rates for the two dyes

the probe, can also affect the overall emission intensity (Fig. 2b). Therefore, normalizing to a second, relatively pH-insensitive, excitation wavelength (440 nm) (Figs. 1a, 2a, and 3a) allows for the measurement of changes in pH, while accounting for other concerns such as photobleaching (Fig. 2b), changes in focal plane (Fig. 2c) and the amount of fluorophore. This technique is referred to as dual-excitation ratio fluorescence imaging. The ratio of the emission intensities recorded with excitation at 490 and 440 nm (ratio 490/440) can then be converted into pH by using an in situ calibration procedure. This entails clamping the pH of the intracellular compartment using ionophore-containing calibration solutions (Figs. 2a and 3a, b) [27]. This technique was promptly adapted to measure phagosomal pH by covalently coupling fluorescein to phagocytic targets [10].

A

B

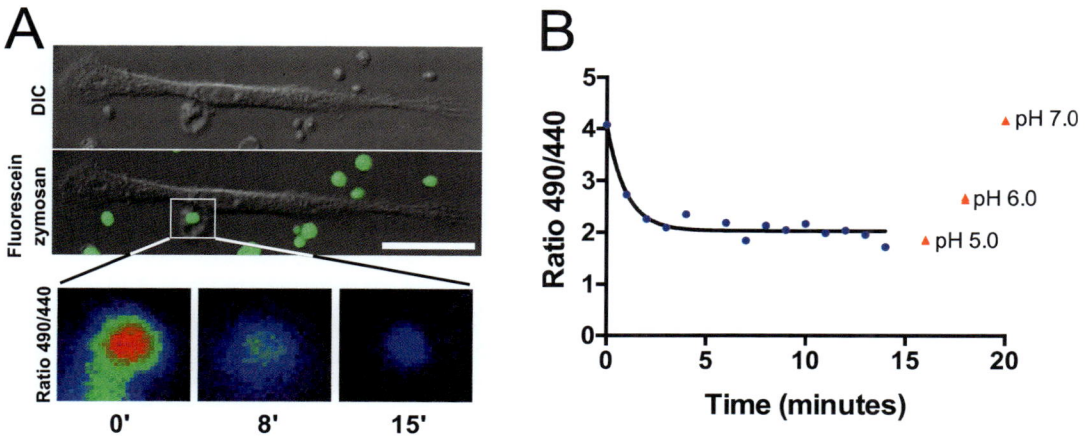

Fig. 3 Live-cell imaging of phagosomal pH. (**a**) Primary human macrophages were allowed to bind fluorescein-labeled zymosan and initiate phagocytosis. A zymosan particle in a phagocytic cup is denoted showing fluorescence with 490 nm excitation. The same particle was observed for 15 min, capturing a picture every 1 min. The background-corrected pseudo-colored 490/440 ratio image is shown at the start of phagocytosis and at 8 and 15 min after engulfment. Scale bar = 10 μm. (**b**) The background-corrected 490/440 ratio for the phagosome depicted in (**a**) was plotted against time (*blue* points). An in situ calibration was then carried out on the same cell (*red* points)

Fluorescence ratio imaging remains the most sensitive method for real-time measurement of the phagosomal pH. Several variations of this technique have emerged, including the measurement of whole populations of cells using a spectrofluorometer [28], individual cells using a flow cytometer [29] and even individual phagosomes using fluorescence microscopy [30]. Each of these approaches requires careful consideration. For example, population-based methods like spectrofluorometry and flow cytometry, although useful for the rapid measurement of large populations of cells, do not distinguish between phagocytic targets that have been internalized and those that are merely bound to the cell surface; therefore, this data represents a composite of the pH in the immediate environment of both internalized and uninternalized particles. Also, recent work has shown that there is a great degree of heterogeneity between cells and even individual phagosomes [31]. Only fluorescence microscopy can resolve the idiosyncrasies of individual phagosomes. In this chapter, we will focus on the application of ratio fluorescence microscopy using several pH-sensitive fluorophores for the dynamic measurement of phagosomal pH.

1.1 Instrumentation

A comprehensive schematic of the required hardware for ratio fluorescence imaging can be found in a recent chapter on the measurement of lysosomal pH (*see* ref. [27]). In short, a high intensity arc lamp is required as a light source. The type of lamp used should be matched to the desired excitation wavelengths. The two most commonly used light sources are xenon and mercury arc lamps. Xenon lamps provide a relatively even output spectrum; whereas, mercury lamps are characterized by distinct, sharp emission peaks [32]. If the discrete emission peaks of the mercury lamp do not match the desired excitation wavelengths for the pH probe, suboptimal excitation of the probe may occur. Xenon lamps also have the added benefit of more stable emission over time, rendering them better suited for real-time or quantitative fluorescence imaging.

From the arc lamp, light is carried through a fiber-optic cable to a computer-controlled, shuttered excitation filter wheel. The filter wheel should allow for a very rapid (on the order of milliseconds) transition between wavelengths. The filtered light then passes through a filter cube containing a dichroic mirror that reflects the light up through the objective lens to the biological sample. The light emitted from the fluorophore is then collected by the objective lens, passes through the dichroic mirror and is filtered through a second computer-controlled, shuttered filter wheel. This emission filter wheel should allow for equally rapid transitions between filters, particularly if performing dual emission wavelength fluorescence imaging. The emitted light can then be collected by a charge-coupled device (CCD)-camera. Ideally, the detector should have a small pixel size and a large imaging area, which allow for a large dynamic range of gray-scale intensity values as well as the capacity for shorter exposure times. These features facilitate the simultaneous measurement of individual cells or phagosomes that may be heterogeneously labeled with the fluorescent probe. Pixel binning can be applied when resolution must be sacrificed to increase sensitivity. Lastly, the appropriate software capable of integrating the exposure times, filter transitions, acquisition intervals and other parameters is required.

1.2 Probe Selection

An initial consideration when selecting a probe to measure organelle pH is appropriate subcellular targeting. Rough estimates of phago-somal pH can be obtained by using fluorescent weak bases that partition into acidic compartments. This is the basis of several commercially available probes such as acridine orange and the LysoTracker/LysoSensor variants. The relatively low membrane permeability of the protonated form of these probes, combined with the much larger permeability of the uncharged species, results in their accumulation in acidic organelles. This is often used to determine whether or not a phagosome has acidified; however, these are purely qualitative probes and cannot be used to estimate pH values accurately. Moreover, weak base probes also label lyso-somes and other acidic compartments and are therefore not phagosome-specific [27]. The accumulation of the protonated form of these acidotropic dyes can also result in alkalizing or osmotic effects in the organelle(s) where they accumulate. Also, any dynamic experiments involving the dissipation of organelle pH with ionophores or V-ATPase inhibitors will result in the irreversible loss of these dyes from the compartment being monitored. Lastly, some LysoTracker probes can photo-convert to species emitting at alternate wavelengths, and are therefore not suitable for two-color imaging [33].

A far more specific way of targeting the pH probe to the phagosome is to covalently link it to the phagocytic target. A number of the pH-sensitive probes come as either an isothiocyanate or a succinimidyl ester. When incubated under slightly basic conditions, these amine-reactive groups allow the probes to attach covalently to primary amines, to yield a thiourea or stable amide bond, respectively. This approach has the advantage of specific phagosomal targeting, no loss of probe upon pH dissipation, and minimal alkalinizing or osmotic effects as only a finite amount of the probe is covalently bound to the phagocytic target.

Upon binding protons, a number of fluorescent molecules undergo either a change in emission intensity or shifts in the emission spectrum. This feature allows for these molecules to be used as sensors of the proton concentration in cellular compartments. After correction for any fluorescence background, the emission intensities at two excitation (dual excitation) or emission (dual emission) wavelengths can be used to generate a ratio that reports the pH [34–36]. As discussed briefly earlier, this is the basis of fluorescein pH sensitivity, but also of other probes such as Oregon Green and SNARF1 (Fig. 1). Covalent coupling of these probes to phagocytic targets allows for ratio fluorescence determinations of phagosomal pH (Figs. 2a and 3a). For these ratiometric probes, an important consideration is their dissociation constant (pK_a). Since the pK_a represents the pH at which half of the probe is protonated, it also represents the pH value at which it is most sensitive to changes in proton concentration. Phagosomal pH spans a very large range of values, starting at near neutral shortly after sealing and rapidly acidifying to near pH 5.0 [37]. In addition, some cell types, such as neutrophils and human M1 macrophages, can exhibit significantly alkaline phagosomal pH values [37, 38]. This represents a challenge when selecting an appropriate pH sensor. As can be noted in Fig. 1, fluorescein ($pK_a = 6.4$) is useful for changes in pH near its pK_a (Fig. 1a, c) but becomes relatively useless at more basic or acidic values. Oregon Green ($pK_a = 4.7$) is, therefore, a more useful probe for measuring subtle changes in an acidified phagosome, while being relatively useless during the early phases of phagosomal acidification (Fig. 1b, c). Neither fluorescein nor Oregon Green are ideal for measurements of the slightly alkaline phagosomal pH values obtained in neutrophils and M1 macrophages; for this a third probe, SNARF1 ($pK_a = 7.5$), is a more suitable choice (Fig. 1d) [37].

Some have recently employed two separate fluorescent probes for measuring phagosomal pH. In this approach, a ratio is generated from the emissions of a pH-sensitive probe (e.g., fluorescein) and a second pH-insensitive probe (e.g., rhodamine). However caution should be taken when using this method, as differential bleaching of the two probes is unavoidable and can result in illumination-dependent changes of the ratio that are not indicative of actual pH changes (Fig. 2d). In view of the above caveats, the stage of phago-some maturation and the desired experimental manipulations should be considered carefully when selecting the appropriate pH sensor.

1.3 Calibration

The ratios obtained by dual wavelength fluorescence imaging are proportional to the pH; however, to make quantitative determina-tions they must be converted to pH by calibrating the sensor in buffers of known pH. Several approaches to calibration have been used over the years. A relatively simple method involves exposing the particles to be used as phagocytic targets that have been labeled with the pH sensor to buffers of known pH (Fig. 2a). The ratios can then be plotted against pH and the experimental values inter-polated. Although this method is quick and comparatively simple, it works on the assumption that the dye behaves similarly inside a phagosome as it does in vitro, and this may not be the case. A pre-ferred method is to collect experimental data and then calibrate the same particles in situ [39]. This method involves bathing the cells in a series of high potassium buffers (approximately the same con-centration as potassium in the cytosol) of various known pH values and adding the K^+/H^+ antiporter nigericin. This manipulation equilibrates the extracellular pH with the cytosolic pH and likely also that of intracellular compartments, including the phagosomal lumen. The result is a "clamping" of the pH of intracellular com-partments at (approximately) the pH of the bathing solution. Dual wavelength ratios can then be obtained at the imposed pH values and this can be used to generate an in situ calibration curve (Fig. 3b). Although this technique makes no assumptions about the behavior of the dye in different physiological settings, it does assume that the concentration of K^+ in the phagosome is the same or similar to the concentration of K^+ in the cytosol, which may not be the case. Moreover, not all cells are equally sensitive to nigericin [40]. Nevertheless, this method has been used with relative success over the years and will be outlined below.

2 Materials

2.1 Cell Lines

1. RAW264.7 murine macrophage-like cells (*see* **Note 1**).

2. Complete RPMI1640 medium: RPMI1640 supplemented with 5% v/v heat-inactivated fetal bovine serum (FBS). Store at 4 °C.

2.2 Reagents

1. 25 mg/mL fluorescein isothiocyanate (FITC) in anhydrous DMSO. Store at –20 °C.

2. 25 mg/mL Oregon Green 488 succinimidyl ester in anhy-drous DMSO. Store at –20 °C.

3. 25 mg/mL SNARF1 succinimidyl ester in anhydrous DMSO. Store at –20 °C.

4. 2 mg/mL Zymosan A bioparticles in PBS (*see* **Note 2**). Store at –20 °C.

5. 1 mg/mL human IgG in PBS. Store at 4 °C.

6. Hank's balanced salt solution (HBSS). Store at 4 °C.

7. Phosphate-buffered saline (PBS). Store at 4 °C.

8. 10 mg/mL nigericin, free acid in ethanol. Store at –20 °C.

9. Potassium-rich calibration buffer: 140 mM KCl, 1 mM $MgCl_2$, 1 mM $CaCl_2$, 5 mM glucose and a buffer suitable for the desired pH (*see* **Note 3**). The solutions are adjusted to desired pH values using 1 M KOH or 1 M HCl for alkalization and acidification, respectively.

2.3 Microscopy System Requirements

1. External high-intensity arc lamp (such as X-Cite 120, EXFO Photonic solutions Inc.)

2. Inverted light microscope (such as the Leica DM4).

3. Computer-controlled excitation and emission filter wheels (such as Lambda 10-2 optical filter changer, Sutter Instrument Company).

4. Appropriate dichroic filter cubes.

5. Electron-multiplied charge-coupled device camera (such as the Cascade II EMCCD camera, Photometrics).

6. Image analysis software (such as MetaFluor fluorescence ratio imaging software, Molecular Devices).

3 Methods

3.1 Labeling of Phagocytic Targets with pH Sensors

1. Centrifuge 500 μL of the 2 mg/mL stock solution of zymosan particles at $6000 \times g$ for 15 s (*see* **Notes 4** and **5**).

2. Aspirate the supernatant, and resuspend the zymosan particles in 1 mL of PBS to wash.

3. Centrifuge the particles once more at $6000 \times g$ for 15 s and repeat the wash a total of three times.

4. Resuspend the zymosan in either 500 μL of 0.5 mg/mL FITC, 0.5 mg/mL Oregon Green succinimidyl ester, or 0.5 mg/mL SNARF1 succinimidyl ester in PBS adjusted to pH 8.3 (*see* **Note 6**).

5. Incubate for 1 h at room temperature, under constant agitation.

6. Wash the cells with PBS five times, or until little or no dye is evident in the supernatant.

7. Resuspend the zymosan in 500 µL of PBS.

8. The labeled zymosan can either be used right away or kept at −20 °C for long-term storage.

3.2 IgG Opsonization of Zymosan

1. Dilute 100 µL of the fluorophore-labeled zymosan in 400 µL of PBS.

2. Centrifuge at $6000 \times g$ for 15 s and resuspend the particles in 200 µL of PBS.

3. Add 4 µL of human IgG from the 1 mg/mL stock and rotate at room temperature for 1 h (*see* **Note 7**).

4. Wash five times with PBS and resuspend the pellet after the final wash in 100 µL of PBS.

5. Opsonized zymosan should be used immediately and not stored long term.

3.3 Phagocytosis Assay

1. One day before performing the phagocytosis assay, seed RAW264.7 cells onto 18 mm coverslips in a 12-well plate containing complete RPMI1640.

2. On the day of the experiment, remove one coverslip from the 12-well plate and place it into a magnetic Chamlide chamber (*see* **Note 8**).

3. Fill the chamber with 500 µL of HBSS pre-warmed to 37 °C.

4. Pipette 5 µL of the opsonized fluorophore-labeled zymosan into the magnetic chamber with the cells.

5. Centrifuge at $500 \times g$ for 1 min, to bring the particles into contact with the RAW264.7 cells.

6. Place the magnetic chamber onto the heated stage (37 °C) of the microscope.

7. Turn the lights off in the imaging room to minimize the contribution of stray light to the fluorescence recordings.

8. Using a 40× objective bring the cells into focus and scan the coverslip for cells that have zymosan particles attached (*see* **Note 9**).

9. Launch the imaging protocol in the software and configure the acquisition parameters, making sure that there are no saturated pixels in the field.

10. Select the regions of interest (individual phagosomes) and a background region (*see* **Note 10**).

11. Acquire a series of images, alternating between either the desired excitation or emission wavelengths (*see* **Note 11**). Keep the number of acquired images to the minimum required for the desired time interval to prevent excessive photobleaching of the probe and phototoxicity to the cells.

**3.4   In Situ
Calibration**

1. Pre-warm all the calibration buffers to 37 °C.

2. Wash the cells 2× with the first calibration buffer (*see* **Note 12**).

3. In a 1.5 mL Eppendorf tube, combine 500 µL of the first calibration buffer with nigericin at a final concentration of 10 µg/mL. Vortex the solution for 5 s and immediately add it to the cells in the Chamlide chamber.

4. Wait for 5 min for the pH of the phagosome to equilibrate to the pH of the calibration buffer. Acquire 3 images 1–2 min apart. If the ratios obtained from each image are very different, wait for an additional 3–5 min to allow the pH to equilibrate.

5. Once equilibrated, acquire five images to determine the wavelength ratio that corresponds to the pH of the calibration solution.

6. Repeat **steps 2–5** for each calibration solution.

7. Obtain a pH calibration curve by plotting the background-corrected wavelength ratios against their corresponding pH values.

8. The experimentally observed phagosomal pH can then be estimated by interpolating the pH values from the calibration curve generated (Fig. 3b).

4   Notes

1. Raw264.7 cells are a macrophage-like cell line and were selected on the basis of them being professional phagocytes. They express several phagocytic receptors including Fc receptors and complement receptors. Other cell lines, such as HeLa cells, are not professional phagocytes, do not express phagocytic receptors and are therefore not ideal for studying phagocytic events.

2. Zymosan was chosen here due to its relative stability. Once dissolved in PBS, zymosan can be stored for several years at −20 °C. Other particles, such as sheep red blood cells, are commonly used for phagocytosis assays and are amenable to labeling with isothiocyanate or succinimidyl ester conjugates, but are not as stable for long-term storage. Moreover, sheep red blood cells must be opsonized, as they do not bear any ligands for mammalian phagocytic receptors. Zymosan, on the other hand, is readily able to bind phagocytic dectin receptors without any form of opsonization.

3. The buffer should be selected based on optimal buffering capacity near the desired pH. Example buffers (shown in parentheses) for select pH values are: pH 4.0 (acetate-acetic acid), pH 4.5 (acetate-acetic acid), pH 5.0 (acetate-acetic

acid), pH 5.5 (2-[N-morpholino]ethanesulfonic acid [MES]),
pH 6.0 (MES), pH 6.5 (MES), pH 7.0 (N-[2-hydroxyethyl]-
piperazine- N-[2-ethanesulfoinc acid] [HEPES]).

4. The phagocytic target chosen requires consideration. Here we
   describe the use of zymosan—a particle composed of protein–
   carbohydrate complexes from yeast cell walls. The presence of
   β-1,3-glycosidic linkages allows this particle to be recognized
   by dectin receptors and internalized by phagocytic cells.

5. Not all cells are equally phagocytic, nor are all targets equally
   stimulatory. The appropriate selection of a phagocytic target
   requires careful consideration. For example, RAW264.7 cells
   do not engulf unopsonized zymosan due to a paucity of dectin
   receptors on their surface. Therefore, unopsonized zymosan
   would not be suitable as a phagocytic target in this experiment.
   Secondly, not all forms of phagocytosis are identical. It should
   be noted that opsonizing targets with different opsonins triggers different phagocytic receptors and hence different internalization and maturation pathways. Therefore, the particle/
   opsonin should be carefully chosen to match the desired experimental question.

6. As discussed above, this protocol takes advantage of the commercial availability of succinimidyl ester and isothiocyanate
   forms of various pH-sensitive fluorophores. Other reagents,
   such as antibodies directly conjugated to pH sensors, can also be
   employed. Antibodies directly conjugated to pH sensors, such
   as fluorescein, are commercially available. Using this approach
   typically requires a primary antibody directed against the phagocytic target itself, followed by labeling with a secondary antibody conjugated to a pH sensor. Of course, this will effectively
   render the target particle immunoglobulin-opsonized.
   Therefore, if other forms of phagocytosis—such as complement-
   mediated phagocytosis or dectin-mediated phagocytosis—are to
   be studied, direct labeling of the target with an isothiocyanate or
   succinimidyl ester conjugate is more suitable.

7. Although primary human and mouse macrophages readily
   internalize unopsonized zymosan through dectin receptors,
   the RAW264.7 macrophage-like cells used in this protocol do
   not internalize unopsonized zymosan efficiently, due to a paucity of dectin receptor expression. For this reason, the zymosan
   must be opsonized with IgG prior to the assay to foster phagocytosis through Fc receptors, which are abundantly expressed
   on RAW264.7 cells.

8. A Chamlide chamber is a coverslip holder consisting of two
   interlocking parts and an O-ring. A coverslip is placed in the
   chamber and a seal is achieved by the magnetic attraction of
   the two interlocking parts. The coverslip can then be bathed in

complete RPMI 1640. The Chamlide fits into a slot on the heated stage of the microscope allowing for continuous, live cell imaging. Finding a cell that is in the process of engulfing a particle can take practice. It is not sufficient to select a particle that is merely in the vicinity of a phagocytic cell. It is helpful to look for cells that have already extended pseudopodia around the phagocytic target to increase the likelihood of capturing a phagocytic event.

9. A background region of interest should be selected in the extracellular space near to the region of interest surrounding the phagosome, and its fluorescence determined at the wavelengths of choice. The background values for both wavelengths should be subtracted separately from the values recorded at the corresponding wavelength in the phagosome region of interest, prior to the generation of the ratio. This allows for the calculation of a properly background-corrected ratio.

10. Bear in mind that some ratiometric probes are dual excitation probes (Fluorescein and Oregon Green) and some are dual emission probes (SNARF1). This means that either the excitation filters (490 and 440 nm for fluorescein and Oregon Green) or the emission filters (640 and 580 nm for SNARF1) will have to be alternated during acquisition.

11. The order of addition of the calibration buffers is theoretically unimportant. However, the further the calibration buffer deviates from neutral pH values does seem to affect the ability of the cell to stay attached to the coverslip. In general, a good practice is to start with the neutral pH buffer and to proceed sequentially to more acidic or alkaline calibration buffers.

12. All in all, the technique of fluorescence ratio imaging represents a powerful, reliable, and relatively simple method for assessing phagosomal pH. Measurements can be made in individual cells and even individual phagosomes. This allows for the observation of cell/phagosome heterogeneity that is inevitably masked by whole-population techniques. By carefully considering the pH sensor, phagocytic target and calibration method chosen, this technique will continue to provide valuable information about the maturation and luminal biochemistry of phagosomes.

Acknowledgments

J.C. was supported by a Cystic Fibrosis Canada postdoctoral fellowship. Research in our laboratory is supported by grants FDN-143202 and MOP-126069 from the Canadian Institutes of Health Research.

References

1. Bloomfield G, Traynor D, Sander SP et al (2015) Neurofibromin controls macropinocytosis and phagocytosis in Dictyostelium. eLife. doi: 10.7554/eLife.04940

2. Arandjelovic S, Ravichandran KS (2015) Phagocytosis of apoptotic cells in homeostasis. Nat Immunol 16:907–917. doi:10.1038/ni.3253

3. Koenigsknecht J, Landreth G (2004) Microglial phagocytosis of fibrillar beta-amyloid through a beta1 integrin-dependent mechanism. J Neurosci 24:9838–9846. doi:10.1523/JNEUROSCI.2557-04.2004

4. Kevany BM, Palczewski K (2010) Phagocytosis of retinal rod and cone photoreceptors. Physiol Bethesda Md 25:8–15. doi:10.1152/physiol.00038.2009

5. Manwani D, Bieker JJ (2008) The erythroblastic island. Curr Top Dev Biol 82:23–53. doi:10.1016/S0070-2153(07)00002-6

6. Blum JS, Wearsch PA, Cresswell P (2013) Pathways of antigen processing. Annu Rev Immunol 31:443–473. doi:10.1146/annurev-immunol-032712-095910

7. Turk V, Stoka V, Vasiljeva O et al (2012) Cysteine cathepsins: from structure, function and regulation to new frontiers. Biochim Biophys Acta 1824:68–88. doi:10.1016/j.bbapap.2011.10.002

8. Canton J (2014) Phagosome maturation in polarized macrophages. J Leukoc Biol 96:729–738. doi:10.1189/jlb.1MR0114-021R

9. Wreden CC, Johnson J, Tran C et al (2003) The H+-coupled electrogenic lysosomal amino acid transporter LYAAT1 localizes to the axon and plasma membrane of hippocampal neurons. J Neurosci 23:1265–1275

10. Geisow MJ, Arcy Hart PD, Young MR (1981) Temporal changes of lysosome and phagosome pH during phagolysosome formation in macrophages: studies by fluorescence spectroscopy. J Cell Biol 89:645–652

11. Claus V, Jahraus A, Tjelle T et al (1998) Lysosomal enzyme trafficking between phagosomes, endosomes, and lysosomes in J774 macrophages enrichment of cathepsin H in early endosomes. J Biol Chem 273:9842–9851. doi:10.1074/jbc.273.16.9842

12. Hart PD, Young MR, Jordan MM et al (1983) Chemical inhibitors of phagosome-lysosome fusion in cultured macrophages also inhibit saltatory lysosomal movements. A combined microscopic and computer study. J Exp Med 158:477–492. doi:10.1084/jem.158.2.477

13. Flannagan RS, Jaumouillé V, Grinstein S (2012) The cell biology of phagocytosis. Annu Rev Pathol 7:61–98. doi:10.1146/annurev-pathol-011811-132445

14. Maxson ME, Grinstein S (2014) The vacuolar-type H+-ATPase at a glance—more than a proton pump. J Cell Sci 127:4987–4993. doi:10.1242/jcs.158550

15. Cuppoletti J, Aures-Fischer D, Sachs G (1987) The lysosomal H+ pump: 8-azido-ATP inhibition and the role of chloride in H+ transport. Biochim Biophys Acta 899:276–284

16. Dell'Antone P (1979) Evidence for an ATP-driven "proton pump" in rat liver lysosomes by basic dyes uptake. Biochem Biophys Res Commun 86:180–189

17. Graves AR, Curran PK, Smith CL, Mindell JA (2008) The Cl-/H+ antiporter ClC-7 is the primary chloride permeation pathway in lysosomes. Nature 453:788–792. doi:10.1038/nature06907

18. Harikumar P, Reeves JP (1983) The lysosomal proton pump is electrogenic. J Biol Chem 258:10403–10410

19. Ohkuma S, Moriyama Y, Takano T (1983) Electrogenic nature of lysosomal proton pump as revealed with a cyanine dye. J Biochem 94:1935–1943

20. Steinberg BE, Huynh KK, Brodovitch A et al (2010) A cation counterflux supports lysosomal acidification. J Cell Biol 189:1171–1186. doi:10.1083/jcb.200911083

21. Metchnikoff E (1968) Lectures on the comparative pathology of inflammation: delivered at the Pasteur Institute in 1891. Dover Publications, New York

22. Sprick MG (1956) Phagocytosis of M. tuberculosis and M. smegmatis stained with indicator dyes. Am Rev Tuberc 74:552–565

23. Pavlov EP, Solov'ev VN (1967) pH changes of cytoplasm in phagocytosis of microbes stained with indicator dyes. Biull Eksp Biol Med 63:78–81

24. Mandell GL (1970) Intraphagosomal pH of human polymorphonuclear neutrophils. Proc Soc Exp Biol Med 134:447–449

25. Jensen MS, Bainton DF (1973) Temporal changes in Ph within the phagocytic vacuole of the polymorphonuclear neutrophilic leukocyte. J Cell Biol 56:379–388. doi:10.1083/jcb.56.2.379

26. Ohkuma S, Poole B (1978) Fluorescence probe measurement of the intralysosomal pH in living cells and the perturbation of pH by various agents. Proc Natl Acad Sci U S A 75:3327–3331

27. Canton J, Grinstein S (2015) Measuring lysosomal pH by fluorescence microscopy. Methods

Cell Biol 126:85–99. doi:10.1016/bs. mcb.2014.10.021

28. Yates RM, Russell DG (2008) Real-time spectrofluorometric assays for the lumenal environment of the maturing phagosome. Methods Mol Biol 445:311–325. doi:10.1007/978-1-59745-157-4_20

29. Vergne I, Constant P, Lanéelle G (1998) Phagosomal pH determination by dual fluorescence flow cytometry. Anal Biochem 255:127–132. doi:10.1006/abio.1997.2466

30. Steinberg B, Grinstein S (2007) Assessment of Phagosome Formation and Maturation by Fluorescence Microscopy. In: DeLeo F, Bokoch G, Quinn M (eds) Neutrophil methods protocols. Humana Press, Totowa, USA, pp 289–300

31. Schlam D, Bohdanowicz M, Chatgilialoglu A et al (2013) Diacylglycerol kinases terminate diacylglycerol signaling during the respiratory burst leading to heterogeneous phagosomal NADPH oxidase activation. J Biol Chem 288:23090–23104. doi:10.1074/jbc.M113.457606

32. Webb DJ, Brown CM (2013) Epi-fluorescence microscopy. Methods Mol Biol 931:29–59. doi:10.1007/978-1-62703-056-4_2

33. Freundt EC, Czapiga M, Lenardo MJ (2007) Photoconversion of Lysotracker Red to a green fluorescent molecule. Cell Res 17:956–958. doi:10.1038/cr.2007.80

34. Grynkiewicz G, Poenie M, Tsien RY (1985) A new generation of Ca2+ indicators with greatly improved fluorescence properties. J Biol Chem 260:3440–3450

35. Tsien RY (1989) Fluorescent indicators of ion concentrations. Methods Cell Biol 30:127–156

36. Tsien RY, Rink TJ, Poenie M (1985) Measurement of cytosolic free Ca2+ in individual small cells using fluorescence microscopy with dual excitation wavelengths. Cell Calcium 6:145–157

37. Canton J, Khezri R, Glogauer M, Grinstein S (2014) Contrasting phagosome pH regulation and maturation in human M1 and M2 macrophages. Mol Biol Cell 25:3330–3341. doi:10.1091/mbc.E14-05-0967

38. Jankowski A, Scott CC, Grinstein S (2002) Determinants of the phagosomal pH in neutrophils. J Biol Chem 277:6059–6066. doi:10.1074/jbc.M110059200

39. Thomas JA, Buchsbaum RN, Zimniak A, Racker E (1979) Intracellular pH measurements in Ehrlich ascites tumor cells utilizing spectroscopic probes generated in situ. Biochemistry (Mosc) 18:2210–2218

40. Chow S, Hedley D, Tannock I (1996) Flow cytometric calibration of intracellular pH measurements in viable cells using mixtures of weak acids and bases. Cytometry 24:360–367. doi:10.1002/(SICI)1097-0320(19960801)24:4<360::AID-CYTO7>3.0.CO;2-J

Chapter 13

Image-Based Analysis of Phagocytosis: Measuring Engulfment and Internalization

Nicholas D. Condon, Adam A. Wall, Jeremy C. Yeo, Nicholas A. Hamilton, and Jennifer L. Stow

Abstract

The process of phagocytosis is crucial for fighting infection in innate immunity, for maintaining homeostasis through clearing cell debris, and for tissue remodeling in development. Here we describe two semi-automated image-based assays for the quantitative characterization of the early stages of phagocytosis and pathogen entry. A feature of these assays is the ability to detect and assess molecules or agents that subtly affect the stages both before and after cup closure or internalization.

Key words Phagocytosis, Quantitative assay, Fluorescent imaging, Macrophage, Image-based, ImageJ, *Salmonella*

1 Introduction

Phagocytosis is a key cellular function for preventing disease, for instance by killing infectious pathogens [1] or preventing the accumulation of aberrant amyloid deposits in the brain [2]. Parenthetically, phagocytosis can also help to harbor disease by being subverted for the entry or sequestration of intracellular bacteria [3–6].

In all these situations, being able to visualize and measure phagocytosis is important for basic scientific discovery and as a focus for the development of new applied biological solutions or clinical treatments. The ability to rapidly and quantitatively assay phagocytosis is also important for experimentally determining the functions of host gene products, pathogen-derived effectors or toxins, drugs or environmental factors that are involved in or can affect the process.

Phagocytosis can be defined as a series of functional steps or stages, each orchestrated by different cellular machinery [7]. These stages can be discerned in macrophages during the phagocytosis of "prey," i.e., pathogens or large particles. The *early* stages of phagocytosis involve membrane trafficking and actin polymerization

Roberto Botelho (ed.), *Phagocytosis and Phagosomes: Methods and Protocols*, Methods in Molecular Biology, vol. 1519, DOI 10.1007/978-1-4939-6581-6_13, © Springer Science+Business Media New York 2017

to extend and engage cell surface projections, such as filopodia and dorsal ruffles, which participate in surveillance and detection of potential prey [8]. Upon encountering prey, these projections can then transition to pseudopods of forming phagocytic cups during the initial stages of engulfment [9–12]. In macrophages, these early steps are usually accompanied by active macropinocytosis that occurs constitutively for sampling of the surrounding milieu and can be enhanced by pathogen contact [11, 13]. The membranes associated with filopodia, ruffles, and phagocytic cups, house cohorts of receptors geared to detecting and deciphering danger signals and signaling from these receptors serves to activate the cells [7, 14, 15]. Thus in the early stages of phagocytosis, membranes and membrane proteins coordinate dual functions of receptor-mediated signaling and prey engulfment. These processes are both also facilitated by membrane lipids, which actively promote receptor clustering, activation, and signaling and also enhance membrane curvature and spreading [16]. The membrane phosphoinositides for instance, transition to species that directly engage the recruitment of adaptors and signaling proteins for receptor mediated cascades [14–16].

Phagosomes are sealed and then internalized in the *middle* phase of phagocytosis, which is accompanied by remodeling and tightening of the phagosome limiting membrane [11]. A series of fusion and fission processes allow membrane exchange between the phagosome, endosomes (early and recycling endosomes and macropinosomes) and the Golgi [17–20]. This active, bidirectional traffic between the phagosome and endocytic networks supports maturation of the phagosome. *Late* stages of phagosome maturation see the formation of a degradative phagolysosome, after its fusion with late endosomes or lysosomes that deliver lysosomal enzymes and complete the conversion of this compartment to an acidified, protease-rich environment for killing pathogens, degrading material and preparing peptides for antigen presentation [1, 18].

Here we present two image-based assays focussed on quantifying the early stages of phagocytosis and pathogen entry. The first, "Phagocytosis quantification assay" is set up to measure the phagocytic uptake of opsonized beads. Although latex beads are the prototype, the methodology can be adapted to measure the uptake of other particles, pathogens, or dead cells. A distinguishing feature of our assay is the capacity to assess molecules or agents that affect the stages both before and after cup closure. This is relevant for deciphering the functions of phosphoinositide modifying kinases and phosphatases, such as the PI3 kinase(s) that are required for cup closure [11, 16]. Other molecular targets for early stage phagocytosis assay include actin polymerizing and depolymerizing proteins, and membrane trafficking and signaling regulators including Rab GTPases [1, 12, 21]. In this assay, beads are defined and quantified according to three successive functional steps diagrammed in Fig. 1a and then depicted in the examples in Fig. 1.

Based on a combination of a bright-field view and immunolabeling with an externally applied IgG antibody, beads are classified as "outside" (bead attached to the cell surface and captured by the cell but engulfment is not yet obvious), "in between" (the bead is semi-surrounded by a phagocytic cup that is still open to the external milieu) or "phagocytosed" (the bead is enclosed within a fully sealed phagosome that is internalized into the cell) (Fig. 1a). The cell depicted in Fig. 1b displays beads in all three stages of early phagocytosis. The image analysis script written to decipher and quantify the beads in here is based on an earlier image-based format intended for high throughput screening (*see* Fig. 2); [22]. It involves thresholding images, creating masks for cell areas and distinguishing antibody-labeled beads.

A second image-based assay called the "*Salmonella* Infection Assay" is presented for the measurement of bacterial entry into macrophages. We have built upon the script for quantifying bead uptake to create methods for quantifying the internalization of bacteria. Several bacteria hijack host pathways to gain entry into the cells, and to create an intracellular replicative niche [3, 5, 6]. It is therefore advantageous to garner an understanding of which host proteins are required for or are susceptible to modulating internalization and colonization by these pathogenic bacteria. To address this, we have developed an assay that measures the internalization of bacteria using fluorescence based imaging. It is demonstrated here for measuring internalization of *Salmonella*. This assay measures the co-localization of phagocytic objects (bacteria in this example) with cells, and it can determine whether bacteria are inside or outside of the cells based on antibody accessibility (Fig. 3). This assay is commonly used in conjunction with colony forming unit (CFU) assays to measure and analyze bacterial infections and to experimentally test the roles of specific pathogen or host genes or proteins (for example, by genetic or pharmacological manipulation of bacteria or host cells). CFUs can vary for different reasons; for instance, fewer colonies may mean higher host-derived killing (for example, normal uptake paired with increased lysosomal killing), or possibly less uptake (reduced internalization of the bacteria resulting in fewer surviving bacteria).

Taken together these image-based assays are versatile, powerful measures for analyzing stage-specific phagocytic processes.

2   Materials

2.1   Phagocytosis Quantification Assay

2.1.1   IgG Opsonization of Latex Beads

1. 3 μm Latex beads.

2. Human IgG antibody.

3. 0.02 % sodium azide.

4. Phosphate buffered saline (PBS).

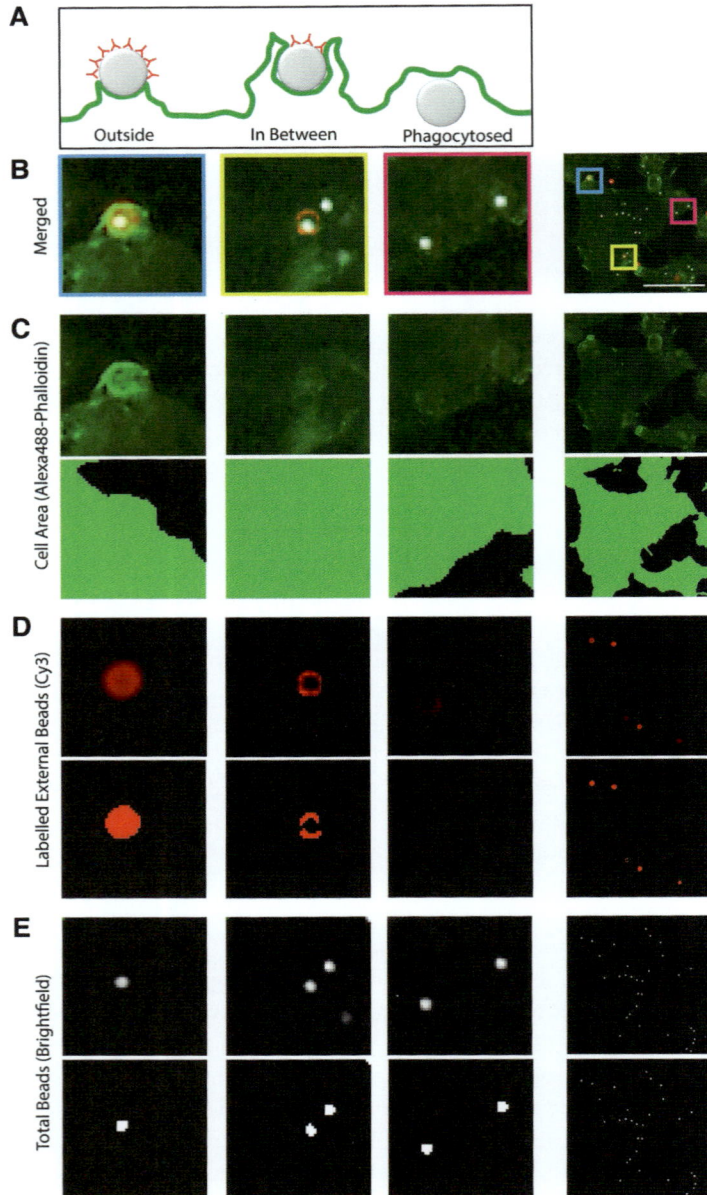

Fig. 1 Stages of phagocytosis. (**a**) Foreign objects or beads are taken up into macrophages by phagocytosis. In the early stages of phagocytosis opsonized beads are captured on the cell surface and in this assay they are defined as "*outside*" beads. The extension of pseudopod membranes form an open phagocytic cup around the bead referred to as the "*in between*" stage, while after complete engulfment, sealing of the phagocytic cup the bead is internalized into a phagosome and is now categorized as "*phagocytosed*." (**b**) A RAW264.7 macrophage was exposed to IgG-coated latex beads for 15 min prior to fixation and was labeled with phalloidin (*green*). External beads (those in the *outside* and *in between* stages) are immunolabeled with a fluorescent Cy3 antibody, labeled red in this image. Examples of *red* outside and in between beads are indicated with *blue* and *yellow boxes* respectively. Unlabeled white beads within the cell perimeter are *phagocytosed* and shown in the *pink box*. These examples are magnified on the *inset* panels to the left. (**c**) Example raw image and binary images of cell labeling to determine the regions of interest for the assay-beads not associated with cells are excluded. (**d**) External bead labeling using the Cy3 antibody. The *upper panels* show full 255 8-bit images and the *lower panels* are the binary output from the macro. (**e**) Total beads detected by brightfield image acquisition. The *upper panels* are the full 255 8 bit images and the *lower panels* are the binary output as determined by the macro

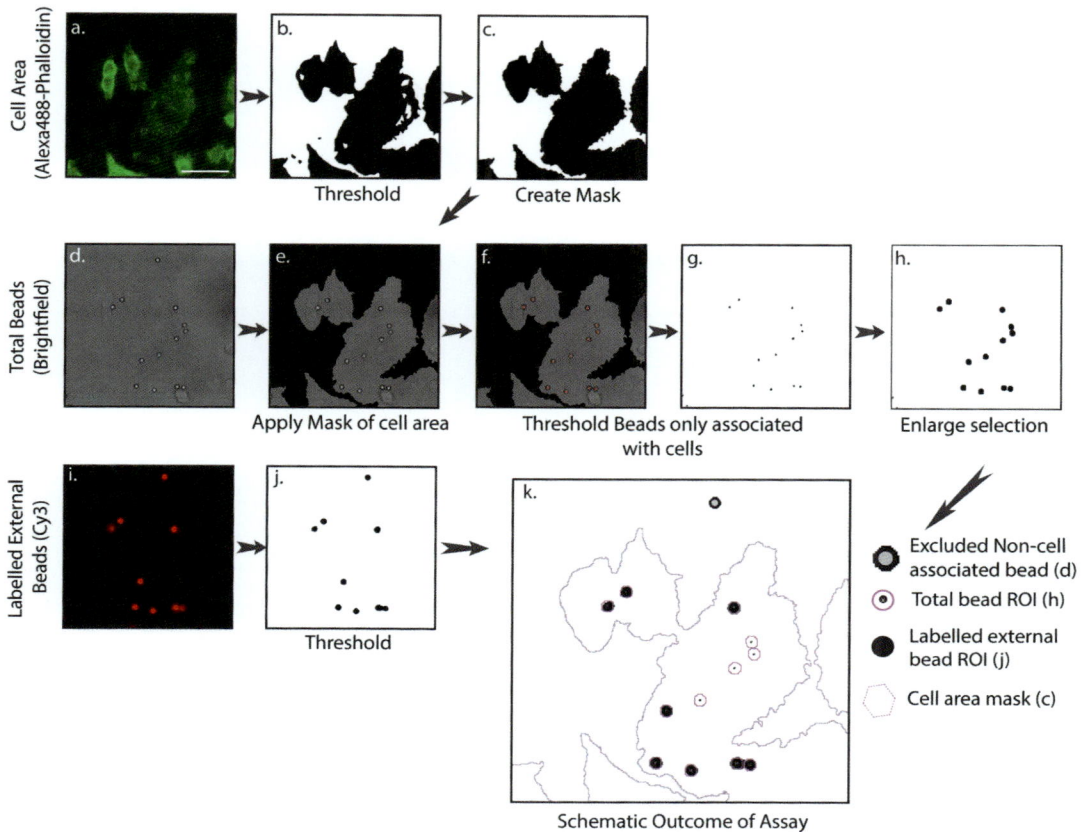

Fig. 2 Phagocytosis assay workflow. Images of the fixed cells are captured in fluorescent and brightfield channels for use in the phagocytosis macro. The images are split into their individual channels, cell area ((**a**) cell outline), brightfield ((**d**) total beads), and external beads ((**i**) all beads outside of the cell). Initially cell area is converted into a mask by thresholding (**b–c**), this is to exclude any beads not attached to cells by overlaying the mask onto the brightfield image (**e**). The individual beads are then detected by thresholding the "masked" brightfield image (**f, g**). As the threshold only detects a small part of the bead, the binary image selection is enlarged (**h**). Next, the external bead image is converted to a binary by user-adjusted thresholding (**j**). Then, every bead ROI is measured in the binary external bead image, the total area of fluorescence is measured as a percentage of the total bead area (**k**). Results are binned into *outside*, *in between* and *phagocytosed* by the percentage labeling of external beads equivalent to 0–5, 5–95, and 95–100 % labeling respectively (*see* Fig. 1d)

2.1.2 Phagocytosis Assay

1. 24-well tissue culture treated plates.
2. #1.5 11 mm round glass coverslips.
3. Complete RPMI culture medium (RPMI culture, 10% fetal bovine serum, 2 mM l-glutamine).
4. PBS.
5. 4% paraformaldehyde in PBS.

2.1.3 Immunolabeling

1. Cy3-anti human secondary antibody (*see* **Note 1**).
2. 0.5% bovine serum albumin (BSA).
3. 0.1% Triton X-100 in PBS.

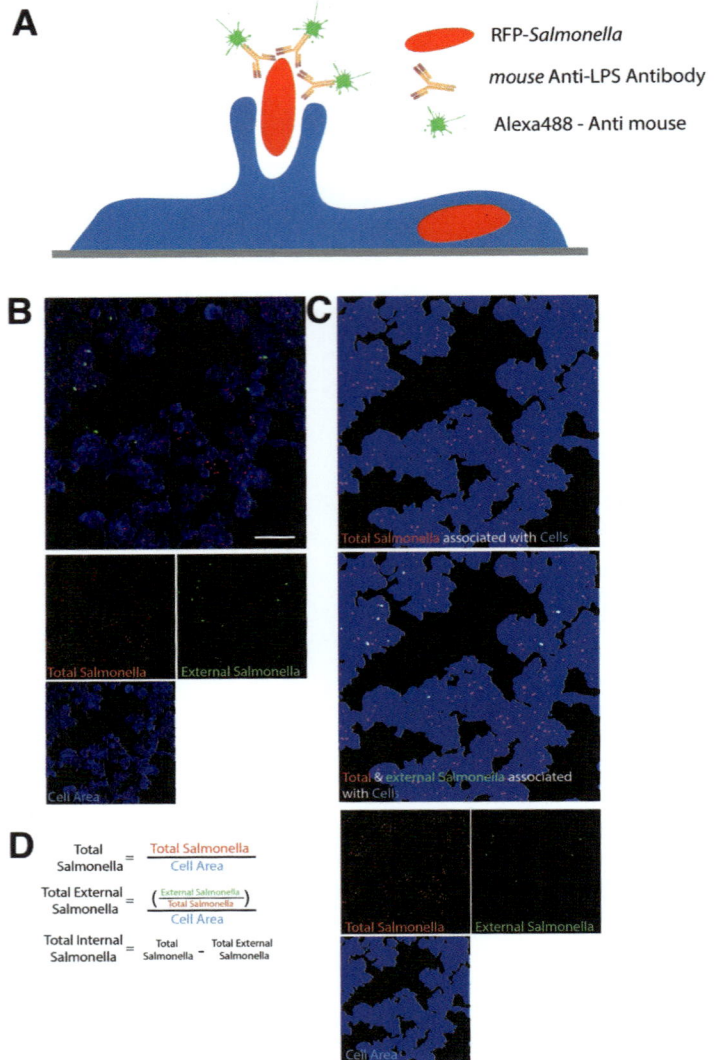

Fig. 3 *Salmonella* internalization assay. Designed as an extension of the phago-cytosis assay, the S*almonella* internalization assay is a wide-field fluorescence based image analysis protocol used to quickly screen for any perturbations in the uptake of bacteria into macrophages. (**a**) A schematic outline of the assay labeling conditions, cells are labeled with Alexa 350-Phalloidin (*blue*) and are infected with RFP-expressing *Salmonella* (*red*), while external bacteria are labeled with an antibody against LPS (*green*). (**b**) Example image of macro-phages (*blue*) total *Salmonella* (*red*) and total external *Salmonella* (*green*). (**c**) Binary images from example captures (**b**) shown as total *Salmonella* (*red*) asso-ciated with cells (*blue, upper*), and total *Salmonella* (*red*) associated with cells (*blue*) as well as externally labeled/outside bacteria (*green*). (**d**) Calculations utilized to determine the percentage of bacteria that are internal and external with respect to the cell outlines

4. Alexa Fluor 488 Phalloidin (*see* **Notes 1** and **2**).

5. Antifade mounting reagent.

6. Clear nail varnish.

7. Microscopy slides.

2.2 Salmonella Infection Assay

1. RFP-*Salmonella enterica* serovar typhimurium expressing *PP-BRD1-RFP* (*see* **Notes 1** and **2**).

2.2.1 Culturing Salmonella

2. Standard LB-agar petri dishes plus 100 μg/mL ampicillin.

3. Standard LB broth plus 100 μg/mL ampicillin.

4. OD_{600} spectrometer and cuvettes.

5. Sodium hypochlorite for decontamination.

2.2.2 Labeling Salmonella and Macrophages

6. Mouse anti-*Salmonella* typhimurium LPS antibody (*see* **Note 2**).

7. Alexa Fluor 350 anti-mouse (*see* **Note 1**).

8. 0.5 % BSA.

9. 0.1 % Triton X-100 in PBS.

10. Alexa Fluor 488 Phalloidin (*see* **Note 1**).

11. Antifade mounting reagent.

12. Clear nail varnish.

3   Methods

3.1 Phagocytosis Quantification Assay

1. Resuspend 100 μL of beads in 1 mL of PBS to make a 10 % suspension and wash thrice with PBS by centrifugation at $10,000 \times g$ for 10 min.

3.1.1 IgG Opsonization of Latex Beads

2. Aspirate excess water and incubate beads in equal volume of human IgG (5 mg/mL), shaking, at room temperature overnight.

3. Beads are washed thrice in PBS by centrifugation at $10,000 \times g$.

4. Finally, resuspend beads in an equal volume of PBS with 0.02 % sodium azide and store at 4 °C for up to 1 month.

3.1.2 Phagocytosis Assay

1. Macrophages are seeded at 0.2×10^6 cells/mL onto glass coverslips and grown overnight (~80 % confluency), in complete RPMI medium.

2. Dilute IgG coated beads for (~1:1000 to achieve ~5–15 beads/cell), into 4 °C complete medium.

3. Replace medium on cells with medium containing beads and synchronize phagocytosis by centrifugation at $300 \times g$ for 2 min at 4 °C using a plate rotor.

4. Return to 37 °C and incubate for desired time-course.

5. Wash coverslips thrice in cold PBS to remove excess beads and fix with 4 % PFA in PBS for 30 min at time points of interest (*see* **Note 3**).

3.1.3   Labeling and Imaging Conditions

1. To label external beads, wash coverslips with fresh PBS, prior to blocking in 0.5 % BSA–PBS for 30 min.

2. Label external beads with an anti-human secondary antibody of your choice (Cy3 was used at 1:400) in 0.5 % BSA–PBS for 30 min (*see* **Note 1**).

3. Then to label the total cell area, wash cells thrice with PBS before permeabilizing membrane with 0.1 % Triton X-100 for 5 min.

4. Wash cells once with PBS before labeling with Alexa 488-Phalloidin for 15 min to demarcate the cell area (*see* **Note 2**).

5. Wash coverslips thrice with PBS to remove excess fluorescent probes.

6. Mount coverslips on a microscope slide with your choice of mounting medium.

3.1.4   Microscope Setup and Image Capture of Cell Area and External Beads

1. Image coverslips using an upright fluorescence widefield microscope capable of bright field imaging. Coverslips should be imaged at a low magnification (10–20×, dependant on camera resolution, *see* **Note 4**).

2. Capture the cell outline image (in this example, Alexa 488-phalloidin is used), whereby the total cell area is in focus.

3. Capture external beads (Cy3-Anti IgG), at a focal level equal to that of the total bead (brightfield) image.

4. Typically ten fields of view each with equivalent and robust cell densities will suffice for quantification. Cells are usually grown to ~70 % confluency for optimal results.

3.1.5   Imaged-Based Assay for Quantification of Phagocytosis

1. Open FIJI imaging software package (fiji.sc/Downloads) and install Phagocytosis macro (link to script: http://doktor-nick.github.io/Phagocytosis-Analysis/) (*see* **Note 5**).

2. Running the macro will prompt you with the order in which image files should be opened, Bright field, beads (Cy3-fluorescence) and "transfected" (phalloidin/overexpressed protein; *see* **Note 6**).

3. Initially the macro will prompt you to define the cell area (this avoids beads not associated with any cell affecting the outcome) based on a threshold of the phalloidin labeled cells. Adjust the slider to best "mask" the entire cell without over sampling into the "background" (For thresholding *see* **Note 4**).

4. The script will prompt you to manually adjust the "Threshold" slider to only highlight the center "white dot" captured in the

bright field image; to designate the total number of beads in the field of view (*see* **Note 7**).

5. Next, the script will automatically remove artifacts and count the total number of beads.

6. External beads are then analyzed for their level of Cy3-labeling. Beads with uneven "crescent" shaped staining are partially internalized and are classified as "inbetween." Adjust the threshold when prompted, to only allow for the selection of the partially stained bead (*see* **Notes 1** and **4**).

7. The macro will then output the count results as a text output stating "of *t* *(total)* there are *x* phagocytosed…Of *t*, there are *y* outside…" and "In between: *z*," where *z* = the total − (outside + inside beads).

8. Re-run the macro for as many field-of-view image sets as required, noting the output values *t*, *x*, *y*, *z*. These can be further analyzed or represented as a percentage using a standalone spreadsheet or statistical software.

3.2 Salmonella Infection Assay

3.2.1 Salmonella Infection and Image Acquisition

1. Macrophages are seeded at 0.2×10^6 cells/mL onto glass coverslips and grown overnight (~80% confluency) in complete medium.

2. Inoculate wild-type RFP-expressing *Salmonella* into a 3 mL culture of LB-ampicillin broth overnight at 37 °C, shaking. Subculture bacteria by diluting at 1:60 LB-ampicillin broth, 4 h prior to experimentation to induce the expression of the SPI-1 invasion genes.

3. Measure the optical density of the subculture of *Salmonella* (OD of $0.1 = \sim 10^8$ bacteria/mL).

4. Calculate the multiplicity of infection (MOI) required for the experiment (*see* **Note 8**), and dilute bacteria into complete medium.

5. Replace medium on coverslips with medium containing *Salmonella*, and spin at $300 \times g$ to synchronize infections.

6. Wash thrice with PBS prior to fixation with 4% PFA for 30 min at the time points required (*see* **Note 9**).

7. Block coverslips with 0.5% BSA–PBS for 30 min.

8. Label external *Salmonella* with mouse-anti-LPS antibody diluted in BSA–PBS for 60 min, followed by washing thrice with BSA–PBS.

9. Label external *anti-LPS antibody* with Alexa 488-anti-mouse secondary for 60 min, followed by washing thrice with BSA–PBS.

10. Permeabilize cells with 0.1% Triton X-100 for 5 min prior to three washes again.

11. Demarcate cells by labeling with Alexa 350-phalloidin (or any blue fluorescent marker, *see* **Note 2**).

12. Mount coverslips on a microscope slide with your choice of mounting medium.

13. Image cells using an upright fluorescence widefield microscope. Coverslips should be imaged at a low magnification. Capture images at consistent exposures for the external *Salmonella* (Alexa 488-anti LPS), total *Salmonella* (RFP, fluorescence), and cell outlines (Alexa 350-phalloidin). Phalloidin binds to F-actin, which is used to demarcate the cell area for this analysis. Typically ten fields of view will suffice for quantification providing approximately 500 bacteria (*see* **Notes 1** and **2**).

3.2.2 Image-Based
Salmonella
Internalization Assay

1. Open FIJI imaging software package (fiji.sc/Downloads) and install *Salmonella* macro (link to script: http://doktor-nick. github.io/Phagocytosis-Analysis/).

2. Merge individual channel images into RGB tiff files (red = total *Salmonella*, green =, externally labeled *Salmonella,* blue = cell outline/phalloidin).

3. Run macro on a small sample of the data to observe if thresholding/expansions are appropriate for the magnification used to capture individual fields of view. Make any modifications required and then alter line 17 of macro from "// setBatchMode(true);" to "setBatchMode(true);" to force the macro to run faster in the background (*see* **Note 10**).

4. Open the Microsoft Excel file located in the "Count_Results" folder within the source directory. The values calculated and tabulated represent the total number of pixels, calculated as follows;

   (a) Total red = *Total Salmonella* associated with cells (red pixels co-localizing with green pixels).

   (b) Internal = *Total Salmonella* within cells (red pixels co-localizing with green pixels, but not with blue pixels).

   (c) External = Total *Salmonella* associated with cells, labeled with the external anti-LPS antibody (red pixels co-localizing with green and blue pixels).

5. Data can be transferred into a statistical analysis software for further analysis such as R.

4 Notes

1. The individual scripts for these assays are based on use of specific fluorophores (Cy3, Alexa 488, and Alexa 350) other fluorophores can be substituted but the script would have to be adjusted accordingly, particularly with regard to thresholds.

2. Fluorescent phalloidin is used here to label the cells (f-actin specifically), however other markers for whole cell labeling could be substituted. RFP-*Salmonella* plasmid labels all bacteria red. LPS antibody is used to label external *Salmonella*, we have tried several commercial antibodies but antibody AB65922 (Abcam) works best in our hands.

3. For preliminarily screening for defects in phagocytosis, a time point of 15 min at 37 °C allows for approximately 50% of the latex beads to be internalized into macrophages. For a typical time course, cells are fixed at 5, 15, and 30 min. When washing coverslips, cold PBS should be directed at the side of the well, gently, to ensure "attached" beads are not dislodged from any cells. Typically cells would be fixed at time points of 15 and 30 min; however, for longer time courses, it is important to complete an initial kill of external bacteria with a high dose antibiotic, followed by thorough washing of the coverslips to remove external dead bacteria. Furthermore, medium for the remainder of the time-course should be dosed with antibiotics to prevent remaining bacteria multiplying and reinfecting cells at an increased MOI.

4. Identifying the best magnification to capture images for the phagocytosis assay depends on two factors, the density of cells visible within the region of interest (ROI), as well as the resolution of the camera used. Initially some parameters of the assay will need to be set to suit each individual imaging system such as the expected bead size (ROI) in the brightfield image (total bead) and the Cy3-labeled bead image (extental beads). For each image type, an example bead will need to be thresholded and the selected ROI measured for its area in pixels. These values will need to be used as a guide to modify the macro to suit your imaging system—see macro line 40 (total bead image) and line 72 (external bead image). Thresholding is the selection of a population of pixels that fall within a given intensity distribution. This selection of pixels can then be used to generate a binary image to allow the selection of objects or particles of interest from pixels of a different intensity. In our case, the total beads captured in the brightfield image are a higher intensity than the surrounding cells. After applying the threshold, the image will be converted to an 8-bit binary image, with the thresholded pixels to be included given a value of 255 and the background to be excluded will 0. The user should select the threshold which best selects for pixels that comprise the cell area, bead/bacteria or external labeling to white, while being careful not to oversample by selecting too many pixels (mostly background noise).

5. FIJI is an image analysis software which stands for FIJI Is Just ImageJ. This version of the software contains extensive libraries of preinstalled macros and plugins.

6. Both macros are designed to only count objects that are associated with cells, specifically excluding any that are not in contact with cells and may be simply attached to the coverslip. The macros achieve this by screening for a cell marker (typically phalloidin staining), which can be thresholded to convert the entire cell region into a binary image, however this can be translated to a number of different applicable markers, e.g., a GFP-tagged protein of interest, or cytosolic GFP co-transfected with shRNA/knockdown treatment. Utilizing a bright fluorescent marker for this step is important as the assay works on binary thresholding of cells and should be "bright" enough to correctly detect the total cell outline. The assay utilizes an ImageJ binary process called "fill holes" which will select and fill encapsulated regions. This has the capability of skewing the results if cells are seeded too densely, resulting in the "gaps" between cells to be "filled" and biasing the "attached beads" category.

7. The macro detects and counts the total number of beads in the field of view by using the bright field image acquired in such a way that the beads refract light to give a higher intensity ROI than the surrounding background. This is only achievable using beads due to the difference in refractive index of the bead compared to the macrophages. Other phagocytic particles such as sheep red blood cells are not appropriate for this script due to their similar refractive index to the fixed phagocytes.

8. In the *Salmonella* assay, for infection with wild-type *Salmonella* into macrophages an MOI of 10 was utilized. Briefly, to calculate the MOI the number of cells for experimentation was estimated from plating densities and bacteria were diluted into complete RPMI immediately prior to infection.

9. The human IgG coating of the beads allows for both their capture and internalization by the macrophage, as well as the detection and labeling of bead surfaces external to the cell with a fluorescent secondary antibody. The level at which this labeling is detected is quantified by the assay and the phagocytic state of a bead is binned into one of three categories, phagocytosed (no more than 5% of the bead is detectable as being external), outside (more than 95% of the bead detectable to external labeling), and in-between (the remaining beads; 5–95% detectable to external labeling). It is critical for the conversion of the immunofluorescence image to a binary output to be as best a representation of the data set as possible to ensure accuracy of the macro output. Over-sampling will create artifacts, likely skewing the data towards decreased phagocytosis and/or in-between particles, while under-sampling can occur when a threshold that is too low is selected, biasing output towards increased number of "phagocytosed" objects.

10. To maximize the speed in which the script runs it is best to operate it in "BatchMode." This mode does not display the images instead freeing up processing power for the running of the script. This modification can be switched on or off, which is required for the initial determination of the threshold values required by the user to input into the script.

Acknowledgments

The authors wish to thank Tatiana Khromykh and Juliana Venturato for technical assistance and Elizabeth Hartland, The University of Melbourne, for providing *Salmonella* strains. Fluorescence imaging was performed in the Australian Cancer Research Foundation funded Cancer Biology Imaging Facility at IMB. This work was funded by National Health and Medical Research Council of Australia (NHMRC) grants (to N.A.H and J.L.S) and by an NHMRC program grant and fellowship (JLS).

References

1. Aderem A, Underhill D (1999) Mechanisms of phagocytosis in macrophages. Annu Rev Immunol 17:593–623

2. Noda M, Suzumura A (2012) Sweepers in the CNS: microglial migration and phagocytosis in the Alzheimer disease pathogenesis. Int J Alzheimers Dis 2012:891087

3. Bueno SM, Wozniak A, Leiva ED, Riquelme SA, Carreño LJ, Hardt W-D, Riedel CA, Kalergis AM (2010) *Salmonella* pathogenicity island 1 differentially modulates bacterial entry to dendritic and non-phagocytic cells. Immunology 130:273–287

4. Schlumberger M, Hardt WD (2005) Triggered phagocytosis by salmonella: bacterial molecular mimicry of RhoGTPase activation/deactivation. In: Boquet P, Lemichez E (eds) Bacterial virulence factors and Rho GTPases, vol 291, Current topics in microbiology and immunology. Springer, Heidelberg, Berlin, pp 29–42. doi:10.1007/3-540-27511-8_3

5. Bakowski MA, Braun V, Brumell JH (2008) *Salmonella*-containing vacuoles: directing traffic and nesting to grow. Traffic 9:2022–2031

6. Sarantis H, Grinstein S (2012) Subversion of phagocytosis for pathogen survival. Cell Host Microbe 12:419–431

7. Flannagan RS, Jaumouille V, Grinstein S (2012) The cell biology of phagocytosis. Annu Rev Pathol 7:61–98

8. Moller J, Luhmann T, Chabria M, Hall H, Vogel V (2013) Macrophages lift off surface-bound bacteria using a filopodium-lamellipodium hook-and-shovel mechanism. Sci Rep 3:2884

9. Lee WL, Mason D, Schreiber AD, Grinstein S (2007) Quantitative analysis of membrane remodeling at the phagocytic cup. Mol Biol Cell 18:2883–2892

10. Huynh KK, Kay JG, Stow JL, Grinstein S (2007) Fusion, fission, and secretion during phagocytosis. Physiology 22:366–372

11. Swanson JA (2008) Shaping cups into phagosomes and macropinosomes. Nat Rev Mol Cell Biol 9:639–649

12. Gerisch G (2010) Self-organizing actin waves that simulate phagocytic cup structures. PMC Biophys 3:7

13. Orth JD, McNiven MA (2006) Get off my back! Rapid receptor internalization through circular dorsal ruffles. Cancer Res 66:11094–11096

14. Luo L, Wall AA, Yeo JC, Condon ND, Norwood SJ, Schoenwaelder S, Chen KW, Jackson S, Jenkins BJ, Hartland EL, Schroder K, Collins BM, Sweet MJ, Stow JL (2014) Rab8a interacts directly with PI3Kγ to modulate TLR4-driven PI3K and mTOR signalling. Nat Commun 5

15. Bohdanowicz M, Schlam D, Hermansson M, Rizzuti D, Fairn GD, Ueyama T, Somerharju P, Du G, Grinstein S (2013) Phosphatidic acid is required for the constitutive ruffling and macropinocytosis of phagocytes. Mol Biol Cell 24:1700–1712

16. Bohdanowicz M, Grinstein S (2013) Role of phospholipids in endocytosis, phagocytosis, and macropinocytosis. 93(1). doi: 10.1152/physrev.00002.2012

17. Murray RZ, Kay JG, Sangermani DG, Stow JL (2005) A role for the phagosome in cytokine secretion. Science 310:1492–1495

18. Kinchen JM, Ravichandran KS (2008) Phagosome maturation: going through the acid test. Nat Rev Mol Cell Biol 9:781–795

19. Roberts RL, Barbieri MA, Ullrich J, Stahl PD (2000) Dynamics of rab5 activation in endocytosis and phagocytosis. J Leukoc Biol 68:627–632

20. Wahe A, Kasmapour B, Schmaderer C, Liebl D, Sandhoff K, Nykjaer A, Griffiths G, Gutierrez MG (2010) Golgi-to-phagosome transport of acid sphingomyelinase and prosaposin is mediated by sortilin. J Cell Sci 123:2502–2511

21. Tse SML, Furuya W, Gold E, Schreiber AD, Sandvig K, Inman RD, Grinstein S (2003) Differential role of actin, clathrin, and dynamin in Fcγ receptor-mediated endocytosis and phagocytosis. J Biol Chem 278:3331–3338

22. Yeo JC, Wall AA, Stow JL, Hamilton NA (2013) High-throughput quantification of early stages of phagocytosis. Biotechniques 55:115–124

Fluorometric Approaches to Measuring Reductive and Oxidative Events in Phagosomes

Dale R. Balce and Robin M. Yates

Abstract

The phagosome is a redox-active organelle. Numerous reductive and oxidative systems play both direct and indirect roles in phagosomal function. With the advent of newer methodologies to study these redox events in live cells, the details of how redox conditions change within the maturing phagosome, how they are regulated, and how they influence other phagosomal functions can be investigated. In this chapter, we detail phagosome-specific, fluorescence-based assays that measure disulfide reduction and the production of reactive oxygen species in live phagocytes such as macrophages and dendritic cells, in real time.

Key words Phagosome, Lysosome, Oxidation, Disulfide reduction, Redox

1 Introduction

Many cellular functions and signaling pathways are modulated by reductive and oxidative (redox) chemistries. The redox environment and its influence on the function of endosomes, lysosomes, and phagosomes have recently garnered much attention [1]. Reductive mechanisms that influence disulfide reduction of protein-based antigens are required for efficient processing of disulfide-containing antigens within the phagosomal lumen [2]. In addition, reductases such as gamma-interferon inducible lysosomal thiol reductase (GILT) influence the proteolytic processing of antigens through modulation of the activities of thiol-based proteases such as the cysteine cathepsins [3, 4]. On the oxidative side, the NADPH oxidase complex (NOX2), which is responsible for the production of reactive oxygen species (ROS) in these cellular compartments, is also critical to many immune functions of phagocytes. In addition to microbial killing, NOX2-mediated ROS production can also control the activity of redox-sensitive proteases in these organelles, resulting in differential processing of antigens [5]. Numerous cellular assays exist to monitor redox events; however not all are

Roberto Botelho (ed.), *Phagocytosis and Phagosomes: Methods and Protocols*, Methods in Molecular Biology, vol. 1519,
DOI 10.1007/978-1-4939-6581-6_14, © Springer Science+Business Media New York 2017

suitable for specific evaluation in endosomes/lysosomes and phagosomes [6].

Phagosome-specific real-time evaluation of a variety of phagosomal chemistries such as proteolysis, acidification, and phagosome-lysosomal fusion has allowed characterization of the phagosomal lumen under various activation states of different phagocytes [7]. The adaptation of these technologies to monitor phagosome-specific redox events has led to the discovery of the characteristics and functions of phagosomal reductive and oxidative mechanisms [4, 8–10]. In this chapter we describe three phagosome-specific fluorescence-based redox assays that measure phagosomal disulfide reduction and ROS production in real-time in live macrophages. These assays evaluate the reductive or oxidative capacity of the phagosome in a population of phagocytes upon internalization of experimental particles bearing various redox reporters. We describe detailed protocols of the generation of phagosome-specific redox reporter-linked beads and the use of these beads to assess phagosome-specific redox measurements in adherent phagocytes in a multi-well format. Compared to other commonly used redox probes, the reagents used in these assays provide high signal to noise ratios and are stable in the presence of other phagosomal chemistries (e.g., pH) [6]. Furthermore, as these experimental protocols require minimal manipulation, they are easily adapted for high-throughput analysis [9, 11].

2   Materials

Cells, Reagents, Buffers, and Instrumentation

1. Phagosomal disulfide bond reduction and ROS production are measured in primary bone marrow-derived macrophages (BMMØs) and dendritic cells (BMDCs); however these approaches can also be employed for monocyte-derived phagocytes and phagocytic cell lines.

2. BMMØ growth media: Dulbecco's modified Eagle's medium (DMEM) supplemented with 10% fetal bovine serum (FBS), 2 mM L-glutamine, 1 mM sodium pyruvate, 100 U/mL penicillin–streptomycin, 20% conditioned media derived from the supernatant of M-CSF producing L929 cells.

3. BMDC growth media: Roswell Park Memorial Institute medium (RPMI) supplemented with 5% FBS, 2 mM L-glutamine, 10 mM HEPES, 0.5 mM β-mercaptoethanol, 100 U/mL penicillin–streptomycin, 20% conditioned media derived from the supernatant of Ag8653 melanoma cells transfected with murine GM-CSF cDNA.

4. Assay buffer: Tissue culture grade PBS supplemented with 1 mM $CaCl_2$, 2.7 mM KCl, 0.5 mM $MgCl_2$, 5 mM dextrose,

and 0.25% gelatin. Filter-sterilize through a 0.22 μm filter. Store at 4 °C. Warm to 37 °C prior to use.

5. MES buffer: ddH$_2$O, 0.1 M 2-[N-Morpholino]ethanesulfonic acid (MES), 0.5 M NaCl, pH 6.0. Store at room temperature.

6. Coupling buffer: ddH$_2$O, 0.1 M sodium borate, pH 8.0. Store at room temperature.

7. Quenching buffer: PBS pH 7.2 containing 250 mM glycine. Filter sterilize through 0.22 μm filter. Store at 4 °C.

8. Sodium azide 2% aqueous solution. Store at room temperature.

9. 96-well assay plates: 96-well μClear black clear bottom (Greiner Bio-One) or equivalent.

10. 3.0 μm carboxylate-modified silica particles/beads 5% suspension (Si-COOH, Kisker Biotech, Steinfurt, Germany) (*see* **Note 1**).

11. Dextran, amino-modified, 70,000 MW (Life Technologies). Store as desiccate.

12. N-hydroxysulfosuccinimide sodium salt (Sulfo-NHS). Store as desiccate. Protect from moisture.

13. N-(3-dimethylaminopropyl)-N'-ethylcarbodiimide hydrochloride (EDC). Store as desiccate. Protect from moisture.

14. Cyanamide. Store as desiccate at 4 °C. Protect from moisture.

15. 5 mg/mL, Alexa Fluor 594 carboxylic acid, succinimidyl ester (AF594-SE, Life Technologies), in DMSO; aliquoted and stored at −20 °C. Protect from light.

16. 5 mg/mL, BODIPY FL L-cystine (Life Technologies), in DMSO; aliquoted and stored at −20 °C. Protect from light.

17. 1 mg/600 μL, OxyBURST Green H$_2$HFF BSA, in coupling buffer; aliquoted and stored at −20 °C. Protect from light. Thaw on ice prior to use.

18. 10 mM, Amplex UltraRed (Life Technologies), in DMSO; aliquoted and stored at −20 °C. Protect from light.

19. 1 U/μL, horseradish peroxidase (HRP), in PBS; aliquoted and stored at −20 °C.

20. 5 mg/mL, (+)-Biotin N-hydroxysuccinimide ester (NHS-Biotin), in DMSO; aliquoted and stored at −20 °C.

21. 2 mg/mL, IgG from human serum, in sterile water; aliquoted and stored at −20 °C.

22. 10 mg/mL, anti-bovine serum albumin IgG, in PBS; aliquoted stored at −20 °C.

23. 10 mg/mL, anti-biotin IgG: Mouse monoclonal (clone BN-34) as ascite fluid, in PBS; aliquoted and stored at −20 °C.

24. 1.5 mL polypropylene tubes with silicone O-rings (VWR International).

25. Fluorescence microplate reader with temperature control. Plate readers may be filter- or monochromator-based. The authors routinely use an Envision microplate reader (Perkin Elmer) to monitor phagosomal oxidative capacity and a Fluostar Optima (BMG Labtech) to monitor phagosomal disulfide reduction.

3   Methods

3.1   Preparation of Redox Reporter Beads

3.1.1   Preparation of Dextran-Linked Reductase-Reporter Beads

1. 50 mg of carboxylate-modified silica beads (1 mL of manufacturer stock solution) is transferred to a low binding 1.5 mL screwcapped polypropylene tube. Beads are pelleted and washed twice with PBS by centrifugation at $6000 \times g$ at room temperature for 30 s using a benchtop microtube centrifuge and vortex.

2. 30 mg of the heterobifunctional crosslinking reagent cyanamide is freshly dissolved in 1 mL PBS and incubated with the beads at room temperature with agitation for 15 min.

3. During the incubation period, 10 mg of amine-modified dextran is dissolved in 1 mL coupling buffer.

4. The beads are washed twice with ice-cold coupling buffer to remove excess cyanamide. The cyanamide-activated beads are then resuspended in the dextran/coupling buffer solution and incubated at room temperature with agitation for 2 h (see Note 2).

5. The beads are washed twice with 1 mL PBS to remove unconjugated amine-modified dextran, and the remaining amine-reactive groups are quenched by incubation with quenching buffer for 5 min at room temperature with agitation. The dextran-coupled beads can be now conjugated with BODIPY FL L-cystine (steps 6–8) or can be stored at 4 °C for later use in 1 mL quenching buffer with 10 µL of 2% sodium azide.

6. Dextran-coupled beads are washed twice with 1 mL MES buffer and resuspended in 100 µL MES buffer in a 1.5 mL screwcapped polypropylene tube.

7. 20 µL of 16 mg/mL EDC in MES (make fresh and use immediately), 27.5 µL of 32 mg/mL sulfo-NHS in MES (make fresh and use immediately) and 20 µL of 5 mg/mL BODIPY-FL L-cystine in DMSO are added (in order as listed with vortexing between additions) to the dextran-coupled bead suspension and incubated with agitation for 2–4 h at room temperature in the dark.

8. The beads are washed twice with 1 mL coupling buffer and resuspended in 500 μL coupling buffer (*see* **Note 3**).

9. 5 μL of 5 mg/mL NHS-Biotin and 2 μL of 5 mg/mL AF594-SE are added to the resuspended beads and incubated with agitation for 1–2 h at room temperature in the dark.

10. Unreacted NHS-Biotin and AF594-SE are quenched and removed by washing beads twice with quenching buffer.

11. The now-completed dextran-linked reductase-reporter beads are resuspended in 1 mL quenching buffer with 10 μL of 2% sodium azide as a preservative, enumerated using a hemocytometer and stored at 4 °C in the dark.

3.1.2 Preparation of ROS-Reactive OxyBURST Beads

1. 5 mg of carboxylate-modified silica beads (0.1 mL of manufacturer stock solution) is transferred to a low binding 1.5 mL screwcapped polypropylene tube. Beads are pelleted and washed twice with PBS by centrifugation at $6000 \times g$ at room temperature (RT) for 30 s using a benchtop microtube centrifuge and vortex (*see* **Notes 4** and **5**).

2. Beads are resuspended in a solution containing 25 mg of the heterobifunctional cross-linker cyanamide in 1 mL PBS and incubated with agitation for 15 min at room temperature.

3. The cyanamide-activated beads are then washed twice with ice cold coupling buffer (to remove excess cyanamide) and resuspended in 100 μL of the OxyBURST/coupling buffer solution (*see* **Note 2**). The bead/OxyBURST/coupling buffer suspension is then incubated with agitation in the dark at room temperature for 1–3 h (*see* **Note 6**).

4. The beads are washed twice with 1 mL coupling buffer then resuspended in 200 μL coupling buffer containing 1 μL of 5 mg/mL AF594-SE and incubated with agitation in the dark at room temperature for an additional 15 min.

5. The now-completed OxyBURST-linked reporter beads are washed twice and resuspended in 1 mL PBS, enumerated using a hemocytometer and stored at 4 °C in the dark. For best results, the beads should be used within 5–7 days.

3.2 Cell Preparation and Handling

1. Bone marrow is flushed from the tibias, femurs and ilia of mice with DMEM (for BMMØs) or RPMI (for BMDCs) and centrifuged at $230 \times g$ at 4 °C for 10 min.

2. To obtain BMMØs freshly isolated marrow is resuspended in BMMØ growth media and plated onto non-treated petri dishes (approximately 8–10 dishes per mouse). 7 days after initial plating, BMMØs are diluted 1:2 and re-plated onto new petri dishes in BMMØ growth media. Approximately 10–14 days after initial plating, BMMØs are ready for use.

3. To obtain BMDCs, freshly isolated marrow is resuspended in RPMI containing 10 % fetal bovine serum, 2 mM L-glutamine, 1 mM sodium pyruvate, and 100 U/mL penicillin–streptomycin and cultured in a 75 cm^2 tissue culture flask overnight. The next day, non-adherent cells are harvested from the flask, resuspended in BMDC growth media and plated onto 100 mm treated tissue culture dishes (approximately 6–8 dishes per mouse). Every 2 days for 10 days, half of the growth media is removed from the culture dish and replenished with new growth media. Approximately 10 days after initial plating, BMDCs are ready for use.

4. To harvest BMMØs, growth media is removed from confluent, fully differentiated BMMØs and replaced with cold PBS. The cells are incubated at 4 °C for 10 min to facilitate BMMØ detachment from the culture dish. BMMØs are gently dislodged with a cell scraper and centrifuged at $230 \times g$ at 4 °C for 10 min.

5. To harvest BMDCs, growth media is removed and 5 mL pre-warmed trypsin–EDTA is added to the culture dish. Following a 5 min incubation at 37 °C, cells are gently dislodged with a cell scraper and centrifuged at $230 \times g$ at 4 °C for 10 min.

6. BMMØs and BMDCs are plated onto 96-well assay plates to 100 % confluency. Approximately 1.2×10^5 cells are required per well in a volume of 100 μL of complete medium to establish a confluent monolayer. Plates are kept at room temperature for 15 min before incubating at 37 °C overnight to ensure an even monolayer will be established. The authors usually plate a minimum of three wells for each experimental condition and additional "test" wells for titrating bead numbers.

3.3 Performing Redox Measurements in Live Cells

3.3.1 Assessment of Phagosome-Specific Disulfide Reduction

1. Shortly prior to phagosomal assessment, the well medium is removed and the adherent cell monolayer is washed twice with 100 μL of pre-warmed assay buffer (37 °C).

2. 50 μL of pre-warmed assay buffer is added to each well prior to the addition of experimental beads. In this volume, cells can be treated with water soluble inhibitors/compounds if desired. In addition, background readings of cells without experimental beads can be obtained at this stage.

3. 100 μL of the reductase bead stock is washed twice with 1 mL PBS (gentle vortex, or by pipetting up and down) and resuspended in 200 μL PBS containing 5 μL anti-biotin IgG (10 mg/mL). Beads are incubated with agitation in the dark at room temperature for 15 min (see **Note 7**).

4. Beads are washed once (gentle vortex, or by pipetting up and down) with 1 mL PBS and resuspended in 1 mL PBS.

5. The appropriate concentration of beads required per well is then determined. Approximately 2–4 beads per cell are opti-

mal. As a starting point, 20 μL of the opsonized reductase beads stock solution is resuspended in 1 mL assay buffer. 50 μL is added to a test well containing a confluent monolayer of cells. After a 5–10 min incubation period, the number of beads per cell is visually inspected under a light microscope. The bead dilution is then adjusted accordingly to achieve approximately 2–4 beads per cell.

6. 50 μL of the reductase beads at the appropriate working dilution in assay buffer is added to each experimental well (*see* **Note 8**).

7. The assay plate is inserted into the plate reader and substrate fluorescence (SF) 488/520 nm and calibration fluorescence (CF) 594/620 nm (excitation λ/emission λ) is recorded. Disulfide reduction is indicated by an increase in fluorescence as the disulfide linker is cleaved resulting in the de-quenching of the disulfide-linked BODIPY fluorophores (*see* **Note 9**).

8. Substrate/calibration fluorescence is recorded every 2–5 min for 1–2 h.

9. Data is exported into a spreadsheet application such as Microsoft Excel. RFU (relative fluorescence unit) = $(SF\text{-}SF_{background})/(CF\text{-}CF_{background})$ is calculated for each time point and plotted against time. An example of analyzed data is shown in Fig. 1.

3.3.2 Assessment of Phagosome-Specific ROS Production

1. Shortly prior to phagosomal assessment, the well medium is removed and the adherent cell monolayer is washed twice with 100 μL of pre-warmed assay buffer (37 °C).

2. 50 μL of pre-warmed assay buffer is added to each well prior to the addition of experimental beads. In this volume, cells can be

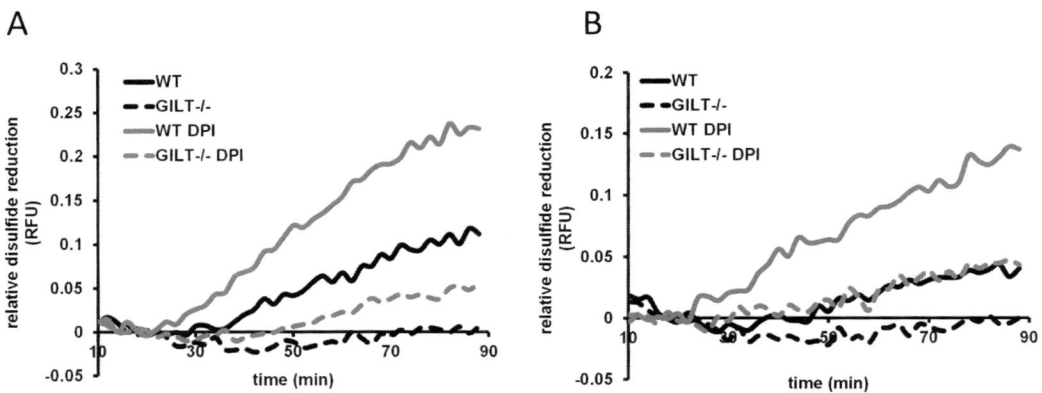

Fig. 1 Assessment of phagosome-specific disulfide reduction in (**a**) BMMØs and (**b**) BMDCs derived from C57Bl6 (WT) or GILT-deficient (GILT−/−) mice. Phagosomal disulfide reduction is significantly decreased in the absence of GILT. As disulfide reduction in the phagosome is sensitive to the presence of ROS, disulfide reduction is increased in the presence of the NOX2 inhibitor diphenyleneiodonium

A

B

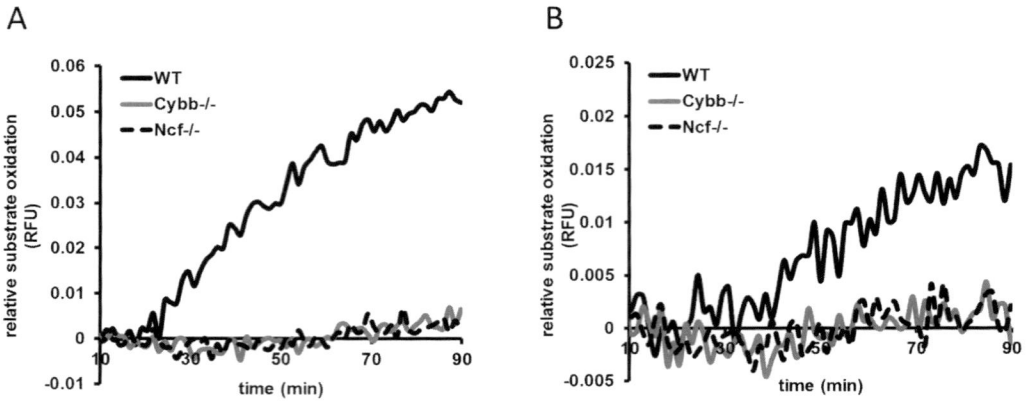

Fig. 2 Assessment of phagosome specific ROS production in (**a**) BMMØs and (**b**) BMDCs derived from C57Bl6 (WT) or NOX2 complex-deficient (Cybb−/− and Ncf−/−) mice. Phagosomal ROS is effectively absent in gp91-deficient (Cybb−/−) and p47-deficient (Ncf−/−) phagocytes

treated with water soluble inhibitors/compounds if desired. In addition, background readings of cells without experimental beads can be obtained at this stage.

3. OxyBURST bead stock is washed once with 1 mL PBS (gentle vortex, or by pipetting up and down) and resuspended in 200 μL PBS containing 5 μL anti-BSA IgG (10 mg/mL). Beads are incubated with agitation in the dark at room temperature for 30 min (*see* **Note 7**).

4. Beads are washed once (gentle vortex, or by pipetting up and down) with 1 mL PBS and resuspended in 1 mL PBS.

5. The appropriate concentration of beads required per well is then determined. Approximately 2–4 beads per cell are optimal. As a starting point, 20 μL of the opsonized OxyBURST beads stock solution is resuspended in 1 mL assay buffer. 50 μL is added to a test well containing a confluent monolayer of cells. After a 5–10 min incubation period, the number of beads per cell is visually inspected under a light microscope. The bead dilution is then adjusted accordingly to achieve approximately 2–4 beads per cell.

6. 50 μL of the OxyBURST bead working solution in assay buffer is then added to each well (*see* **Note 8**).

7. The assay plate is inserted into the plate reader and substrate fluorescence (SF) 488/520 nm and calibration fluorescence (CF) 594/620 nm (excitation λ/emission λ) is recorded. Oxidation of the OxyBURST substrate by phagosomal ROS leads to increased fluorescence of the fluorophore (*see* **Note 10**).

8. Data is exported into a spreadsheet application such as Microsoft Excel. $RFU = (SF\text{-}SF_{background})/(CF\text{-}CF_{background})$ is calculated for each data point and plotted against time. An example of analyzed data is shown in Fig. 2.

To complement the OxyBURST bead assay described above, a simple assay to measure extracellular release of H_2O_2 in response to the phagocytosis of opsonized zymosan particles can be used. Amplex UltraRed reacts with H_2O_2 in a stoichiometry of 1:1 to produce the fluorescent resorufin; thus this assay can be used to quantify extracellular concentrations of H_2O_2 when regressed to a standard curve.

1. Shortly prior to phagosomal assessment, the well medium is removed and the adherent cell monolayer is washed twice with 100 μL of pre-warmed assay buffer (37 °C).

2. 50 μL of pre-warmed assay buffer is added to each well prior to the addition of experimental beads. In this volume, cells can be treated with water soluble inhibitors/compounds if desired. In addition, background readings of cells without experimental beads can be obtained at this stage.

3. Serum-opsonized zymosan particles (100 mg zymosan/mL fetal bovine serum) are resuspended in assay buffer to a final concentration of 0.1–1 mg zymosan/mL. 50 μL is added to each well (*see* **Note 11**).

4. The assay plate is incubated for 1–2 h in a 37 °C culture incubator.

5. The well-supernatant is transferred to a fresh 96-well assay plate.

6. In a clean 15 mL conical tube Amplex UltraRed and HRP are added to freshly degassed PBS at a concentration of 20 μg/mL and 20 U/mL respectively. 50 μL of the Amplex UltraRed/HRP/PBS solution is immediately added to each well supernatant. The assay plate is incubated for 15 min in the dark at room temperature.

7. Amplex Red fluorescence ~571/585 nm (excitation λ/emission λ) is then measured using a microplate reader. 3–5 measurements from each well and from each well replicate are collected and averaged.

8. Data is exported into a spreadsheet application such as Microsoft Excel. Background values from no zymosan or other negative control samples (e.g., NOX2-deficient) are deducted from each data set. An example of analyzed data is shown in Fig. 3.

4   Notes

1. 1.5–5.0 μM silica or latex beads can also be used.

2. Washing at this step should be done quickly and dextran/OxyBURST solution added immediately after washing.

3. Beads at this stage should be a deep orange in color, indicating successful conjugation of the BODIPY FL L-cystine substrate.

A

B

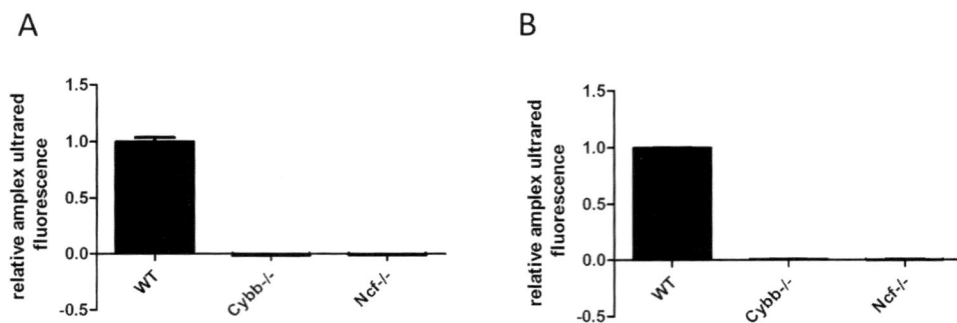

Fig. 3 Assessment of extracellular H_2O_2 produced by (**a**) BMMØs and (**b**) BMDCs derived from C57Bl6 (WT) or NOX2 complex-deficient (Cybb–/– and Ncf–/–) mice upon internalization of serum-opsonized zymosan particles. H_2O_2 production is compromised in the absence of the NOX2 complex subunits gp91 (Cybb–/–) and p47 (Ncf–/–)

4. As OxyBURST beads must be made fresh, quantities have been adjusted to yield a small batch of beads (approximately enough for 100 wells of a 96-well assay plate).

5. PBS and coupling buffers are first degassed in a water bath sonicator prior to use.

6. Depending on the objectives of the experiment, the surface of the beads can be covalently coupled to IgG in lieu of (noncovalent) opsonization with anti-BSA IgG prior to the addition of the beads to cells (Subheading 3.3.2, **step 3**). Hence, if desired, 1 μL of human IgG (2 mg/mL) is added to the bead/OxyBURST solution during the 1–3 h incubation.

7. IgG opsonization of the experimental beads is most efficient when performed immediately prior to use.

8. Vortex working bead solution vigorously before applying to cells to avoid clumping of beads.

9. A variety of fluorescence plate reader platforms can be used to monitor monolayer phagosomal activity; however a few specific functionalities are required. Fluorescence plate readers with bottom read and height optimization are required as BMMØs and BMDCs establish adherent monolayers. Temperature control is required for kinetic measurements in live cells.

10. The same instrument protocol used to monitor BODIPY FL L-cystine de-quenching can be used to monitor OxyBURST substrate oxidation.

11. Wells without zymosan or cell treated with the NOX2 inhibitor diphenyleneiodonium can be included as background or negative controls respectively.

References

1. Yates RM (2013) Redox considerations in the phagosome: current concepts, controversies, and future challenges. Antioxid Redox Signal 18:628–629

2. Collins DS, Unanue ER, Harding CV (1991) Reduction of disulfide bonds within lysosomes is a key step in antigen processing. J Immunol 147:4054–4059

3. Phan UT, Arunachalam B, Cresswell P (2000) Gamma-interferon-inducible lysosomal thiol reductase (GILT). Maturation, activity, and mechanism of action. J Biol Chem 275: 25907–25914

4. Balce DR, Allan ER, McKenna N et al (2014) Gamma-interferon-inducible lysosomal thiol reductase (GILT) maintains phagosomal proteolysis in alternatively activated macrophages. J Biol Chem 289:31891–31904

5. Allan ER, Tailor P, Balce DR et al (2014) NADPH oxidase modifies patterns of MHC class II-restricted epitopic repertoires through redox control of antigen processing. J Immunol 192:4989–5001

6. Balce DR, Yates RM (2013) Redox-sensitive probes for the measurement of redox chemis-tries within phagosomes of macrophages and dendritic cells. Redox Biol 1:467–474

7. Yates RM, Russell DG (2008) Real-time spec-trofluorometric assays for the lumenal environment of the maturing phagosome. Methods Mol Biol 445:311–325

8. Balce DR, Li B, Allan ER et al (2011) Alternative activation of macrophages by IL-4 enhances the proteolytic capacity of their phagosomes through synergistic mechanisms. Blood 118:4199–4208

9. Balce DR, Greene CJ, Tailor P et al (2015) Endogenous and exogenous pathways maintain the reductive capacity of the phagosome. J Leukocyte Biol 100(1):17–26

10. VanderVen BC, Yates RM, Russell DG (2009) Intraphagosomal measurement of the magnitude and duration of the oxidative burst. Traffic 10:372–378

11. Rybicka JM, Balce DR, Khan MF et al (2010) NADPH oxidase activity controls phagosomal proteolysis in macrophages through modulation of the lumenal redox environment of phagosomes. Proc Natl Acad Sci U S A 107: 10496–10501

Chapter 15

Simultaneous Analysis of Multiple Lumenal Parameters of Individual Phagosomes Using High-Content Imaging

Samuel Cheung*, Catherine Greene*, and Robin M. Yates

Abstract

The biochemical processes within the phagosomes of macrophages and dendritic cells are essential to immunity and homeostasis. Measurable properties of the phagosomal lumen include: assessment of various hydrolytic activities, reduction and oxidation events, pH, ion concentrations, and electrochemical gradients. These often-interdependent phagosomal features are commonly evaluated individually, hindering the analysis of the biochemical relationship between these factors within the same phagosome. In addition, the ability of phagosomes within the same cell to behave autonomously is becoming more evident, thus highlighting the need for a technique capable of multiplex analyses of phagosomal lumenal chemistries at the single phagosome level. In this chapter, we outline an approach that is capable of simultaneously measuring multiple phagosomal parameters of individual phagosomes by utilizing specifically designed experimental particles with multiple reporter fluors, in combination with real-time fluorometric measurement via automated microscopy. Subsequent analysis using high-content image analysis software enables each phagosomal parameter to be evaluated in a quantitative or semi-quantitative manner. This approach facilitates investigation of the complex relationship between phagosomal properties in a population of macrophages in real time, at the level of individual phagosomes.

Key words Phagosome, Lysosome, Macrophage, Proteolysis, Acidification, Endosome

1 Introduction

Phagosomal maturation encompasses a complex series of membrane fusion and fission events within the endo-lysosomal system, enabling the phagosome to recruit enzymes and signaling complexes for the generation of reactive oxygen species (ROS), lumenal acidification and the hydrolysis of phagocytosed macromolecules [1–3]. In addition to the study of signaling pathways and fusion machinery that orchestrate phagosomal maturation, exploration of the microenvironmental and biochemical changes that occur within the lumen of the maturing phagosome provides

---

*These authors are contributed equally.

Roberto Botelho (ed.), *Phagocytosis and Phagosomes: Methods and Protocols*, Methods in Molecular Biology, vol. 1519,
DOI 10.1007/978-1-4939-6581-6_15, © Springer Science+Business Media New York 2017

insight into how phagosomes perform their multitudinous functions. Considering the complex series of events that are involved in phagocytosis and phagosomal maturation, in addition to the multiplicity of phagosomal functions, it is unsurprising that phagosomes behave in a heterogeneous fashion [4, 5]. Utilizing fluorometric evaluation of lumenal properties of phagosomes across a population of phagocytes offers a convenient and robust approach to phagosomal assessment [6, 7]. However, heterogeneity between phagosomes and phagocytes cannot be evaluated using a population-based approach. Additionally, the measurement of single parameters in isolation limits the exploration of the complex interplay between phagosomal chemistries [8, 9].

Fluorometric assessments of phagosome-specific lumenal parameters have previously been used in our laboratory by utilizing customized reporter beads phagocytosed by populations of phagocytes in a synchronous fashion, with measurements taken by simple plate- or cuvette-based fluorometers [6, 7]. More recently we have adapted this technology to permit the simultaneous evaluation of multiple parameters of *individual* phagosomes in a population of live macrophages. This approach requires the use of automated high-content microscopic imaging and image analysis that automates the identification and tracking of phagosomes and records relative fluorescent intensities of reporter fluors. Whilst considerably more laborious than population-based assays, evaluation of multiple phagosomal properties within individual phagosomes enables the dissection of phagosomal lumenal chemistry as well as the exploration of phagosomal autonomy and heterogeneity.

When designing multiplexed phagosomal assays it is important to carefully consider the limitations of using multiple fluorophores or substrates, to ensure that each measurement taken accurately reflects its experimental objective and is not overtly influenced by other phagosomal parameters or proximal fluorophores. In addition to choosing fluor combinations that do not have overlapping emission spectra, the possible unintended resonance energy transfer between fluor combinations should be considered and controlled. It is also important to consider each fluor's suitability for and stability in the phagosomal environment. Generally a reference or reporter fluor (that is intended to measure attributes other than pH) should be stable at low pH and its fluorescent quantum yield should be unaffected by pH changes in the maturing phagosome (pH ~7.5–4.0). Many fluors are also affected by phagosomal ROS. For example, the newer-generation pH reporter fluors pHrodo (Invitrogen) and cypHer (GE Healthcare) are irreversibly inactivated by the levels of phagosomal ROS generated by cultured macrophages and dendritic cells [9, 10]. In certain cases, it is not possible to assemble an ideal set of reporter fluors that can be multiplexed to measure the desired combination of phagosomal parameters. In such scenarios the limitation of each fluor needs be taken into account during the analysis and evaluation of results.

In this chapter we describe one of the multiplexed assays used in our laboratory, which is designed to measure proteolytic efficiency in concert with the acidification of individual phagosomes in bone marrow derived macrophages (BMMØs). Although the protocol is specific for this particular assay and for the instrumentation and analysis software used, the experimental platform and principles described can be applied to the design of any assay intended to simultaneously evaluate multiple lumenal parameters in individual phagosomes.

2 Materials

2.1 Cells, Reagents, and Buffers

1. BMMØ media: Dulbecco's modified Eagle's medium (DMEM) supplemented with 10% fetal bovine serum (FBS), 2 mM l-glutamine, 1 mM sodium pyruvate, 100 U/mL penicillin–streptomycin, and 20% conditioned medium from the supernatant of macrophage colony-stimulating factor (M-CSF) producing L929 cells.

2. 96-well assay plates: 96-well μClear black clear bottom (Greiner Bio-One) (see **Note 1**).

3. Dulbecco's phosphate-buffered saline, tissue culture grade, sterile, pH 7.2 (DPBS).

4. Assaying buffer: DPBS adjusted to contain 1 mM $CaCl_2$, 2.7 mM KCl, 0.5 mM $MgCl_2$, 5 mM glucose, 0.1% calf skin gelatin (see **Note 2**) and 10 mM HEPES. Filter sterilize through 0.22 μm filter. Store at 4 °C. Warm to 37 °C before use.

5. 3.0 μm carboxylate-modified silica particles/beads (Si-COOH) 5% suspension (Kisker Biotech, Steinfurt, Germany) (see **Note 3**).

6. IgG from human serum. Store as desiccate at 4 °C.

7. Cyanamide. Store as desiccate at 4 °C. Protect from moisture.

8. Coupling buffer: 0.1 M sodium borate in ddH_2O. Adjust pH to 8.0 with 10 M NaOH. Filter sterilize through 0.22 μm filter. Store at room temperature.

9. Quenching buffer: 250 mM glycine in DPBS pH 7.2. Filter sterilize through 0.22 μm filter. Store at 4 °C.

10. 2% sodium azide, aqueous solution. Toxic. Store at room temperature.

11. 1 mg/mL DQ Green BSA, dissolved in coupling buffer. Store at –20 °C. Protect from light and moisture (Life Technologies, Part of Thermo Fisher Scientific, Burlington, ON).

12. 5 mg/mL pHrodo Red succinimidyl ester (pHrodo-SE), dissolved in high quality anhydrous dimethylsulfoxide (DMSO).

Store at -20 °C. Protect from light and moisture (Life Technologies, Part of Thermo Fisher Scientific, Burlington, ON).

13. 5 mg/mL Alexa Fluor 647 succinimidyl ester (Alexa 647-SE), dissolved in high quality anhydrous dimethylsulfoxide (DMSO). Store at -20 °C. Protect from light and moisture (Life Technologies, Part of Thermo Fisher Scientific, Burlington, ON).

2.2 Imaging Instrument and Analysis Software

1. IN Cell Analyzer 2000 Live C TEMP/LH/EC High-Content Analysis system (GE Healthcare Life Sciences) (*see* **Note 4**). The imaging chamber in this system has temperature and atmospheric control (humidity and CO_2). The instrument is equipped with polychroic QUAD band mirrors that have been optimized for the installed excitation filters ($350 \pm 50, 490 \pm 20, 579 \pm 24$, and 645 ± 30 nm) and emission filters ($455 \pm 50, 525 \pm 20, 624 \pm 40$, and 705 ± 72 nm).

2. IN Cell Analyzer 2000 Acquisition Software (Version 4.6) (GE Healthcare Life Sciences).

3. Analysis software: IN Cell Developer Toolbox 1.9.3(x64) as a component of the IN Cell Investigator image analysis software package (Version 1.6.3) (GE Healthcare Life Sciences).

3 Methods

3.1 Preparation of Experimental Particles

1. 50 mg of 3.0 μm carboxylate-modified silica particles (1 mL of manufacturer stock) is added to a low binding polypropylene 1.5 mL screw-cap microtube. Brief centrifugation at $6000 \times g$ at room temperature using a bench-top centrifuge and three DPBS washes allows the removal of the manufacturer's storage buffer and the pelleting of beads (*see* **Note 5**).

2. 30 mg of the heterobifunctional crosslinker, cyanamide is freshly dissolved in 1 mL DPBS and incubated with the silica particles at room temperature for 30 min with agitation.

3. During the incubation, 1 mg of DQ Green BSA and 0.01 mg of human IgG (optional) are dissolved in 1 mL of coupling buffer (*see* **Note 6**).

4. After the incubation, the silica particles are washed three times with ice-cold coupling buffer to remove excess cyanamide. The silica particles are resuspended in the DQ Green BSA/IgG solution, covered and incubated at room temperature for 3–6 h with agitation, allowing the DQ Green BSA and IgG to be covalently linked to the silica particles through the cyanamide crosslinker (*see* **Note 7**).

5. The silica particles are washed twice with 500 μL coupling buffer to remove unbound protein and incubated with 500 μL of

quenching buffer for 10 min at room temperature with agitation to quench any remaining amine-reactive moieties.

6. DQ green BSA-coupled silica particles are washed twice with 500 μL coupling buffer and resuspended in 1 mL of coupling buffer containing 1 μL of 5 mg/mL pHrodo-SE and 1 μL of 5 mg/mL Alexa 647-SE. The particles are incubated at room temperature for 30 min with agitation, allowing labeling of the particle bound protein with the amine-reactive fluors (*see* **Note 8**).

7. The particles are washed three times with 1 mL DPBS to remove excess fluor, and incubated with 500 μL of quenching buffer for 10 min at room temperature with agitation to quench residual unreacted succinimidyl esters.

8. The labeled silica particles are washed three times with 1 mL DPBS and can be stored at 4 °C for later use in 1 mL quenching buffer with 10 μL of 2% sodium azide as a preservative.

3.2 Macrophage Monolayer Preparation and Handling

1. Phagosomal chemistries are assessed in primary murine BMMØs (*see* **Note 9**). BMMØs are derived from bone marrow flushed from the femurs, tibias, and ilia of euthanized mice with DMEM and centrifuged at $230 \times g$ at 4 °C for 10 min.

2. The freshly isolated bone marrow is resuspended in BMMØ media and plated onto untreated 10 cm petri dishes. After 7 days, the BMMØs are subcultured 1:2 and plated on fresh 10 cm petri dishes in BMMØ media. The cells are allowed to differentiate for 10–14 days after bone marrow extraction before use.

3. Growth media is removed from confluent, fully differentiated BMMØs in untreated 10 cm petri dishes, and replaced with cold DPBS. The cells are incubated at 4 °C for 10 min to facilitate BMMØ detachment from the culture dish. BMMØs are gently dislodged with a cell scraper and centrifuged at $230 \times g$ at 4 °C for 10 min.

4. The BMMØ cell pellet is resuspended in 1 mL of BMMØ media and a cell count is performed using trypan blue exclusion on a hemocytometer. An appropriate working dilution of the cells is made to achieve ~5.0×10^6 BMMØs/mL.

5. The counted BMMØs are plated at 25% confluency (approximately 30,000 cells in 100 μL complete BMMØ medium) per well in a maximum of 48-wells in 96-well assay plates (*see* **Note 10**). The cells are then incubated at room temperature for 15 min to establish an even monolayer before the plates are transferred to a 37 °C incubator at 7% CO_2 overnight (*see* **Note 11**).

6. Prior to assay commencement, the medium is removed from the pre-plated BMMØ monolayer and the adhered BMMØs are gently washed twice with 100 μL of pre-warmed assaying buffer (37 °C).

7. 50 μL of pre-warmed assay buffer is added to each well before the addition of beads (Subheading 3.4).

3.3 Imaging System Setup

Here we describe high-content multiplex imaging using the IN Cell Analyzer 2000 Live C TEMP/LH/EC High-Content Analysis system (GE Healthcare Life Sciences) (*see* **Note 12**). The instrument is equipped with an imaging chamber with environmental control (temperature, humidity, and CO_2) and is capable of accommodating different plate formats. The acquisition software contains predefined coordinates for compatible plates (the authors prefer 96-well μClear black clear bottom (Greiner Bio-One)) (*see* **Note 13**). Image acquisition protocols generally require customization with respect to plate/wells used, the kinetics of acquisition, objectives to be used and the requisite filter sets for the fluors being imaged.

1. The system is pre-warmed to 37 °C for at least an hour prior to assay commencement (*see* **Note 14**).

2. The three required channels are assigned (corresponding to the DQ Green BSA, pHrodo-SE, and Alexa 647-SE fluors) as well as inclusion of a brightfield channel.

3. The 40× magnification objective is selected (*see* **Note 15**).

4. The correct excitation emissions are set for each set of images; FITC (DQ Green BSA), Texas Red (pHrodo-SE), Cy5 (Alexa 647-SE) and the brightfield channel will default to DAPI.

5. It is recommended that the exposure value is set to 0.02 s for all channels at 10 % lamp intensity to avoid overt phototoxicity. All images are acquired in a *2-D mode*.

6. The instrument has an offset adjustment for each channel which can be optimized based on the assay being carried out and the experimental plate used. We have found that an offset of 7 for FITC, 6 for Texas Red, and 5 for Cy5 works best for our system.

7. The real-time function of this live-cell imaging system is engaged by defining the desired length of assay and time interval. To maximize sample size and minimize phototoxicity, the authors typically use 180 s as the acquisition time interval over 1–3 h. Under these experimental conditions minimal cell death is observed for up to 10 h.

8. Laser-guided autofocus is utilized to refocus at the start of each time point.

9. For *Image Processing*, the *High pass filter* is set at a value of 10 with the *Invert phase polarity* disabled. A defined *contrast angle* of 135°, with the *combiner prism* at 0, and *intensity modulation* disabled, provides the sharpest images in these experiments.

3.4 Acquisition of Images

1. Prior to assay commencement, the medium is removed from the pre-plated BMMØ monolayer and the adhered BMMØs are gently washed twice with 100 µL of pre-warmed assaying buffer (37 °C).

2. 50 µL of pre-warmed assay buffer is added to each well before the addition of beads.

3. 100 µL of the fluorescently labeled bead stock (prepared in Subheading 3.1) is washed twice in 1 mL of PBS, resuspended in 1 mL PBS and enumerated using a hemocytometer.

4. The appropriate concentration of beads required per well is then determined. Approximately 1–2 beads per cell are optimal. Initially, 20 µL of the fluorescently labeled beads is resuspended in 1 mL assay buffer and 50 µL of the bead suspension in assay buffer is added to a test well containing cells. The number of beads per cell is counted under a light microscope after a 5–10 min incubation period. The working dilution of beads is then adjusted to achieve approximately 1 to 2 beads per cell, as a higher number of beads than this becomes difficult to track (*see* **Note 16**).

5. Once the number of required beads per well has been determined, 50 µL of the appropriate working dilution of bead stock in assaying buffer is transferred into each well. The assay plate is incubated at room temperature for 10 min to allow the particles to evenly settle for phagocytosis to occur (*see* **Note 17**). An additional 100 µL of pre-warmed assay buffer can be added per well for assays longer than 2 h.

6. The sample plate is loaded into the pre-warmed imaging chamber of the IN Cell Analyzer 2000 and the previously developed acquisition protocol (Subheading 3.3) is loaded, wells selected, and the acquisition protocol is commenced.

7. After image acquisition is completed, the computer will automatically process the images to generate the image stack file in *xdce* format. The generation of the image stack file may take several minutes depending on the file size. Fluorescent tracking described in the next step using IN Cell Investigator software cannot be performed without the *xdce* file.

3.5 Analysis

Analysis of the acquired image data is performed using the IN Cell Investigator software and Spotfire DescisionSite data visualization software. The IN Cell Investigator requires an *xdce* file generated during the acquisition using the IN Cell Developer software. The analysis is underpinned by customized algorithms that identify the fluorescent beads in each field of view based on the dimensions and characteristics of the object. The following section will describe how to identify, track and record fluorescent intensities of the 3.0 μm silica particles bearing DQ green BSA, pHrodo-SE, and Alexa 647-SE.

1. Once the *xdce* file has been loaded to the IN Cell Investigator software, the individual image sets (*New target type*) corresponding to the individual channels are selected (*Source Images*) and named (e.g., enter DQ green BSA for FITC channel).

2. From the options under the channel name, the appropriate representative color is assigned (e.g., green for DQ green BSA). To enhance the accuracy of the automated detection of the phagocytosed beads, the *Filled* targets, and *Remove single pixel targets* boxes are selected under *Target Detail*.

3. Under *Segmentation, nuclear* detection is selected as this predefined algorithm accurately identifies 3.0 μm beads (*see* **Note 18**).

4. The *Minimum target area* is changed to match the bead size of 3 μm (mean diameter). The bead position and area that will be subject to analysis can be previewed with the *Preview selected operation* option. This allows manual validation that all the beads have been correctly identified and should be performed for at least two fields of view.

5. The *sensitivity* is adjusted depending on the fluorescent signal of the bead. The authors use a value between 90 and 95 to yield maximal coverage of bead fluorescence without interference from background noise.

6. Under *Postprocessing, Fill Holes* is selected as it is useful if the core of the particle is nonfluorescent (e.g., solid-core polystyrene beads).

7. To remove background signal from the edge of the images, *Border Object Removal* is selected and adjusted to a 10 pixel distance from all borders.

8. To obtain a numerical value for fluorescent intensity for each bead at each time point, the *Dens-Levels* option is selected under the *Measures* option for each channel (*see* **Note 19**).

9. Repeat **steps 1–8** for DQ Green BSA and pHrodo-SE by creating two new target sets.

10. Over time, cells and phagosomes will move within each field of view so in order to quantify bead-associated fluorescence over time the cell (bead) tracking function is required. The calibration fluor, Alexa 647-SE is selected as the *Track Target Set* (*see* **Note 20**). For the 3.0 μm beads, *Use Intensity as extra tracking* is enabled as an additional tracking parameter. By default the *Outlier detection* has a value of 4 and the *Tracking Method* is preset as proximity.

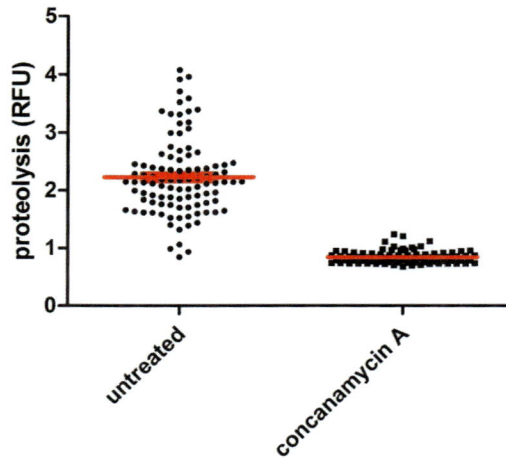

Fig. 1 Assessment of phagosomal proteolysis in BMMØs at 60 min following phagocytosis of silica particles coupled to DQ green BSA. The V-ATPase inhibitor concanamycin A is used to demonstrate the dependence of phagosomal proteolysis on acidification of the phagosomal lumen. Each point represents the degree of substrate hydrolysis within an individual phagosome at 60 min

11. Once all the parameters are set and the protocol is saved, analysis is commenced through selection of the *New Analysis Wizard* and application of the preprepared protocol. The data can be exported in Microsoft Excel format to allow the analysis program to generate a copy of all quantified parameters in a Microsoft Excel file.

3.6 Using Spotfire DecisionSite to Visualize Data in Graphical Format

Once the IN Cell Developer analysis is complete, the software automatically exports the data as a Microsoft Excel file in addition to IN Cell Analyzer 2000 specific files. At this point, basic statistical analysis of multiple or individual parameters can be performed. For instance, the degree of phagosomal heterogeneity in a parameter can be quantified at one point in time, or over a period of time, using basic functions in Microsoft Excel (Fig. 1).

1. The Spotfire DecisionSite, graphical software embedded in the IN Cell Developer software, is selected, allowing conversion of the data generated in the IN Cell Developer software to a graphical format.

2. Once connected to the Spotfire DecisionSite software, the fluorescence change in each particle over time is viewed as a 3D plot.

3. When plotting the 3D graph, the left axis is set to display the fluorescence channel of interest, the top axis for *Linked TrackID* and the bottom axis to show *This Time Index*.

4. Line connection is selected between *Linked TrackID* and *This Time Index*, so each fluorescent bead can be cross-referenced on the 3D plot and on the original image by selecting any data point or area on an image. An example of 3D plots generated in Spotfire DecisionSite and the corresponding images from the IN Cell Developer software is shown in Fig. 2.

4   Notes

1. 96 Well µClear black clear bottom (Greiner Bio-One) plates are used for live-cell imaging in this protocol. Alternatively, other IN Cell Analyzer 2000 compatible microplates could be used.

2. Gelatin may be slow to dissolve at room temperature. It can be heated up gently in a small volume of DPBS before transferring to a final 1 L of DPBS and filtering. FBS can be used as a substitute for gelatin but will result in a higher background in many fluorescent channels.

3. The authors prefer the 3.0 µm COOH-modified silica particles due to their ease of handling and low autofluorescence. Alternatively, 1–5 µm silica or polystyrene particles may be used. It should be noted that other macrophage-like cell lines do not efficiently phagocytose particles with diameters over 3 µm. Therefore particles larger than 3 µm are not advised for those cell types. The IN Cell Investigator analysis software is capable of measuring fluorescence in particles of various sizes but a higher degree of error was found when using particles smaller than 2.0 µm.

4. Equivalent automated live cell imaging systems are offered by PerkinElmer, Leica, Olympus, and Molecular Devices. We have specifically chosen the IN Cell Analyzer 2000 system (GE Healthcare Life Sciences) because of its ability to track individual phagosomes in its platform software in a high-content format.

5. When working with the silica beads, it is imperative to use low binding polypropylene tubes, otherwise the beads may adhere to the tube.

6. The DQ Green BSA/ IgG mix can be reused for the labeling of 2–3 other batches of beads. Store mix at −20 °C.

7. After DQ Green BSA conjugation, the beads should be deep orange in color.

8. DQ Green BSA, once de-quenched has excitation λ/emission λ of 505/515 nm. The spectral characteristics of pH rodo and Alexa 647 are 560/585 nm and 650/665 nm (excitation λ/ emission λ), respectively.

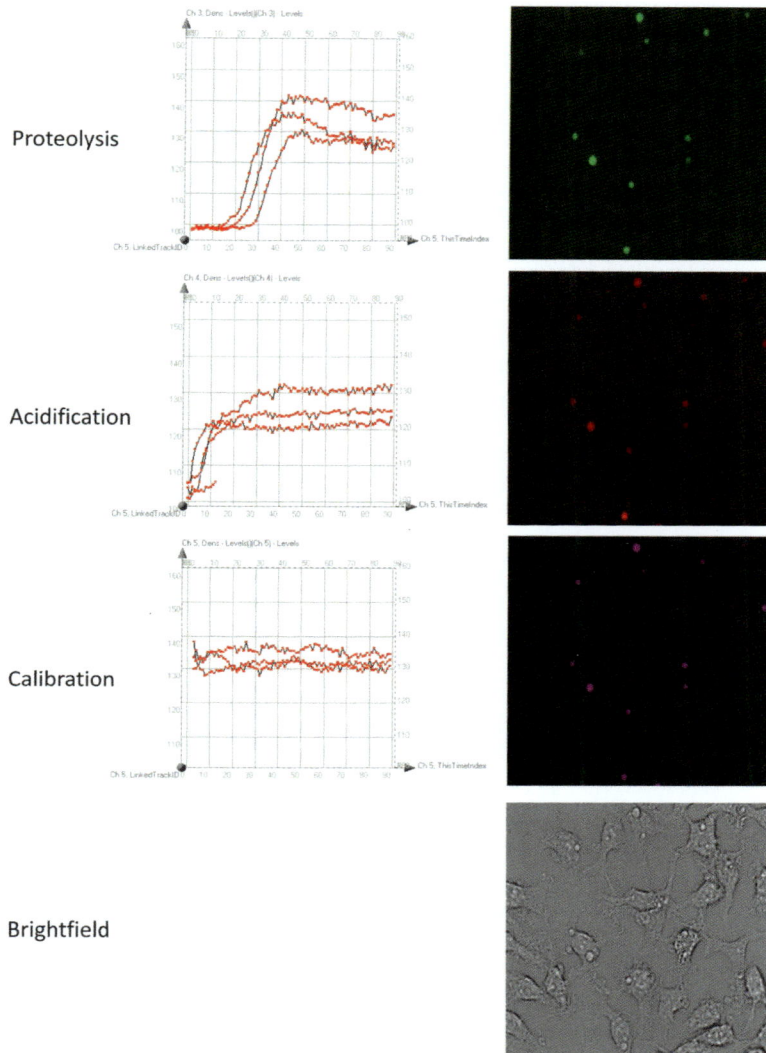

Fig. 2 Assessment of proteolytic efficiency and acidification of individual phagosomes in BMMØs over a 90 min period following phagocytosis. Relative fluorescent intensities of DQ Green BSA (proteolysis), pHrodo-SE (acidification), and Alexa 647-SE (calibration) of individual phagosomes are shown alongside a corresponding fluorescent image (at 90 min). The Spotfire DecisionSite software package was used to generate the graphs

9. Although macrophage-like cell lines can be used, primary BMMØs are generally preferred for their enhanced phagocytic proficiency and adhesion.

10. The sample size is limited by a number of factors. In our experimental platform, a 3-min interval with acquisition of four channels imposes a maximum 48 wells per assay.

11. Additional BMMØ medium and/or cytokines/treatments can be added at this stage.

12. Image acquisition can be performed using the IN Cell Analyzer 6000 system or other automated microscopes. Although the IN Cell Analyzer 6000 generates higher resolution images, we found that the acquisition time for each field of view is much longer compared to the IN Cell Analyzer 2000, decreasing the resolution of time, or the number of samples that can be included in one assay.

13. A list of *Plate/Slide Selection* is preprogrammed in the acquisition software. However, plates/slides that are not on the list may be used, but the focal plane and well coordinates must be entered manually.

14. It is important to pre-warm the imaging chamber to avoid undesirable movement of the plate due to thermal changes during acquisition.

15. The authors found that the 40× objective yields the best results for the study of phagosomes containing 3 μm beads. Using lower-power objectives will result in the loss of a significant amount of cellular detail, while higher-magnification objectives will only cover a small population of cells.

16. The number of beads required for each well can be empirically determined through the serial dilution method.

17. Particles may settle faster with a smaller volume. The rate of settling is also dependent on the particle size and density.

18. Since the 3 μm silica particles are uniformly spherical, the *nuclear* function is an appropriate predefined algorithm to identify the beads in this assay.

19. The *measure option* also allows the quantification of total fluorescence in each image by choosing the *Mean* function, if desired.

20. It is important to perform the tracking calculations using a constant fluorescence signal (i.e., the reference fluor Alexa 647-SE).

References

1. Vieira OV, Botelho RJ, Grinstein S (2002) Phagosome maturation: aging gracefully. Biochem J 366:689–704

2. Kinchen JM, Ravichandran KS (2008) Phagosome maturation: going through the acid test. Nat Rev Mol Cell Biol 9:781–795

3. Desjardins M, Huber LA, Parton RG et al (1994) Biogenesis of phagolysosomes proceeds through a sequential series of interactions with the endocytic apparatus. J Cell Biol 124:677–688

4. Schlam D, Bohdanowicz M, Chatgilialoglu A et al (2013) Diacylglycerol kinases terminate diacylglycerol signaling during the respiratory burst leading to heterogeneous phagosomal NADPH oxidase activation. J Biol Chem 288: 23090–23104

5. VanderVen BC, Yates RM, Russell DG (2009) Intraphagosomal measurement of the magnitude and duration of the oxidative burst. Traffic 10:372–378

6. Yates RM, Hermetter A, Russell DG (2009) Recording phagosome maturation through the real-time, spectrofluorometric measurement of hydrolytic activities. Methods Mol Biol 531: 157–171

7. Yates RM, Russell DG (2008) Real-time spectrofluorometric assays for the lumenal environment of the maturing phagosome. Methods Mol Biol 445:311–325

8. Rybicka JM, Balce DR, Khan MF et al (2010) NADPH oxidase activity controls phagosomal proteolysis in macrophages through modulation of the lumenal redox environment of phagosomes. Proc Natl Acad Sci U S A 107: 10496–10501

9. Rybicka JM, Balce DR, Chaudhuri S et al (2012) Phagosomal proteolysis in dendritic cells is modulated by NADPH oxidase in a pH-independent manner. EMBO J 31: 932–944

10. Balce DR, Yates RM (2013) Redox-sensitive probes for the measurement of redox chemistries within phagosomes of macrophages and dendritic cells. Redox Biol 1:467–474

Chapter 16

Isolation and Western Blotting of Latex-Bead Phagosomes to Track Phagosome Maturation

Anetta Härtlova, Julien Peltier, Orsolya Bilkei-Gorzo, and Matthias Trost

Abstract

Phagocytosis plays an essential role in the immune system for the defense against invading microorganisms and the clearing of apoptotic cells. After internalization, the newly formed phagosome is constantly remodeled by fusion with early endosomes, late endosomes, and lysosomes. These changes ultimately deliver the engulfed material into the terminal degradative compartments known as phagolysosomes. However, defective phagosome maturation can result in inflammatory or autoimmune disease depending on the type of phagosome cargo. Therefore, characterization of the components involved in phagosome formation and maturation is important for a better understanding of macrophage physiological and pathological functions. In this chapter we describe a step-by-step protocol for the isolation of large-scale latex/polystyrene bead phagosome preparations with high degrees of purity for Western blotting analysis of phagosome maturation.

Key words Phagosomes, Organelle purification, Phagocytosis, Western blotting

1 Introduction

Phagocytosis is the essential component of the innate immunity. It is the process by which foreign particles (>0.5 μm) including microbial pathogens, apoptotic cells, and cellular debris are internalized and eliminated by phagocytic cells [1]. It is an important cellular mechanism for almost all eukaryotes, highly conserved in evolution [2], and, in mammals, a key part of the innate immune response to invading microorganisms. During homeostasis, development, and wound healing, macrophages phagocytose apoptotic cells and cell debris to recycle cellular building blocks. Phagocytosis is induced through binding of particles to cell surface receptors [3] such as scavenger receptors, which induces cytoskeletal changes for the uptake of the particle. After internalization, newly formed phagosomes engage in a maturation process that involves fusion with various organelles, including endosomes and ultimately lysosomes [4, 5]. In these phagolysosomes the foreign matter is degraded. In case of exogenous particles, antigens are presented

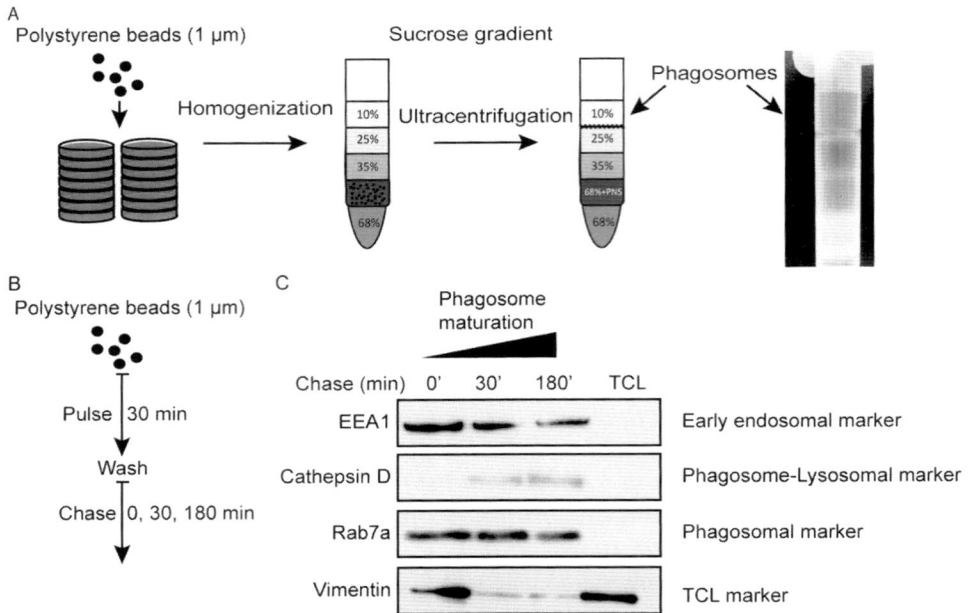

Fig. 1 Isolation and western blot analysis of latex-bead containing phagosomes. (**a**) Schematic illustration of latex-bead phagosome preparation isolated by flotation on a sucrose step gradient. (**b**) Scheme of phagocytosis and phagosome preparation experimental setup. (**c**) Immunoblot analysis of EEA1, Cathepsin D, Rab7a, and vimentin characterizing the particular stages of phagosome maturation

via MHC class I and II molecules, bridging innate and adaptive immunity [6].

To better understand phagosome biology, a method for isolation of intact and highly pure phagosomes is required. Phagocytosis of low-density latex or polystyrene beads enables us to purify bead-containing phagosomes via flotation in sucrose-step gradients [7, 8]. These phagosome preparations give us highly pure phagosomes reflecting the high level complexity of the process and avoiding contaminations from other organelles. By isolating latex/polystyrene bead phagosomes from different time-points, different stages of phagosomal maturation can be tracked by Western blotting analysis of the endosomal and lysosomal markers such as early endosomal marker 1 (EEA1), lysosomal hydrolases such as cathepsins and Rab7 [9] (Fig. 1).

2   Materials

2.1   Cell Culture

1. Dulbecco's modified Eagle's medium with glutamine (DMEM).

2. DMEM supplemented with 10% heat-inactivated fetal bovine serum, 1% (v/v) L-glutamine, and 1% (v/v) penicillin–streptomycin.

3. Bone marrow derived macrophages or cultured macrophage-like cell lines such as RAW264.7 or J774 [10] grown in 21 cell culture dishes (150 mm) in a 37 °C incubator with 5 % CO_2.

4. 1× phosphate buffered saline (PBS), without calcium or magnesium, pH 7.2.

2.2 Induction of Phagocytosis and Cell Lysis

1. 1× PBS.

2. Latex beads: 0.8–1 μm diameter, non-modified, -COOH- or -NH_2-surface (*see* **Note 1**), dyed (*see* **Note 2**).

3. Dounce homogenizer.

4. Protease and Phosphatase Inhibitor Cocktail Tablets.

5. 300 mM imidazole solution, pH 7.4.

6. Hypotonic buffer (HB): 250 mM sucrose and 3 mM imidazole solution in deionized water.

2.3 Phagosome Purification

1. 500 ml of 10 % (w/v) sucrose solution in water. Add 5 ml of 300 mM imidazole solution. Filter through 0.45 μm filter. Store at 4 °C.

2. 500 ml of 25 % (w/v) sucrose solution in water. Add 5 ml of 300 mM imidazole solution. Filter through 0.45 μm filter. Store at 4 °C.

3. 500 ml of 35 % (w/v) sucrose solution in water. Add 5 ml of 300 mM imidazole solution. Filter through 0.45 μm filter. Store at 4 °C.

4. 500 ml of 68 % (w/v) sucrose solution in water. Add 5 ml of 300 mM imidazole solution. Store at 4 °C.

5. Optima™ L-90 K ultracentrifuge (Beckman Coulter) or equivalent model.

6. SW 41Ti swinging bucket rotor (Beckman Coulter) or equivalent model.

7. Ultra-clear centrifuge tubes, 13.2 ml capacity (catalog number: 344059) (Beckman Coulter).

8. Polyallomer centrifuge tubes, 13.2 ml capacity (catalog number: 344059) (Beckman Coulter).

2.4 SDS Polyacrylamide Gel Electrophoresis (SDS-PAGE)/ Components

This chapter describes SDS PAGE components using Nu-PAGE gel system that gives optimal and reproducible separation of a broad molecular weight range of proteins under denaturing conditions during gel electrophoresis but other systems are also acceptable to use.

1. Nu-PAGE™ Novex™ 4–12 % Bis-Tris Protein Gels, 1.0 mm, 10-well (catalog number: NP0301) (Invitrogen) (*see* **Note 3**).

2. 1× Running buffer.

3. 2× NuPAGE® LDS lysis buffer: mix the equal amount of NuPAGE® LDS lysis buffer (4×) (catalog number: NP0007) (Invitrogen) and deionized water.

4. 100 µl of 500 mM dithiothreitol (DTT) solution in water.

2.5 Immunoblotting

1. Immobilon-P membrane PVDF, 0.45 µm.

2. Western blot transfer buffer: 25 mM Tris, 0.192 M glycine, and 20% (v/v) methanol.

3. Tris buffered saline (TBS, 10×): 1.5 M NaCl, 0.1 M Tris–HCl, pH 7.4.

4. 1× TBS containing 0.1% (v/v) Tween 20 (1× TBST): mix 100 ml of 10× TBS, 900 ml deionized water and 1 ml Tween 20.

5. Ponceau S red solution.

6. Blocking solution: 5% (w/v) milk in 1× TBST: dissolve 5 g of nonfat dried-milk powder in 100 ml 1× TBST.

7. Mini-Cell Electrophoresis system or equivalent.

8. Mini Trans-Blot® Cell or equivalent.

2.6 Immuno-detection

1. Diluent solution (primary antibody): 3% (w/v) BSA in 1× TBST; Store are at 4 °C.

2. Diluent solution (secondary antibody): 5% (w/v) milk in TBST.

3. Prepare ECL substrate according to the manufacturer's instructions.

4. Autoradiography films or a camera-based gel-documentation system.

3 Methods

3.1 Induction of Phagocytosis

1. Grow adherent macrophages to 80% confluency (21 cell culture dishes, 150 mm) in complete DMEM medium.

2. Sonicate the polystyrene polymeric microspheres (beads) in their original bottle in a sonicator bath for 20 s.

3. Transfer 2.1 ml of beads into 2.0 ml microcentrifuge tubes.

4. Centrifuge beads ($7000 \times g$, 5 min) and wash beads 3 times with 1× PBS. Sonicate beads 3 min in a sonicator bath before the experiment.

5. Add 2.1 ml of beads in 105 ml of pre-warmed complete DMEM (37 °C) in a sterile bottle to get a final dilution of beads 1:50 (v/v) and mix it.

6. Remove old DMEM medium from the culture dishes by aspiration, wash cells with 5 ml of pre-warmed 1× PBS and then add 5 ml of the beads–medium mixture to the cells, always

proceed with seven plates together. Afterwards, put the cells back into the incubator and incubate them at 37 °C for 30 min to internalize the beads (*see* **Note 4**).

3.2 Cell Lysis

All the processes in this section, including centrifugation step should be performed on ice or at 4 °C.

1. After 30 min incubation, place the dishes on ice, remove the medium by aspiration, and then wash cells with 5 ml of ice-cold 1× PBS to wash non-internalized beads. During washes, the dishes on ice buckets are placed on a shaker (*see* **Note 5**).

2. Scrape the cells with a rubber policeman, collect the cells into 50 ml conical centrifuge tubes (7 dishes per 50 ml tube), and centrifuge them at ~500×g for 3 min at 4 °C, followed by removing the supernatant by aspiration.

3. Resuspend cells with 10 ml 1× PBS and transfer them into 15 ml conical centrifuge tubes and centrifuge as before. Repeat the wash step until supernatant is clear (*see* **Note 6**).

4. Resuspend cells carefully with 3.0 ml of ice-cold HB buffer containing protease and phosphatase inhibitors, and centrifuge them at ~500×g for 3 min at 4 °C, followed by removing the supernatant by aspiration (*see* **Note 7**).

5. Resuspend the cells carefully with 3.0 ml of ice-cold HB buffer containing protease and phosphatase inhibitors, and then transfer the suspension into a Dounce homogenizer precooled on ice.

6. Homogenize the cell suspension by Dounce homogenization (start with 12 strokes). Verify the disruption efficiency of the cells by trypan blue staining and light microscopy. Carry out more strokes, if necessary, until about 90 % of cells are broken. Nuclei should remain intact.

7. Centrifuge the cell homogenate by centrifugation at ~1000×g for 5 min at 4 °C to remove nuclei and unbroken cells.

3.3 Phagosome Purification

All the processes in this section, including centrifugation steps should be performed on ice or at 4 °C.

1. Collect carefully the supernatant (the post-nuclear fraction containing phagosomes) (*see* **Note 8**) and divide into three 2-ml microfuge tubes (0.8 ml/tube). The phagosome suspension is brought to 38 % sucrose by adding 1 ml of 68 % sucrose + cocktails of protease and phosphatase inhibitors into each 2-ml microcentrifuge tube. Mix the suspension by inverting the tube.

2. This phagosome containing mixture is layered on top of a 1 ml cushion of 68 % of sucrose into a SW41 Ultra-clear centrifuge

tube, followed by addition of 2 ml of 35 % sucrose, 2 ml of 25 % sucrose, and 2 ml of 10 % sucrose solutions (*see* **Note 9**).

3. Centrifugation is performed at 4 °C in a swinging bucket (SW41, Beckman Instruments) at $72,300 \times g$ (24,000 rpm) for 1 h.

4. Collect phagosomes (a band colored in blue) carefully from the interface of the 10 and 25 % sucrose layers and resuspend them in 12 ml ice-cold 1× PBS containing protease and phosphatase inhibitors. Phagosomes from two tubes can be pooled together for the next centrifugation (*see* **Note 10**).

5. Pellet phagosomes by centrifugation at $28,400 \times g$ (15,000 rpm) in a SW41 swinging bucket rotor at 4 °C for 15 min.

6. Remove the supernatant and resuspend the phagosome pellet in 50 µl of 2× LDS buffer containing DTT. It is important to remove beads sticking to the side of the tube and wipe dry the wall of the tube before adding the lysis buffer (*see* **Notes 11** and **12**).

3.4 Phagosome Solubilization and SDS-PAGE Electrophoresis

1. Mix the sample, and then heat it to 70 °C for 10 min. Perform a quick centrifugation before protein concentration determination.

2. Perform an EZQ protein assay (Thermo Fisher Scientific) or similar according to manufacturer's instructions to determine protein concentration (*see* **Note 13**).

3. Set up an electrophoresis apparatus and immerse in 1× running buffer. Remove gel combs and clean wells of any residual stacking gel.

4. Load samples and protein markers into the gel wells using gel-loading tips. Set electrophoresis power pack to 80 V (through the stacking gel), before increasing it to 120 V when the protein front reaches the separation gel.

3.5 Protein Transfer

1. Cut the PVDF membranes to the size of the gel and soak them in methanol for 30 s before moving to transfer buffer. Soak the filter papers and sponges in transfer buffer as well.

2. Sequentially assemble the transfer constituents according to the manufacturer's instructions and ensure no bubbles lie between any of the layers. Apply 90 V for 2 h at 4 °C.

3.6 Immunoblotting

1. After transfer, wash the membrane twice with 1× TBST, and stain the membrane with commercial Ponceau S red solution for 1 min to visualize protein bands, then wash any Ponceau S red staining with copious amounts of 1× TBST.

2. Block the membrane with blocking solution for 1 h.

3. After blocking, wash the membranes with 1× TBST for 5 min.

3. Dilute primary antibody in dilution solution with a starting dilution ratio of 1:1000 (optimal dilutions should be determined experimentally). Incubate the membrane with primary antibody overnight at 4 °C.

4. Wash membrane three times with 1× TBST for 10 min each.

5. Incubate the membrane with a suitable HRP-conjugated secondary antibody (recognizing the host species of the primary antibody), diluted at 1:5000–1:50,000 in blocking solution. Incubate for 1 h with constant rocking.

6. Wash membrane three times with 1× TBST for 10 min each.

3.7 ***Signal Detection***

1. Prepare ECL substrate according to the manufacturer's instructions.

2. Incubate the membrane completely with substrate for 1–5 min (adjust time for more sensitive ECL).

3. Expose the membrane to autoradiography film in a dark room or read using a chemiluminescence imaging system.

4   Notes

1. Different bead surfaces will trigger different phagocytic receptors. In our hands, carboxylated beads which likely trigger scavenger receptors (Guo et al. submitted) are taken up most efficiently.

2. Dyed beads will make it considerably easier to follow and to collect from sucrose gradients.

3. NuPAGE® Bis-Tris Gels precast polyacrylamide gels can be replaced by homemade SDS-PAGE gels.

4. Timing needs to be very accurate as small differences in time can result in substantial changes of the phagosome proteome.

5. To obtain more matured phagosome/phagolysosomes, cells are further incubated for a chase period in fresh medium. Chase periods depend on the protein target of interest, which might be recruited to phagosomes at early or later time points after initiation of phagocytosis.

6. Usually wash cells three times with 1× PBS.

7. This buffer is hypotonic and thus, cells will swell and break easily.

8. Use a small Pasteur pipette with a long, fine tip. Make sure not to take up any of the soft pellet of unbroken cells/nuclei.

9. The sucrose gradient is gently loaded with 3-ml Pasteur pipette. Avoid air bubbles.

10. The phagosomes are collected with a small Pasteur pipette with a long, fine tip.

11. The phagosome pellet can be frozen at −80 °C for storage.

12. In case of homemade SDS-PAGE gels, use 2× Laemmli buffer instead of 2× NuPAGE® LDS lysis buffer.

13. Protein amounts need to be quantified by reducing agent compatible assay kits because of the presence of DTT.

Acknowledgments

This work was funded by the Medical Research Council UK and the pharmaceutical companies supporting the Division of Signal Transduction Therapy (DSTT) (AstraZeneca, Boehringer-Ingelheim, GlaxoSmithKline, Janssen Pharmaceutica, Merck KGaA, and Pfizer).

References

1. Underhill DM, Goodridge HS (2012) Information processing during phagocytosis. Nat Rev Immunol 12(7):492–502. doi:10.1038/nri3244

2. Boulais J, Trost M, Landry CR, Dieckmann R, Levy ED, Soldati T et al (2010) Molecular characterization of the evolution of phago-somes. Mol Syst Biol 6:423. doi:10.1038/msb.2010.80, msb201080 [pii]

3. Dill BD, Gierlinski M, Hartlova A, Arandilla AG, Guo M, Clarke RG et al (2015) Quantitative proteome analysis of temporally resolved phago-somes following uptake via key phagocytic recep-tors. Mol Cell Proteomics 14(5):1334–1349. doi:10.1074/mcp.M114.044594

4. Desjardins M, Huber LA, Parton RG, Griffiths G (1994) Biogenesis of phagolysosomes pro-ceeds through a sequential series of interac-tions with the endocytic apparatus. J Cell Biol 124(5):677–688

5. Campbell-Valois FX, Trost M, Chemali M, Dill BD, Laplante A, Duclos S, et al. Quantitative proteomics reveals that only a subset of the endoplasmic reticulum contributes to the phagosome. Mol Cell Proteomics 2012;11(7):M111 016378. doi: 10.1074/mcp.M111.016378.

6. Botelho RJ, Grinstein S (2011) Phagocytosis. Curr Biol 21(14):R533–R538. doi:10.1016/j.cub.2011.05.053

7. Desjardins M, Griffiths G (2003) Phagocytosis: latex leads the way. Curr Opin Cell Biol 15(4):498–503

8. Trost M, English L, Lemieux S, Courcelles M, Desjardins M, Thibault P (2009) The phago-somal proteome in interferon-gamma-activated macrophages. Immunity 30(1):143–154, doi: S1074-7613(08)00548-7 [pii] 10.1016/j.immuni.2008.11.006

9. Kinchen JM, Ravichandran KS (2008) Phagosome maturation: going through the acid test. Nat Rev Mol Cell Biol 9(10):781–795. doi:10.1038/nrm2515

10. Guo M, Hartlova A, Dill BD, Prescott AR, Gierlinski M, Trost M (2015) High-resolution quantitative proteome analysis reveals substan-tial differences between phagosomes of RAW 264.7 and bone marrow derived macrophages. Proteomics 15(18):3169–3174. doi:10.1002/pmic.201400431

Chapter 17

Assessing the Phagosome Proteome by Quantitative Mass Spectrometry

Julien Peltier, Anetta Härtlova, and Matthias Trost

Abstract

Phagocytosis is the process that engulfs particles in vesicles called phagosomes that are trafficked through a series of maturation steps, culminating in the destruction of the internalized cargo. Because phagosomes are in direct contact with the particle and undergo constant fusion and fission events with other organelles, characterization of the phagosomal proteome is a powerful tool to understand mechanisms controlling innate immunity as well as vesicle trafficking. The ability to isolate highly pure phagosomes through the use of latex beads led to an extensive use of proteomics to study phagosomes under different stimuli. Thousands of different proteins have been identified and quantified, revealing new properties and shedding new light on the dynamics and composition of maturing phagosomes and innate immunity mechanisms. In this chapter, we describe how quantitative-based proteomic methods such as label-free, dimethyl labeling or Tandem Mass Tag (TMT) labeling can be applied for the characterization of protein composition and translocation during maturation of phagosomes in macrophages.

Key words Maturation of phagosomes, Macrophages, Label free, Isotope labeling, Quantitative proteomics, Phagosome dynamics, Dimethyl, TMT, Mass spectrometry

1 Introduction

Phagocytosis, the mechanism by which particles are internalized, leads to the formation of a specialized organelle called the phagosome. After uptake, phagosomes mature by the fusion of the nascent organelle with early endosomes, late endosomes and ultimately lysosomes forming a phagolysosome in which the engulfed material is degraded [1]. This process is important for tissue homeostasis, innate immunity and adaptive immunity through antigen presentation from the phagosome. Numerous studies have contributed to the understanding of the importance of many factors in phagosome maturation including small GTPases, phosphoinositol kinases, signaling, and actin dynamics [2, 3]. However, the mechanisms of uptake and maturation as well as signaling from the phagosome are still relatively poorly understood. In recent

Roberto Botelho (ed.), *Phagocytosis and Phagosomes: Methods and Protocols*, Methods in Molecular Biology, vol. 1519,
DOI 10.1007/978-1-4939-6581-6_17, © Springer Science+Business Media New York 2017

years, proteomics approaches of phagosomes led to a better understanding of these molecular mechanisms occurring during phagosome maturation. For example, proteomic analysis of phagosomes showed for the first time translocation of the endoplasmic reticulum (ER) to the phagosome membrane [4, 5], providing new potential mechanisms of antigen cross-presentation [6, 7]. It further provided clues how macrophage activation changed phagosome dynamics for the gain of antigen presentation [8]. Moreover, proteomics of phagosomes from Dictyostelium, Drosophila, and mouse showed strong conservation of a core phagosome proteome over evolution [9].

Historically, two-dimensional gel electrophoresis (2DE) coupled to matrix-assisted laser desorption/ionization time-of-flight (MALDI-TOF) or electro-spray ionization mass spectrometry (ESI-MS) were the first methods to study complex protein mixtures like phagosomes [10, 11]. However, 2DE-MS had inherent limitations such as limited sample capacity and sensitivity, restricting the identification of low abundance proteins as well as an incompatibility for transmembrane proteins. To overcome these critical limitations, the last decade has seen the development of reproducible quantitative proteomic approaches using sensitive data-dependent tandem mass spectrometry analysis (LC-MS/MS) that now allow the systematic identification and quantification of thousands of proteins in complex proteomes.

Quantitative data from proteomics come in two forms, the absolute quantity of the protein or the relative change of the protein in two samples. Relative quantification is mostly used in the discovery phase, comparing levels of a protein in different states with results being expressed as a relative fold change of protein abundance. Absolute quantification refers to the exact determination of the amount of a protein in question (e.g. ng/ml) and is mostly used in targeted proteomics approaches. Differential analysis is generated from LC-MS/MS experiments and can be carried out using either label-free or isotope labeling approaches such as stable isotope labeling by amino acids in cell culture (SILAC) [12], dimethyl labeling [13], or Tandem Mass Tag™ (TMT) reagents. Exciting progress in multiplexing allows us to compare currently up to 10 experimental conditions in one single LC-MS/MS analysis. Application of such analytical methods could resolve fundamental questions about molecular mechanisms in phagosome biogenesis [8, 14, 15].

In this chapter, we describe reliable quantitative proteomic strategies for studying the phagosome proteome. Important considerations in every step of the sample preparation, dimethyl/TMT labeling, and mass spectrometry analysis are described and discussed in further details. We do not discuss SILAC labeling strategies for phagosome proteomics as growth of macrophage cell lines such as RAW264.7 in dialyzed serum seems to induce autophagy and thus affecting phagosome proteome composition (data not shown).

2 Materials

2.1 Stock Solutions

1. HPLC grade water.
2. HPLC grade acetonitrile.
3. Lysis buffer: 1 % Sodium 3-[(2-methyl-2-undecyl-1,3-dioxolan-4-yl)methoxy]-1-propanesulfonate (commercially available as RapiGest, Waters) 50 mM Tris pH 8.0 and 1 mM TCEP (Tris(2-carboxyethyl)phosphine hydrochloride) in HPLC grade water.
4. 100 mM Triethylammonium bicarbonate buffer (TEAB) in HPLC grade water, store at 4 °C.
5. 1 % Ammonium hydroxide solution (NH$_4$OH, 28.0–30.0 % NH$_3$ basis) in HPLC grade water.
6. 5 % Hydroxylamine solution (HDA, NH$_2$OH) in HPLC grade water.
7. Trifluoroacetic Acid solution (TFA).
8. Formic Acid solution (FA).
9. 500 mM Iodoacetamide in HPLC grade water (*see* **Note 1**).
10. 500 mM Dithiothreitol (DTT) in HPLC grade water.
11. 50 mM acetic acid in HPLC grade water.
12. Formaldehyde, CH$_2$O (36.5–38 % in H$_2$O).
13. Formaldehyde-d2, CD$_2$O (~20 wt% in D$_2$O, 98 atom % D).
14. Solubilization Buffer for mass spectrometry analysis: 2 % HPLC grade acetonitrile, 0.1 % TFA in HPLC grade water, store at 4 °C.

2.2 Material Preparation

1. Protein LoBind tubes 1.5 ml.
2. 100 µg of MS grade porcine trypsin in 100 µl of 50 mM acetic acid. Aliquot trypsin and store at –80 °C.
3. Sodium cyanoborohydride, NaBH$_3$CN (M = 62.84 g/mol) (*see* **Note 2**).
4. Isobaric Tandem Mass Tag™ reagents (TMT, Thermo Scientific); duplex TMT, 6plex TMT, or 10plex TMT.
5. Vacuum centrifuge with cold trap.
6. C18 Solid Phase Extraction columns (C18 SPE).
7. Washing buffer: 90 % (v/v) acetonitrile + 0.1 % (v/v) TFA in HPLC grade water.
8. Loading buffer: 0.1 % (v/v) TFA in HPLC grade water.
9. Elution buffer: 50 % (v/v) acetonitrile + 0.1 % (v/v) TFA in HPLC grade water.

2.3 LC-MS/MS Analysis

1. Orbitrap Fusion Tribrid Mass Spectrometer (Thermo Scientific) with an EASY-Spray Ion Source (Thermo Scientific) (*see* **Note 3**).

2. Liquid chromatography; Dionex Ultimate 3000 RSLC-nano System (Thermo Scientific).

3. Buffer A: HPLC grade water + 0.1% formic acid.

4. Buffer B: 80% HPLC grade acetonitrile + 0.1% formic acid in HPLC grade water.

5. C18 trap column; Acclaim PepMap® 100, 100 μm × 2 cm nanoViper C18 trap column, 5 μm, 100 Å (Thermo Scientific).

6. C18 analytical column; PepMap RSLC, 75 μm × 50 cm C18 column, 2 μm, 100 Å (Thermo Scientific).

3 Methods

An overview of the common quantitative mass spectrometry workflows used for the study of phagosomes is outlined in Fig. 1.

3.1 Sample Preparation for Quantitative Proteomic Analysis

1. Isolate intact and highly pure phagosomes as described in Chap. 16 or elsewhere [8, 16]. In short, incubate macrophages with 1 μm sized latex/polystyrene beads which are phagocytosed by the cells. Disrupt cells then mechanically and put the cytosolic supernatant at the bottom of a sucrose gradient. After ultracentrifugation, the beads float due to the low density enabling the purification of highly pure phagosomes. In order to remove the sucrose, pellet phagosomes by an additional centrifugation step.

2. Lyse phagosome pellets by the addition of 50–100 μl of lysis buffer. Incubate samples for 30 min at 55 °C on a thermomixer, 1000 rpm.

3. Perform a protein assay according to manufacturer's instructions to determine protein concentration (*see* **Note 4**).

4. Cool down samples on the bench at room temperature and add 500 mM iodoacetamide (1:100 v/v). Incubate again for 30 min at room temperature, protected from light.

5. Quench iodoacetamide by addition of 500 mM DTT (1:100 v/v).

6. Immediately before use, resuspend 100 μg of MS grade porcine trypsin in 100 μl of 50 mM acetic acid.

7. For protein digestion, dilute sample 1:10 (v/v) with 100 mM TEAB pH 8.0, add trypsin to each sample as a ratio of enzyme and substrate 1:100 (w/w) and incubate at 37 °C for 3 h. After 3 h, add another aliquot of trypsin 1:100 (w/w) and incubate overnight at 37 °C (*see* **Note 5**).

A

B

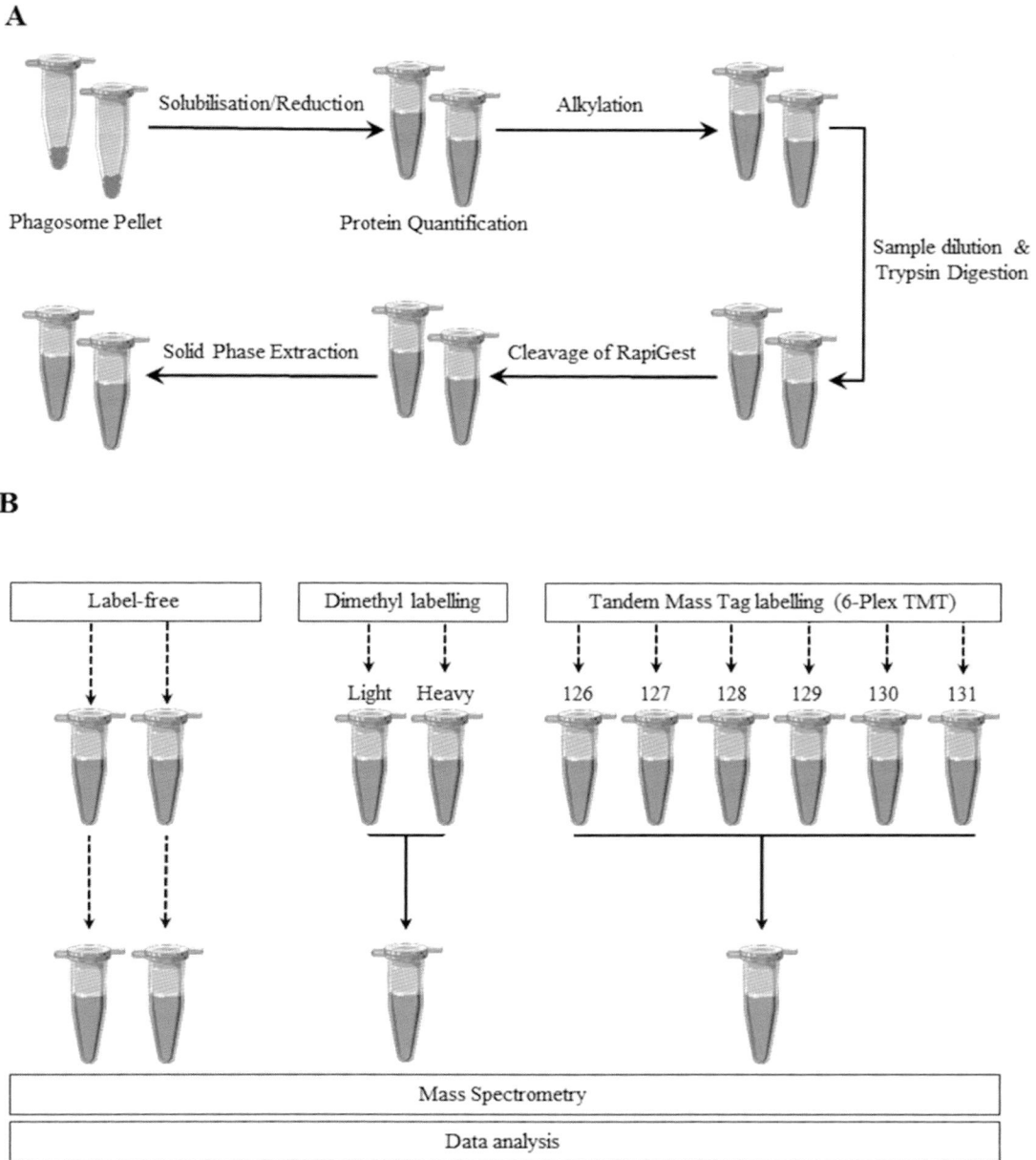

Fig. 1 Schematic of phagosome preparation and proteomics. (**a**) Sample preparation scheme for phagosome proteomics. After solubilization of phagosome pellets, the lysate is reduced by 1 mM TCEP, alkylated with 500 mM iodoacetamide and digested with trypsin. Cleavage of Rapigest is performed by the addition of TFA. Cleanup of the samples is performed at the final stage of the procedure in order to remove all potential contaminants for peptide labeling and mass spectrometry analysis. (**b**) Quantitative mass spectrometry work-flows used for studying phagosome proteomes. Horizontal lines indicate when samples are combined and dashed lines indicate at which point experimental variation and quantification errors may occur. For label-free mass spectrometry analysis, all experimental procedures are processed in parallel until data analysis. In Dimethyl and TMT mass spectrometry experiments, experimental procedures are performed in parallel until proteins have been digested. Each experimental condition is individually labeled by a different chemical stable isotope. From this point, samples are combined for mass spectrometry and data analyses

8. After digestion, acidify samples to 1 % TFA as a final concentration in order to cleave Rapigest. Incubate for 1 h at 37 °C in a thermomixer. Pellet Rapigest and any insoluble material by centrifugation at $14,000 \times g$ for 30 min at 4 °C.

9. In parallel, equilibrate a C18 SPE column (*see* **Note 6**) with the following solutions, using a benchtop centrifuge at $0.1 \times g$ for 3 min at room temperature (*see* **Note 7**). Wash the C18 SPE column by adding 300 μl of the washing buffer. Remove the flow-through and equilibrate the C18 SPE column with 300 μl of the loading buffer. Again, remove the flow-through, load peptides onto the C18 SPE column and collect column eluent into 1.5 ml tubes. Wash the C18 SPE column two times with the loading buffer.

10. Peptides are eluted from the C18 SPE column with 300 μl of the elution buffer. Freeze and dry the fractions using a vacuum concentrator.

11. Store samples at −80 °C (*see* **Note 8**).

12. For label free-based quantitative proteomic strategy, resuspend samples and perform mass spectrometry analysis as described in Subheading 3.4, **step 1**.

3.2 Quantitative Proteomic of Phagosomes Using Dimethyl Labeling

1. Resuspend peptide samples (from Subheading 3.1, **step 10**) in 100 μl of 100 mM TEAB.

2. Prepare fresh working solutions as follow: 108 μl of formaldehyde (CH_2O) is added to 892 μl of HPLC grade water whereas 20 μl of d-formaldehyde (CD_2O) is added to 80 μl of HPLC grade water. From these two stock solutions, 4 μl of the diluted formaldehyde (CH_2O) or d-formaldehyde (CD_2O) are added to the corresponding sample. Vortex briefly (*see* **Note 9**).

3. From this point, all experiments are performed in a chemical hood. Take 37 mg of Sodium cyanoborohydride ($NaBH_3CN$) in a tube and measure on a precision balance. Dissolve the sodium cyanoborohydride in 1 ml of HPLC grade water ($[NaBH_3CN] = 0.59$ M) and keep solution on ice. Add 4 μl of sodium cyanoborohydride to each sample and vortex briefly. Incubate for 30 min using a thermomixer at room temperature, 1000 rpm.

4. Repeat **steps 2** and **3** by using fresh aliquots of formaldehyde, d-formaldehyde, and sodium cyanoborohydride. Again, incubate for 30 min at room temperature using a thermomixer at 1000 rpm.

5. Quench the dimethyl labeling reaction by the addition of 16 μl of 1 % ammonium hydroxide, vortex briefly and place samples at 4 °C.

6. Add 8 μl of 100 % TFA to neutralize ammonium hydroxide.

7. Perform a C18 SPE cleanup with the following solutions, using a benchtop centrifuge at $0.1 \times g$ for 3 min at room temperature. Wash the C18 SPE column by adding 300 μl of the washing buffer. Remove the flow through and equilibrate the C18 SPE column with 300 μl of the loading buffer. Again, remove the flow-through, load peptides onto the C18 SPE column and collect column eluent into 1.5 ml tubes. Wash the C18 SPE column two times with the loading buffer.

8. Dry the fractions using a vacuum concentrator.

9. Store samples at −80 °C.

3.3 Quantitative Proteomic of Phagosomes Using Tandem-Mass-Tag™ Labeling (TMT)

1. Immediately before use, equilibrate one set of TMT Label Reagents to room temperature before opening (*see* **Note 10**).

2. Resuspend peptides samples (from Subheading 3.1, **step 10**) with 30 μl with 100 mM TEAB (pH 8.0). Add 70 μl acetonitrile to each tube, vortex briefly and collect the solution at the bottom of the tube with a microfuge pulse.

3. Transfer each peptide samples into a different TMT labeling vial. Close and vortex the labeling vials. Place on a shaker at room temperature for 1 h to allow labeling of peptides by the isobaric TMT reagents (*see* **Note 11**).

4. Quench the labeling reaction by the addition of 8 μl 5% hydroxylamine to the sample. Vortex, collect the solution at the bottom of the tube with a microfuge pulse and incubate for 30 min (*see* **Note 12**).

5. Pool the two, six or ten differentially labeled samples into 1.5 ml LoBind tube and reduce to dryness by vacuum centrifugation. Resuspend the peptides in each tube in 300 μl water 0.1% TFA and check pH (pH≈2) and perform C18 SPE cleanup with the following solutions, using a benchtop centrifuge at $0.1 \times g$ for 3 min at room temperature. Wash the C18 SPE column by adding 300 μl of the washing buffer. Remove the flow-through and equilibrate the C18 SPE column with 300 μl of the loading buffer. Again, remove the flow-through, load peptides onto the C18 SPE column and collect column eluent into 1.5 ml tubes. Wash the C18 SPE column two times with the loading buffer.

6. Dry the fractions using a vacuum concentrator (*see* **Note 13**).

7. Store samples at −80 °C.

3.4 Mass Spectrometry Analysis of Dimethyl and Label-Free Samples

1. Resuspend labeled fractions in a small volume (~20 μl) of solubilization buffer (2% HPLC grade acetonitrile, 0.1% TFA in HPLC grade water). Load 500 ng–2 μg of peptide digest of each fraction for LC-MS/MS analysis.

2. Set up Dionex Ultimate 3000 RSLC-nano method as follow: Load peptides onto a C18 trap column with 2% acetonitrile at a

flow rate of 5 μl/min, with the post-column flow directed to waste. After 5 min loading time, switch the solvent flow onto the C18 analytical column at 0.3 μl/min flow rate. Peptides will be resolved with a linear gradient of 25% of buffer B over 270 min, followed by a step from 25 to 35% of buffer B over 60 min, 10 min of high organic wash (99% of buffer B) and 15 min re-equilibration at 3% of buffer B. Peptides are eluted from the analytical column, ionized and analyzed in an Orbitrap Fusion mass spectrometer.

3. Set up the Orbitrap Fusion as follows (*see* **Note 14**): set up the instrument in "Top Speed" data dependant mode, operated in positive ion mode. FullScan spectra are acquired in a range from 400 to 1600 m/z, at a resolution of 120,000 (at 200 m/z), with an automated gain control (AGC) (*see* **Note 15**) of 3.0×10^5 and a maximum injection time of 50 ms. Charge state screening is enabled to exclude precursors with a charge state of 1. The intensity threshold for a MS/MS fragmentation is set to 10^4 counts. The most intense precursor ions are isolated with a quadrupole mass filter width of 1.6 m/z and Collision-induced dissociation (CID fragmentation) (*see* **Note 16**) is performed in one-step collision energy of 32% and activation Q of 0.25. MS/MS fragments ions are analyzed in the segmented linear ion trap with a normal scan range, in a rapid mode. The detection of MS/MS fragments is set up as the "Universal Method," using a maximum injection time of 300 ms and a maximum AGC of 10^3 ions.

3.5 Mass Spectrometry Analysis of TMT Samples

1. Samples preparation and Liquid Chromatography for Mass Spectrometry analysis are the same as described in Subheading 3.4, **steps 1** and **2**.

2. For the analysis of TMT-tagged peptides, the Orbitrap Fusion is operated as follow (*see* **Note 17**): the instrument is turn on "Top N" data dependant mode, operated in positive ion mode. FullScan spectra are acquired in a range from 400 to 1600 m/z, at a resolution of 120,000 (at 200 m/z), with an automated gain control (AGC) of 3.0×10^5 and a maximum injection time of 50 ms. Charge state screening is enabled to exclude precursors with a charge state of 1. The intensity threshold for a MS/MS fragmentation is set to 10^4 counts. The 12 most intense precursor ions are isolated with a quadrupole mass filter width of 1.6 m/z and CID fragmentation is performed in one-step collision energy of 32% and 0.25 activation Q. MS/MS fragments ions are analyzed in the segmented linear ion trap with a normal scan range, in a rapid mode. The detection of MS/MS fragments is acquired with an AGC of 10^4 ions and a maximum injection time of 40 ms. Quantitative analysis of TMT-tagged peptides occurs in the last step of analysis by selecting

the Top 10 MS/MS fragment ions for SPS-MS^3 fragmentation and analysis in the Orbitrap mass analyzer. Higher-energy C-trap dissociation (HCD fragmentation) on MS/MS fragments is performed in one-step collision energy of 55% to ensure maximal TMT reporter ion yield. MS^3 spectra are acquired in the Orbitrap mass analyzer with a range from 100 to 500 m/z, at a resolution of 60,000 (at 200 m/z), with an automated gain control (AGC) of 10^5 ions and a maximum injection time of 120 ms.

3.6   Data Processing and Analysis

The primary requirement of data processing is to convert raw files from mass spectrometry analysis into a list of identified peptides, coupled with quantitative information given by the precursor intensity of the identified peptides (current method used for label-free and dimethyl labeling) or given by the reporter masses for TMT labeling. Many software packages have been developed to handle both identification and quantification of proteins such as Proteome Discoverer (Thermo Scientific), Mascot (Matrix Science), Scaffold Q+ (Proteome software), and MaxQuant [17]. The final outcome of these proteomic approaches is a list of proteins and their quantitative information. A brief overview of expected phagosome proteins will be displayed in summary tables.

3.7   Quality Control

It is important to check phagosome purity and check for marker proteins. From experience, latex bead phagosome proteomes usually contain at least 2500–3000 proteins. We have listed the 20 most abundant proteins according to peptide count (Table 1) and intensity based absolute quantitation (iBAQ) (Table 2) from a recent paper describing the phagosome proteomes of RAW264.7 cells and bone marrow-derived macrophages (*see* **Note 18**) [15]. Latex bead phagosomes are generally rich in endosomal, lysosomal, plasma membrane proteins and contain subsets of ER and Golgi proteins and usually few proteins with described mitochondrial localization [5]. Common potential contaminants found in phagosome proteome analyses are: fetal bovine serum proteins, keratins, histones, and ribosomes.

4   Notes

1. Prepare iodoacetamide solution always fresh just before use.

2. Sodium cyanoborohydride is toxic and must be prepared and used in a fume hood. See the Safety Data Sheet (SDS) for further information.

3. While the pipeline described here utilizes the Orbitrap Fusion Tribrid Mass Spectrometer, experiments can also be performed on other high-resolution mass spectrometers such as

Table 1
Identification of the 20 most abundant phagosome proteins according to peptide count

Uniprot ID	Uniprot accession	Protein name	Peptides	Mol. weight (kDa)
VIME_MOUSE	P20152	Vimentin	82	53.7
GRP78_MOUSE	P20029	78 kDa glucose-regulated protein	57	72.4
VATA_MOUSE	P50516	V-type proton A TPase catalytic subunit A	53	68.3
VATB2_MOUSE	P62814	V-type proton A TPase subunit B	48	56.6
ITB2_MOUSE	P11835	Integrin beta-2	46	84.9
PDIA3_MOUSE	P27773	Protein disulfide-isomerase A 3	41	56.7
G3P_MOUSE	P16858	Glyceraldehyde-3-phosphate dehydrogenase	38	35.8
Q9JHF5_MOUSE	Q9JHF5	A 3 subunit of vacuolar-adenosine triphosphatase	38	93.5
CALR_MOUSE	P14211	Calreticulin	36	48.0
ACTB_MOUSE	P60710	Actin, cytoplasmic 1	32	41.7
GNAI2_MOUSE	P08752	Guanine nucleotide-binding protein G(i) subunit alpha-2	31	40.5
STOM_MOUSE	P54116	Erythrocyte band 7 integral membrane protein	29	31.4
RAB7A_MOUSE	P51150	Ras-related protein Rab-7a	27	23.5
RAB14_MOUSE	Q91V41	Ras-related protein Rab-14	27	23.9
VA0D1_MOUSE	P51863	V-type proton ATPase subunit d 1	24	40.3
GPNMB_MOUSE	Q99P91	Transmembrane glycoprotein NMB	23	63.7
CATD_MOUSE	P18242	Cathepsin D	23	45.0
EF1A1_MOUSE	P10126	Elongation factor 1-alpha 1	23	50.1
GBB2_MOUSE	P62880	Guanine nucleotide-binding protein G(I)/G(S)/G(T) subunit beta-2	21	37.3
RAPIB_MOUSE	Q99JI6	Ras-related protein Rap-1b	21	20.8

quadrupole-time of flight or quadrupole-Orbitrap instruments. The major advantages of the hybrid Quadrupole-Orbitrap such as the Fusion Tribrid mass spectrometer are the ability to select ion for fragmentation with high specificity and perform MS^3 experiments to improve quantitative performance (*see* **Note 13**).

4. We recommend EZQ protein assay (Thermo Fisher Scientific) as it is insensitive to reducing agents and detergents.

Table 2
Identification of the 20 most abundant phagosome proteins according to the intensity based absolute quantitation (iBAQ), i.e., estimated absolute abundance

Uniprot ID	Uniprot accession	Protein name	Peptides	Mol. weight (kDa)	iBAQ
CRIP_MOUSE	P63254	Cysteine-rich protein 1	9	8.5	8.02E+09
G3P_MOUSE	P16858	Glyceraldehyde-3-phosphate dehydrogenase	38	35.8	5.62 E+09
RL38_MOUSE	Q9JJI8	60S ribosomal protein L38	8	8.2	4.41E+09
LYZ1_MOUSE	P17897	Lysozyme C-1	15	16.8	3.33E+09
LYZ2_MOUSE	P08905	Lysozyme C-2	5	16.7	3.11 E+09
LTOR5_MOUSE	Q9D1L9	Regulator complex protein LAMTOR5	6	9.6	2.57 E+09
IFM3_MOUSE	Q9CQW9	Interferon-induced transmembrane protein 3	6	15.0	2.35 E+09
VIME_MOUSE	P20152	Vimentin	82	53.7	2.21 E+09
STOM_MOUSE	P54116	Erythrocyte band 7 integral membrane protein	29	31.4	2.00 E+09
E9Q9C5_MOUSE	E9Q9C5	V-type proton ATPase 16 kDa proteolipid subunit	3	15.3	1.83 E+09
ACTB_MOUSE	P60710	Actin, cytoplasmic 1	32	41.7	1.62 E+09
RL40_MOUSE	P62984	Ubiquitin-60S ribosomal protein L40	12	14.7	1.55 E+09
RAB7A_MOUSE	P51150	Ras-related protein Rab-7a	27	23.5	1.51 E+09
VATB2_MOUSE	P62814	V-type proton ATPase subunit B	48	56.6	1.40 E+09
CATZ_MOUSE	Q9WUU7	Cathepsin Z	19	34.0	1.30 E+09
CD68_MOUSE	P31996	Macrosialin	7	34.8	1.20 E+09
GNAI2_MOUSE	P08752	Guanine nucleotide-binding protein G(i) subunit alpha-2	31	40.5	1.06 E+09
GPNMB_MOUSE	Q99P91	Transmembrane glycoprotein NMB	23	63.7	9.41 E+08
VATA_MOUSE	P50516	V-type proton ATPase catalytic subunit A	53	68.3	9.40 E+08
GBB2_MOUSE	P62880	Guanine nucleotide-binding protein G(I)/G(S)/G(T) subunit beta-2	21	37.3	8.99 E+08

5. The efficiency of digestion is notably decreased by the presence of the acid cleavable detergent Rapigest in the phagosome lysate. Samples must be diluted in 1:10 (v/v) in order to provide a final concentration of 0.1% Rapigest.

6. Make sure to choose C18 SPE columns with the appropriate the binding capacity.

7. Solid phase extraction (SPE) is a technique designed for rapid, selective sample purification prior to mass spectrometric analysis. Normally, reversed-phase liquid chromatography principles are used to control selectivity. SPE provides the sample cleanup, recovery, and concentration necessary for accurate quantitative analysis. For first time users it is recommended to keep all washes and flow-throughs as peptides can easily be lost during this step.

8. For long-term storage, we recommend storing the peptides samples in a deep freezer at –80 °C. As the stability and solubility of any given peptide is dependent on its sequence, repeated freeze-thaw cycles (3–4 thaws) may result in increasing changes during LC-MS/MS quantification.

9. Dimethyl labeling can be used as a triplex reagent by the combination of deuterated and ^{13}C-labeled formaldehyde with cyanoborodeuteride. The use of the heavy label makes it possible to quantitatively analyze three different samples in a single mass spectrometry analysis. Labeling with the "light" reagent (Formaldehyde with sodium cyanoborohydride) generates a mass increase of 28 Da per primary amine on a peptide. Labeling with the "intermediate" reagent (d-formaldehyde with sodium cyanoborohydride) generates a mass increase of 32 Da per primary amine. Incorporation of the "heavy" label (^{13}C-d-formaldehyde with cyanoborodeuteride) results a mass increase of 36 Da.

10. Reagents dissolved in acetonitrile are stable for 1 week when stored at –20 °C and warmed to room temperature before opening. Anhydrous ethanol can be used as an alternative solvent to dissolve reagents.

11. The volumes used should be sufficient to resuspend peptides. If the sample cannot be fully resuspended, then the volume can be increased from 100 to 200 μl, provided that the final concentration of acetonitrile remains 70% (v/v).

12. This quenching step can also be performed by the addition of 200 μl of HPLC grade water to each tube. Place peptide samples on a shaker at room temperature for 2 h. Dilution with water increases the rate of hydrolysis of unreacted TMT reagents. After 2 h the remaining proportion of unquenched tag is negligible.

13. If the pH is not as expected, TEAB buffer, used in previous steps, was not completely evaporated during the vacuum centrifugation step. In such cases, the sample might be resuspended again in 200 μl water 0.1 % TFA and dried once more by vacuum centrifugation.

14. Incorporation of deuterium at the peptide level for quantitative proteomic analysis will create a small but a significant retention time difference during reversed-phased chromatography, compared to their non-deuterated counterparts. The observed retention time difference will not have an impact on the identification of the peptides but possibly on the protein quantification, leading to an incorrect quantitative value or missing quantitative values. This effect is dramatically increased by the use of the "heavy" reagent. Therefore we do not recommend the use of the "heavy" reagent for long LC gradients as proposed in this protocol.

15. Automatic gain control (AGC) is a software feature in mass spectrometers that predicts the time (in ms) required to fill an ion trap with a certain number of ions using the intensities of the ions from the precursor scan. This is important as underfilling of the ion trap could result in low sensitivity while overfilling could lead to space-charge effects.

16. The most commonly used fragmentation techniques are collision-induced dissociation (CID) and higher-energy collisional dissociation (HCD). HCD is a CID technique specific to Orbitrap and Thermo Scientific ion trap instruments that allows making use of the high resolution of the Orbitrap mass spectrometer or—when used in ion traps—allows the detection of lower m/z fragments that are otherwise usually lost in CID MS/MS experiments in ion traps. TMT-based quantification requires the detection of low masses (126–131), therefore, HCD is essential for these types of experiments in ion-trap/Orbitrap hybrid instruments.

17. An established limitation of isobaric tagging strategies occurs during mass spectrometry analyses of samples. Precursor ions selected for MS/MS fragmentation in complex peptide mixtures are typically co-selected with a population of interfering ions. Co-selection of such ion background contributes to the isobaric tag reporter ion ratios [18]. Setting the ion precursor selection window to 0.5 m/z notably improves the accuracy and precision of quantification, but lowers spectral quality and sensitivity. Reciprocally, applying a higher precursor ion selection window of 1.6 m/z will allow higher rate of identification but the robustness of quantification is typically poorer. Similar approaches are also applicable to other mass spectrometers used for isobaric tag experiments, such as hybrid quadrupole-time of flight and quadrupole-Orbitrap mass spectrometers.

A recent alternative approach using MS^3 on Orbitrap mass spectrometers have been proposed and indicates that the precursor co-isolation effect in complex samples can be addressed to a large extent by relying on the reporter ion intensities extracted from MS^3 spectra [19, 20]. A novel Fourier Transform MS^3 Synchronous Precursor Selection method (SPS-MS^3 fragmentation) on the Orbitrap Fusion showed around 20 % reduction in the number of identified proteins but improved quantitation accuracy and quantitation of 95 % of the identified proteins [21].

18. The iBAQ (intensity-based absolute quantification) method is an algorithm implemented in MaxQuant search engine that estimates the absolute abundance of a protein by taking the sum of intensities of identified peptides of proteins, normalized to the length and theoretical number of possible peptides.

Acknowledgments

We would like to thank the mass spectrometry team of the MRC Protein Phosphorylation and Ubiquitylation Unit for their support. We would like to thank Natalia Shpiro for the synthesis of sodium 3-[(2-methyl-2-undecyl-1,3-dioxolan-4-yl)methoxy]-1-propane-sulfonate. This work was funded by the Medical Research Council UK and the pharmaceutical companies supporting the Division of Signal Transduction Therapy (DSTT) (AstraZeneca, Boehringer-Ingelheim, GlaxoSmithKline, Janssen Pharmaceutica, Merck KGaA and Pfizer).

References

1. Kinchen JM, Ravichandran KS (2008) Phagosome maturation: going through the acid test. Nat Rev Mol Cell Biol 9(10):781–795. doi:10.1038/nrm2515

2. Gutierrez MG (2013) Functional role(s) of phagosomal Rab GTPases. Small GTPases 4(3):148–158. doi:10.4161/sgtp.25604

3. Levin R, Grinstein S, Schlam D (2015) Phosphoinositides in phagocytosis and macropinocytosis. Biochim Biophys Acta 1851(6):805–823. doi:10.1016/j.bbalip.2014.09.005

4. Gagnon E, Duclos S, Rondeau C, Chevet E, Cameron PH, Steele-Mortimer O et al (2002) Endoplasmic reticulum-mediated phagocytosis is a mechanism of entry into macrophages. Cell 110(1):119–131

5. Campbell-Valois FX, Trost M, Chemali M, Dill BD, Laplante A, Duclos S et al. Quantitative proteomics reveals that only a subset of the endoplasmic reticulum contributes to the phagosome. Mol Cell Proteomics. 2012; 11(7):M111 016378. doi:10.1074/mcp.M111.016378.

6. Houde M, Bertholet S, Gagnon E, Brunet S, Goyette G, Laplante A et al (2003) Phagosomes are competent organelles for antigen cross-presentation. Nature 425(6956):402–406. doi:10.1038/nature01912

7. Guermonprez P, Saveanu L, Kleijmeer M, Davoust J, Van Endert P, Amigorena S (2003) ER-phagosome fusion defines an MHC class I cross-presentation compartment in dendritic cells. Nature 425(6956):397–402

8. Trost M, English L, Lemieux S, Courcelles M, Desjardins M, Thibault P (2009) The phagosomal proteome in interferon-gamma-activated macrophages. Immunity 30(1):143–154. doi:10.1016/j.immuni.2008.11.006

9. Boulais J, Trost M, Landry CR, Dieckmann R, Levy ED, Soldati T et al (2010) Molecular characterization of the evolution of phagosomes. Mol Syst Biol 6:423. doi:10.1038/msb.2010.80

10. Desjardins M, Huber LA, Parton RG, Griffiths G (1994) Biogenesis of phagolysosomes proceeds through a sequential series of interactions with the endocytic apparatus. J Cell Biol 124(5):677–688

11. Garin J, Diez R, Kieffer S, Dermine JF, Duclos S, Gagnon E et al (2001) The phagosome proteome: insight into phagosome functions. J Cell Biol 152(1):165–180

12. Ong SE, Blagoev B, Kratchmarova I, Kristensen DB, Steen H, Pandey A et al (2002) Stable isotope labeling by amino acids in cell culture, SILAC, as a simple and accurate approach to expression proteomics. Mol Cell Proteomics 1(5):376–386

13. Boersema PJ, Raijmakers R, Lemeer S, Mohammed S, Heck AJ (2009) Multiplex peptide stable isotope dimethyl labeling for quantitative proteomics. Nat Protoc 4(4):484–494. doi:10.1038/nprot.2009.21

14. Dill BD, Gierlinski M, Hartlova A, Arandilla AG, Guo M, Clarke RG et al (2015) Quantitative proteome analysis of temporally resolved phagosomes following uptake via key phagocytic receptors. Mol Cell Proteomics 14(5):1334–1349. doi:10.1074/mcp.M114.044594

15. Guo M, Hartlova A, Dill BD, Prescott AR, Gierlinski M, Trost M (2015) High-resolution quantitative proteome analysis reveals substantial differences between phagosomes of RAW 264.7 and bone marrow derived macrophages.

Proteomics 15(18):3169–3174. doi:10.1002/pmic.201400431

16. Desjardins M, Griffiths G (2003) Phagocytosis: latex leads the way. Curr Opin Cell Biol 15(4):498–503

17. Cox J, Mann M (2008) MaxQuant enables high peptide identification rates, individualized p.p.b.-range mass accuracies and proteome-wide protein quantification. Nat Biotechnol 26(12):1367–1372, doi:nbt.1511 [pii] 10.1038/nbt.1511

18. Savitski MM, Sweetman G, Askenazi M, Marto JA, Lang M, Zinn N et al (2011) Delayed fragmentation and optimized isolation width settings for improvement of protein identification and accuracy of isobaric mass tag quantification on Orbitrap-type mass spectrometers. Anal Chem 83(23):8959–8967. doi:10.1021/ac201760x

19. McAlister GC, Nusinow DP, Jedrychowski MP, Wuhr M, Huttlin EL, Erickson BK et al (2014) MultiNotch MS3 enables accurate, sensitive, and multiplexed detection of differential expression across cancer cell line proteomes. Anal Chem 86(14):7150–7158. doi:10.1021/ac502040v

20. Ting L, Rad R, Gygi SP, Haas W (2011) MS3 eliminates ratio distortion in isobaric multiplexed quantitative proteomics. Nat Methods 8(11):937–940. doi:10.1038/nmeth.1714

21. Erickson BK, Jedrychowski MP, McAlister GC, Everley RA, Kunz R, Gygi SP (2015) Evaluating multiplexed quantitative phosphopeptide analysis on a hybrid quadrupole mass filter/linear ion trap/orbitrap mass spectrometer. Anal Chem 87(2):1241–1249. doi:10.1021/ac503934f

Chapter 18

Dissecting Phagocytic Removal of Apoptotic Cells in *Caenorhabditis elegans*

Shiya Cheng, Kai Liu, Chonglin Yang, and Xiaochen Wang

Abstract

The unique features of programmed cell death during *C. elegans* development provide an outstanding system to decipher the mechanisms governing phagocytic removal of apoptotic cells. Like in many other organisms, phagocytosis in *C. elegans* involves several essential events, including exposure of eat-me signals on the cell corpse surface, cell corpse recognition and engulfment by phagocytes, and maturation of phagosomes for cell corpse destruction. Forward or reverse genetic approaches, microscopy-based cell biological methods, and biochemical assays have successfully been employed to identify key factors that control different steps of phagocytosis and to understand their functions in these cellular events. In this chapter, we mainly describe how to apply genetic and cell biological approaches to dissect cell corpse removal by phagocytosis in *C. elegans*.

Key words *C. elegans*, Phagocytosis, Cell corpse, Engulfment, Phagosome formation, Phagosome maturation, Microscopy, Fluorescent reporters

1 Introduction

Developmental cell death in *C. elegans* provides a unique and excellent system for deciphering the mechanisms underlying phagocytosis of apoptotic cells using genetic and cell biological approaches. During the lifetime of a *C. elegans* hermaphrodite, 131 out of its 1090 somatic cells die at embryonic and early larval stages, and nearly half the germ cells undergo apoptosis during oogenesis in the adult gonad [1]. Apoptotic cells (also termed cell corpses) are visualized as raised button-like structures by DIC (Differential Interference Contrast) microscopy. Cell corpses can also be identified with fluorescently tagged phagocytic factors, depending on the stage at which the factor functions during cell corpse clearance. In general, phagocytic removal of cell corpses involves several distinct stages [2]. Firstly, apoptotic cells expose on their surface phosphatidylserine (PtdSer) as an eat-me signal, which is recognized by phagocytic receptors on phagocytes (neighboring

Roberto Botelho (ed.), *Phagocytosis and Phagosomes: Methods and Protocols*, Methods in Molecular Biology, vol. 1519, DOI 10.1007/978-1-4939-6581-6_18, © Springer Science+Business Media New York 2017

cells in the soma or sheath cells in the gonad). This triggers cytoskeleton rearrangement and membrane remodeling of the phagocytes, leading to cell corpse engulfment and phagosome sealing. Secondly, the cell corpse-containing phagosomes undergo a maturation process by sequentially fusing with early endosomes, late endosomes and lysosomes, leading to formation of phagolysosomes in which cell corpses are ultimately digested by lysosomal hydrolases.

The PtdSer eat-me signal exposed on the apoptotic cell surface is recognized by the phagocytic receptor CED-1 with the help of TTR-52, a transthyretin-like protein secreted from intestine cells. TTR-52 binds both PtdSer and the extracellular domain of CED-1, thus acting as a bridging molecule for cell corpse recognition [3, 4]. In addition, the ABC transporter CED-7 cooperates with TTR-52 and NRF-5, another PtdSer-binding protein secreted from phagocytes, to promote the transfer of PtdSer from cell corpses to phagocytes, thus facilitating phagocyte-cell corpse recognition [5, 6]. Other phagocytic receptors, including PSR-1, PAT-3/INA-1, PAT-3/PAT-2, and MOM-5, may act in parallel to CED-1/CED-7/TTR-52 to mediate cell corpse recognition (see below) [7–11]. These phagocytic receptors orchestrate two partially redundant signaling pathways to mediate the engulfment of cell corpses [12]. In one pathway, CED-1 interacts with an adaptor protein, CED-6 (GULP), to transduce engulfment signals to downstream effectors, including clathrin and its adaptors (AP2 or EPN-1/epsin), leading to actin rearrangement for engulfment [13, 14]. In the other pathway, engulfment signals mediated by PSR-1 or other receptors are transduced to the intracellular proteins CED-2 (CrKII), CED-5 (DOCK180), and CED-12 (ELMO). CED-5 and CED-12 form a complex and act as a bipartite GEF (guanine exchange factor) to activate the Rac GTPase CED-10, while CED-2 (CrKII) can modulate the activity of the CED-12/CED-5 complex. Activation of CED-10 in turn triggers actin rearrangement, which is required for cell corpse internalization [2].

Sealing of cell corpse-containing phagosomes is necessary for completion of the engulfment process. Two phosphoinositides, $PtdIns(4,5)P_2$ and PtdIns3P, play important roles in phagosome sealing [15]. $PtdIns(4,5)P_2$ is generated by the PtdIns4P 5-kinase PPK-1 and mainly accumulates on unsealed phagosomes. The PtdIns3P phosphatase MTM-1 (myotubularin) acts as a $PtdIns(4,5)P_2$ effector, and cooperates with the Class II PI 3-kinase PIKI-1 to control PtdIns3P levels on unsealed phagosomes [16–18]. The SNX9 family protein LST-4 is recruited to phagosomes through coincident detection of $PtdIns(4,5)P_2$, PtdIns3P, and MTM-1, and it further recruits DYN-1 (dynamin) to accomplish phagosome scission, probably by a mechanism similar to scission of endocytic vesicles. In addition, phagosome sealing requires timely depletion of PtdIns(4,5)P_2 by the inositol-5-phosphatase OCRL-1 [19]. OCRL-1-dependent elimination of phagosomal PtdIns(4,5)P_2 releases MTM-1, LST-4, and DYN-1, allowing subsequent accumulation of PtdIns3P on fully sealed phagosomes [15].

Phagosome maturation involves a Rab GTPase cascade and V-ATPase-dependent acidification of phagosomes [20]. RAB-5 associates with early-stage phagosomes, and promotes PtdIns3P generation probably by activating the class III PI3 kinase VPS-34 [21]. PtdIns3P effectors, such as SNX-1 and SNX-6, are recruited to phagosomes and mediate phagocytic receptor recycling [22, 23]. Phagosomes mature from a RAB-5-positive early stage to a RAB-7-positive late stage by a Rab switching mechanism with the help of SAND-1 (Mon1) and CCZ-1 (Ccz1) [24]. These factors form a complex, which activates RAB-7 on phagosomes and releases RAB-5. The release of RAB-5 from phagosomes also requires TBC-2, a GTPase activating protein (GAP) that inactivates RAB-5 [25]. Two other Rab GTPases, RAB-14 and UNC-108, act earlier than RAB-7 during phagosome maturation [26–28]. RAB-14 and UNC-108 probably promote lysosome recruitment to phagosomes, while RAB-7 is responsible for docking and fusion of lysosomes with phagosomes [28]. In addition, the HOPS complex and the Arf-like small GTPase ARL-8 are required for phagolysosome formation [29, 30]. Finally, lysosomal membrane proteins such as the lysine/arginine transporter LAAT-1 and lysosomal hydrolases including cathepsin L (CPL-1) and DNase II (NUC-1) are important for digestion of phagosomal contents [31, 32].

Both somatic and germ cell corpses can be characterized by their raised button-like morphology under DIC microscope (*see* Subheading 3.2.1). If cell corpse numbers are significantly higher in a given strain than in wild type at all stages, defective clearance of cell corpses or increased apoptosis should be considered (*see* Subheading 3.2.2). To distinguish between these two possibilities, time-lapse analysis can be used to analyze cell death events and the duration of cell corpses (*see* Subheading 3.2.3). Transmission electron microscopy (TEM) can be used to ultrastructurally characterize apoptotic cells in germ line and examine whether they are engulfed by phagocytes (*see* Subheading 3.3). Phagocytosis and maturation of apoptotic cell-containing phagosomes can be followed in *C. elegans* by expressing various reporters that label apoptotic cells or phagosomes at overlapping or distinct stages (*see* Subheading 3.4). Apoptotic cells can be labeled by the PtdSer-binding protein Annexin V (*see* Subheading 3.5). The cell corpse engulfment process can be followed by time-lapse analysis of the fluorescently tagged engulfment receptor CED-1 or cytoskeleton protein ACT-1 in engulfing cells (*see* Subheading 3.6.1), and the *ced-1* promoter is usually used to drive expression of reporters in engulfing cells [33]. PtdIns(4,5)P_2, which is labeled by mCherry/GFP::PLCδ1-PH, and PtdIns3P, which is indicated by YFP/mCherry::2xFYVE, accumulate transiently and sequentially on unsealed and fully sealed phagosomes, respectively. The phosphoinositide switch from PtdIns(4,5)P$_2$ to PtdIns3P is thus a gold standard to judge whether a phagosome is sealed [15] (*see* Subheading 3.6.2). FM4-64, a membrane-impermeable dye, can also be used to detect phagosome sealing because only unsealed phagosomes and the outer leaflet of the plasma membrane are accessible to FM4-64 staining (*see* Subheading 3.6.3). Phagosome maturation requires gradual acidification, which occurs before phagosomal fusion with lysosomes. LysoSensor Green DND-189 staining can be used to detect phagosome acidification in the germ line [28] (*see* Subheading 3.7.1). In addition, phagosome maturation involves dynamic interaction with early endosomes, late endosomes, and lysosomes, which correlates with sequential recruitment of RAB-5 and RAB-7, and the lysosomal membrane protein LAAT-1 (*see* Subheadings 3.7.2 and 3.7.3). Finally, phagolysosome formation can be assessed by monitoring the attachment, clustering, and incorporation of NUC-1- or CPL-1-positive vesicles with phagosomes [28, 32] (*see* Subheading 3.7.3). Nearly all the genes required for phagocytic removal of apoptotic cells are identified by forward (e.g. EMS mutagenesis) or reverse (e.g. RNA interference) genetic screens, which are described in Subheading 3.8.

2 Materials

2.1 Worm Culture and Handling

1. Worm incubators with different temperature (10, 16, 20, and 25 °C).

2. Nematode growth medium (NGM) agar plates seeded with *Escherichia coli* OP50.

3. Worm picker. Platinum wire (32 gauge) is flattened at one end, and the other end is inserted into the tip of a glass pipette and anchored by melting the tip in a Bunsen burner.

2.2 Strains

1. Worm strains are available from the *C. elegans* Genetics Center or individual research labs upon request. The *N2* Bristol strain is used as the wild-type strain.

2. *gla-3 RNAi* animals have elevated germ cell apoptosis with normal clearance of cell corpses, and thus are used as a control to study phagocytosis in the germ line [34].

3. Mutations are generated by EMS mutagenesis or by CRISPR/Cas9 editing [35, 36].

4. Transgenic animals carrying extrachromosomal arrays are generated by microinjection. Integrated arrays are generated by γ–irradiation or bombardment [37].

5. Constructs for building transgenic animals: genes of interest are cloned into worm expression vectors (pPD49.26, pPD49.78/83, pPD95.75/77/79, etc.).

6. Microinjection system (*see* **Note 1**).

7. [137]Cesium Gammacell 1000 irradiator (MDS Nordion, Ottawa, ON, Canada).

8. Gene bombardment system (*see* **Note 2**).

2.3 Fluorescent Dyes

1. Alexa Fluor 488-conjugated Annexin V in a solution containing 25 mM HEPES pH 7.4, 140 mM NaCl, 1 mM EDTA, and 0.1% bovine serum albumin (BSA). Store at 4 °C (Invitrogen, Carlsbad, CA, USA).

2. 200 μg/mL FM4-64 in distilled water, aliquoted and stored at −20 °C (Invitrogen, Carlsbad, CA, USA).

3. 1 mM LysoSensor green DND-189 in DMSO, aliquoted and stored at −20 °C (Invitrogen, Carlsbad, CA, USA).

2.4 DIC and Fluorescent Microscopy

1. Microscope slides and coverslips.

2. Agar pads for visualizing embryos and worms under the microscope.

3. M9 buffer: For 1 L buffer, distilled water (ddH$_2$O) is added to 3 g KH$_2$PO$_4$, 6 g Na$_2$HPO4, 5 g NaCl to a final volume of 999 mL. Sterilize the buffer by autoclaving. Add 1 mL sterilized 1 M MgSO$_4$ after the buffer is cooled to 60 °C.

4. 3–10 mM levamisole in M9.

5. Gonadal buffer: 60 mM NaCl, 32 mM KCl, 3 mM Na$_2$HPO$_4$, 2 mM MgCl$_2$, 20 mM HEPES (pH 7.4), 2 mM CaCl$_2$, 50 μg/mL penicillin, 50 μg/mL streptomycin, 100 μg/mL neomycin, 10 mM glucose, 33% fetal calf serum (FCS).

6. Vaseline and Petrolin (melted and mixed in a 1:1 ratio before use).

7. Concave microscope slide.

8. Surgical knife.

9. Syringe with needle.

10. Mouth pipette.

11. 20-μm Polybeads (Polysciences Inc., Warrington, PA, USA) diluted in the same volume of M9 buffer.

12. Epifluorescent microscope, confocal laser scanning microscope or spinning disk confocal microscope.

13. Image scanning and processing software.

2.5 Transmission Electron Microscopy

1. Pre-fixation buffer: 2.5% glutaraldehyde, 1% paraformaldehyde, 0.1 M sucrose, 0.05 M cacodylate in distilled water. This should be prepared freshly before use.

2. Resin mixture: 20 mL Epon-812 Substitute (EMbed 812), 16 mL dodecenyl succinic anhydride (DDSA), 8 mL nadic methyl anhydride (NMA), 770 µL 2,4,6-Tris(dimethylamino methylphenol) (DMP-30). All are available from Electron Microscopy Sciences (Hatfield, PA).

3. JEM-1400 Transmission electron microscope (*see* **Note 3**).

2.6 Genetic Screens

1. 47 mM Ethyl methanesulfonate (EMS) in M9, freshly prepared.

2. Bleach buffer: For 10 mL buffer, 1.2 mL NaClO, 2 mL NaOH (5 M), and 6.8 mL ddH$_2$O are mixed together.

3. *C. elegans* RNAi library (constructed by Julie Ahringer's group at The Wellcome CRC Institute, University of Cambridge, Cambridge, UK and is distributed by Source BioScience).

4. 1 M Isopropyl β-d-1-thiogalactopyranoside (IPTG) in ddH$_2$O, aliquoted and stored at –20 °C.

3 Methods

3.1 Preparing Buffers and Worms

1. Prepare Nematode growth medium (NGM) agar plates seeded with *Escherichia coli* OP50. For 1 L medium, add ddH$_2$O to 20 g agar, 3.0 g NaCl, and 2.5 g peptone to a final volume of 970 mL. Autoclave, and after the buffer is cooled to 60 °C, add the following sterilized solutions sequentially: 25 mL 1 M K$_2$HPO$_4$/KH$_2$PO$_4$ pH 6.0, 1 mL 1 M CaCl$_2$, 1 mL 1 M MgSO$_4$, and 3.2 mL 5 mg/mL cholesterol. After thorough mixing, distribute the medium to plastic plates (60 × 15 mm). *E. coli* OP50 is a uracil auxotrophic strain whose growth is limited on NGM plates. Spot 300 µL OP50 cultured in Luria–Bertani (LB) medium overnight at 37 °C on each NGM agar plate, and dry at room temperature overnight before use.

2. Make agar pads for visualizing embryos and worms under the microscope. Place a microscope slide (A) between two slides (B, C), which are parallel to slide A and carry self-adhesive labels (Fig. 1). Place a drop of 3% agarose in the middle of slide A, and immediately cover it with another slide as indicated in Fig. 1. Remove the cover slide before use to expose the agar pad.

3.2 Evaluation of Cell Death and Cell Corpse Duration

3.2.1 Visualizing Apoptotic Cells with DIC

1. Culture worms on OP50-seeded NGM plates at 20 °C (*see* **Note 4**).

2. To visualize somatic cell corpses at embryonic stages, mount embryos on agar pads in a drop of M9 buffer, and cover the samples with coverslips immediately. To observe germ cell corpses, mount synchronized worms of different ages (e.g., 24/36/48 h post L4 molt) on agar pads in 3–10 mM

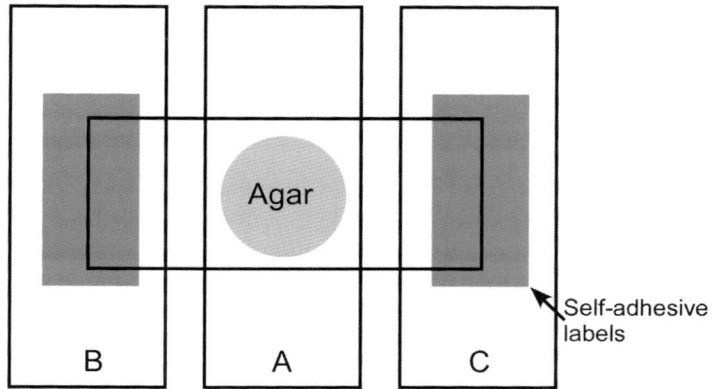

Fig. 1 Diagram showing preparation of an agar pad on a slide. A microscope slide (**A**) is placed between two slides (**B**, **C**) carrying self-adhesive labels. A drop of 3 % agarose is distributed in the middle of slide (**A**), and immediately covered with another slide. Self-adhesive labels help to support the slide on top

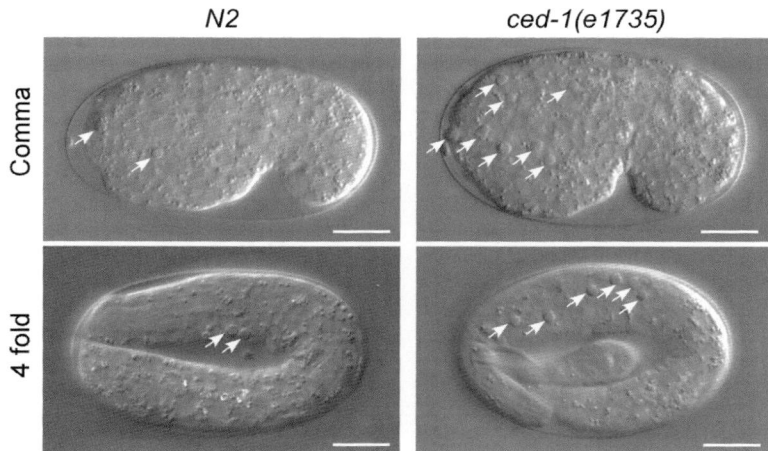

Fig. 2 DIC images showing apoptotic cells in embryos of *N2* (wild type) and *ced-1(e1735)* mutant. Raised button-like apoptotic cells in comma- and fourfold stage embryos are indicated by *arrows*. Bars, 10 μm

levamisole, and cover the samples with coverslips immediately. For time-lapse imaging, seal the slides with Vaseline/ Petrolin mixture.

3. Mount slide on microscope stage, and observe the apoptotic cells under a DIC microscope (*see* **Note 5**).

3.2.2 Quantification of Cell Corpses

1. Quantify somatic cell corpses in the head region of living embryos at six different embryonic stages (comma/bean, 1.5-fold, 2-fold, 2.5-fold, 3-fold, and 4-fold) under DIC microscope (Fig. 2). Analyze at least 15 embryos at each embryonic stage for each genotype (*see* **Note 6**).

2. Quantify germ cell corpses in the U-turn region of one gonad arm at various adult ages (24/36/48 h after L4 molt). Examine at least 15 animals at each time point for each genotype (*see* **Note 6**).

3.2.3 Examining Cell Death Occurrence and Cell Corpse Duration

1. Mount several two-cell stage embryos on agar pads in M9 buffer (*see* **Note** 7). Seal the samples with Vaseline–Petrolin mixture after removing agar outside the coverslip.

2. Capture images in 30–40 Z-series (0.5 μm/section) every 1 min for 8 h at 20 °C using a DIC microscope (e.g., Carl Zeiss Axio Imager M1 equipped with DIC module).

3. Analyze images using software coupled with the microscope (e.g., Carl Zeiss AxioVision Rel. 4.7). Record cell deaths occurring between 200 and 380 min after the first embryonic cell division, and quantify the duration of cell corpses at the same time (*see* **Note 8**).

3.3 Transmission Electron Microscopy (TEM)

The following protocol is adapted from Hall et al. [38].

3.3.1 Pre-fixation

1. Pick healthy worms at 36–48 h post L4 into M9 and anesthetize in 8 % alcohol in M9 for 4 min.

2. Rinse worms three times in M9, 5 min each.

3. Fix worms in pre-fixation buffer on ice for 2 h.

4. Rinse worms three times in 0.2 M cacodylate on ice, 10 min each.

5. Cut worms open as close to the tissue of interest as possible, usually at the pharynx.

3.3.2 Post-fixation

1. Fix samples in 0.1 M cacodylate containing 1 % OsO_4 (osmium tetroxide) for 90 min.

2. Wash samples 3 times in 0.1 M cacodylate, 10 min each.

3.3.3 Sample Embedding

1. Rinse samples three times with ddH_2O, and leave overnight in ddH_2O.

2. Embed samples in 3 % agarose.

3. Cut out small blocks of worm-containing agarose to fit into the embedding mold.

3.3.4 Dehydration

Treat samples sequentially in the following solutions at room temperature:

1. 5 min in 30 % EtOH.

2. 5 min in 50 % EtOH.

3. 5 min in 70% EtOH.

4. 10 min in 90% EtOH, three times in total.

5. 10 min in 100% EtOH, three times in total.

6. 10 min in 100% propylene oxide, three times in total.

3.3.5   Infiltration

1. Resin preparation: Mix resin mixture well and then place resin in a 4 °C freezer to remove gas before use.

2. Place samples in resin solution (resin–propylene oxide, 1:2); shake for 3 h.

3. Place samples in resin solution (resin–propylene oxide, 1:1); shake for 5 h.

4. Place samples in fresh resin mixture for 4 h; repeat 1 more time.

3.3.6   Embedding Samples in Resin

1. Place samples in fresh resin mixture in an embedding mold, close to the edge.

2. Incubate samples in a 60 °C oven for 3 days.

3.3.7   Cutting Blocks

1. Trim blocks with a blade to remove excessive resin.

2. Using a diamond (or glass) knife, cut 2 μm sections from the pharynx until both distal and proximal arms are seen.

3. Cut 70 nm-thick sections with a diamond knife and hold the ribbon of sections with the coated grid.

3.3.8   Staining

1. Stain sample sections in 2% UAc for 15 min; wash three times with ddH$_2$O.

2. Stain sample sections in citric acid lead for 8 min; wash three times with ddH$_2$O (*see* **Note 9**).

3. Observe with TEM (*see* **Note 10**) (Fig. 3) [22].

3.4   Using Reporter Arrays to Monitor Phagocytosis

3.4.1   Generation of Transgenic Animals Expressing Phagocytic Reporters

DNA transformation and integration of extrachromosomal arrays are performed essentially as follows (*see* **Note 1**).

1. Construct genes of interest in frame with fluorescence proteins into worm-expressing vectors, such as pPD49.26, pPD49.78/ 83, pPD95.75/77/79 (*see* **Note 11**) (Tables 1–3).

2. Co-inject vectors carrying genes of interest with a transformation marker into the gonads of young hermaphrodites (~30 animals). Put 1–3 injected animals onto a plate to generate F1 progeny.

3. Single out F1 transgenic progeny to obtain F2 worms. F2 transgenic worms from the same mother are kept as one transgenic line.

4. Irradiate 100 L4-stage animals with 3800 Rad (*see* **Note 12**).

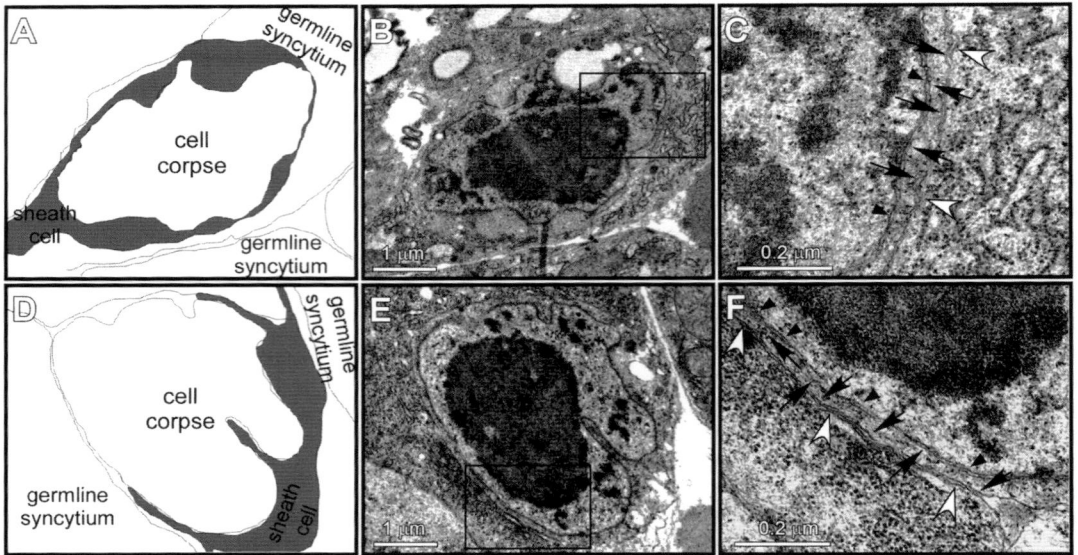

Fig. 3 TEM analysis of germ cell corpse engulfment. (**a–c**) TEM micrographs of an engulfed germ cell corpse in an *N2* animal. (**d–f**) TEM micrographs of an unengulfed germ cell corpse in a *snx-1(tm847)* animal. Membranes are traced in diagrams (**a**) and (**d**). Boxed regions in (**b**) and (**e**) are magnified in (**c**) and (**f**), respectively. *Black arrows* indicate gonadal sheath cell membranes. *Black* and *white arrowheads* indicate cell corpse membranes and germ line syncytium membranes, respectively

Table 1
Reporters for monitoring Ptdser exposure

Function	Fusion protein	Promoter	Description	Localization	Reference
PtdSer exposure	Annexin V::GFP	*hsp-16.2* and *hsp-16.41*	PtdSer binding	Apoptotic cell surface	[5]
	TTR-52::mCHERRY	*hsp-16.2* and *hsp-16.41*	Bridging molecule	Apoptotic cell surface	[3]
	ssGFP::mNRF-5	*hsp-16.2* and *hsp-16.41*	PtdSer transfer	Apoptotic cell surface	[5]
	ssGFP::Lact-C2	*hsp-16.2* and *hsp-16.41*	PtdSer binding	Apoptotic cell surface	[5]

5. Transfer worms to new NGM plates seeded with *E. coli* OP50 for recovery for several hours or overnight. Then transfer 5 worms to a new NGM plate (20 in total). Culture worms until the food is exhausted.

6. Transfer a large number of worms from each of the 20 plates to 20 new plates. After 1–2 days, pick 15 transgenic animals from each of the 20 plates to 15 new plates. Animals from a plate that produces 100% transgenic offspring are kept as an integration line.

Table 2
Reporters for monitoring engulfment and phagosome formation

Function	Fusion protein	Promoter	Description	Localization	Reference
Engulfment and phagosome formation	CED-1::GFP	*ced-1*	Phagocytic receptor	Plasma membrane of engulfing cell	[33]
	GFP::ACT-1	*ced-1*	Cytoskeleton, actin	Unsealed phagosome	[15]
	mCHERRY::PLCδ1-PH	*ced-1*	PtdIns(4,5)P$_2$ binding	Unsealed phagosome	[15]
	mCHERRY::MTM-1 (C378S)	*ced-1*	PtdIns3P phosphatase, PtdIns(4,5)P$_2$ effector	Unsealed phagosome	[15]
	GFP::PIKI-1	*ced-1*	Class II PI3 kinase	Unsealed phagosome	[15, 17]
	LST-4::GFP	*ced-1* or *lst-4*	SH3-PX-BAR domain protein	Unsealed phagosome	[13, 15]
	DYN-1::GFP (dynamin)	*ced-1*	Large GTPase	Unsealed phagosome	[13, 15]

Table 3
Reporters for monitoring phagosome maturation

Function	Fusion protein	Promoter	Description	Localization	Reference
Phagosome maturation	GFP::RAB-5	*ced-1*	Small GTPase	Early phagosome	[25]
	YFP::2xFYVE	*ced-1*	PtdIns3P binding	Early phagosome	[21]
	mCHERRY::RAB-14	*ced-1*	Small GTPase	Early phagosome	[28]
	GFP::UNC-108	*ced-1*	Small GTPase	Late phagosome	[26]
	GFP::RAB-7	*ced-1*	Small GTPase	Late phagosome	[28]
	LAAT-1::GFP	*ced-1*	Lysosomal membrane protein	Phagolysosome	[31]
	NUC-1::nCHERRY	*ced-1*	Lysosomal enzyme	Phagolysosome	[28]
	CPL-1::mChOint	*cpl-1*	Lysosomal enzyme	Phagolysosome	[32]

7. Back cross the integration lines with wild-type worms at least 4 times to eliminate background mutations before further analysis.

3.4.2 Time-Lapse Analysis of Phagocytosis and Phagosome Maturation Using Fluorescent Reporters

1. For time-lapse analysis of embryonic cell corpses, mount pre-comma stage embryos on agar pads in M9 buffer. To analyze germ cell corpse clearance, mount adult worms at 24 h or 36 h post L4 molt on agar pads in M9 buffer with 3–10 mM levamisole.

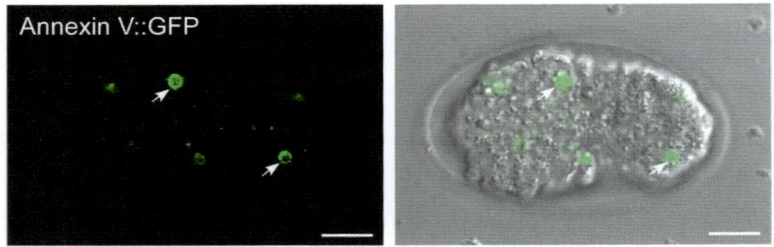

Fig. 4 Apoptotic cells in a wild-type embryo are labeled by Annexin V::GFP. *Arrows* indicate Annexin V::GFP rings around apoptotic cells. Bars, 10 μm

2. Cover the samples with coverslips. Use Vaseline/Petrolin to seal the samples after cutting off agar outside the coverslips.

3. Capture fluorescence images using a confocal laser scanning microscope or spinning disk confocal microscope (e.g., Nikon Eclipse Ti-E equipped with PerkinElmer UltraView spinning-disk confocal scanner unit). In general, images in 20–25 Z-series (1.0 μm/section) are captured every 1 or 2 min for 1–2 h (embryos) or 2–6 h (adult worms) at 20 °C (*see* **Note 13**).

4. Review and analyze images using commercial software packaged with the microscope (e.g., PerkinElmer Velocity) or other public domain software (e.g., Image J).

3.5 Examination of the PtdSer Eat-Me Signal on Apoptotic Cells

3.5.1 Monitoring PtdSer-Exposing Apoptotic Cells in Embryos with Annexin V and TTR-52

1. Seal plates containing worms expressing secreted Annexin V::GFP or TTR-52::mCHERRY under the control of heat-shock promoters (Table 1). Incubate the sealed plates in a water bath at 33 °C for 0.5–1 h to induce expression and secretion of Annexin V::GFP or TTR-52::mCHERRY.

2. Return plates to a 20 °C incubator and culture for 2.5–3 h.

3. Pick up embryos and mount them on agar pads in M9 buffer. Observe embryos under an epifluorescent or confocal laser-scanning microscope (Fig. 4).

4. Capture DIC and fluorescence images of at least 15 embryos at 1.5-fold stage. Quantify the percentage of cell corpses labeled by Annexin V::GFP or TTR-52::mCHERRY by dividing the total number of raised button-like cell corpses by the number of labeled ones.

3.5.2 Annexin V Staining in the Germ Line

1. Put worms aged for 24/36/48 h after the L4 molt in a concave microscope slide in 200 μL gonadal buffer containing 2 μL Annexin V dye (*see* **Note 14**). Cut worms at the head or tail region using a surgical knife to expose the gonads.

2. Keep samples in the dark and incubate for 20 min at room temperature.

3. Remove the dye with a mouth pipette, and wash three times with gonadal buffer (5 min each time).

4. Transfer worms with exposed gonads onto a microscope slide with a mouth pipette, and make sure that the total volume of liquid on the slide is less than 20 μL.

5. Cover the samples with a coverslip after adding 20-μm Polybeads (20–50 beads) to the medium.

6. Capture DIC and fluorescent images under an epifluorescent or confocal microscope. Calculate the ratio of Annexin V-positive cell corpses by dividing the total number of raised button-like cell corpses by the number of Annexin V-labeled ones. At least 15 animals should be scored in each strain.

3.6 Evaluation of Engulfment and Phagosome Formation

1. Perform time-lapse recording in strains expressing CED-1::GFP or GFP::ACT-1 as described in Subheading 3.4.2 (Table 2).

2. Examine the recruitment of CED-1 (ACT-1) on at least 10 cell corpses.

3.6.1 Using CED-1::GFP or GFP::ACT-1 to Evaluate Engulfment

3. Define the time point when CED-1 (ACT-1) is first detected on the cell corpse as "0 min," and measure the time required for the cell corpse to be fully encircled by CED-1 (ACT-1). If significantly longer time is required for CED-1 (ACT-1) "ring" formation, defects in engulfment should be considered.

3.6.2 Examining Phagosome Sealing by Monitoring Phosphoinositide Switch

1. Perform time-lapse recording in strains expressing both mCHERRY::PLCδ1-PH and YFP::2xFYVE as described in Subheading 3.4.2 (*see* **Note 15**) (Tables 2 and 3).

2. Follow the persistence of PLCδ1-PH and the appearance of 2xFYVE on at least 10 phagosomes (Fig. 5). If PLCδ1-PH persists longer in the mutant than in wild type, a phagosome sealing defect may be considered.

3.6.3 Detecting Phagosome Sealing with FM4-64 and Annexin V

1. Dissect adult worms (36 or 48 h after L4 molt) to expose gonads in 200 μL gonadal buffer with or without 2 μL Alexa Fluor 488-conjugated Annexin V as described in Subheading 3.5.2.

2. Remove 155 μL gonadal dissection buffer, and then add 5 μL FM4-64 (200 μg/mL) to make a final concentration of 20 μg/mL.

3. Keep the samples in the dark for 5 min at 4 °C (which prevents endocytosis) before examination by fluorescent microscopy.

4. Transfer worms together with 10–20 μL medium onto a microscope slide, and cover the samples with a coverslip after adding 20-μm Polybeads (20–50 beads).

5. Capture DIC and fluorescent images. Calculate the percentage of cell corpses labeled by FM4-64 or Annexin V by dividing the total number of raised button-like cell corpses by the number of labeled ones. At least 10 animals should be scored in each strain (*see* **Note 16**).

Somatic cell corpse

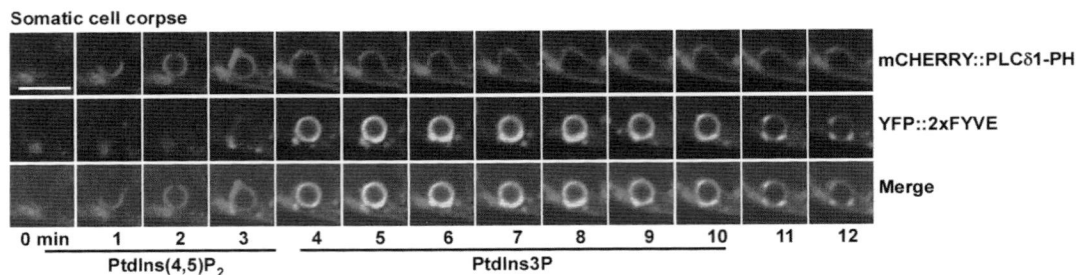

Fig. 5 PtdIns(4,5)P$_2$ and PtdIns3P accumulate sequentially on phagosomal membranes. Time-lapse images of a cell corpse in a wild-type embryo expressing mCHERRY::PLCδ1-PH and YFP::2xFYVE are shown. "0 min" represents the time point when PLCδ1-PH was first detected on the cell corpse. Bars, 5 μm

3.7 Evaluation of Phagosome Maturation

3.7.1 Using LysoSensor Green DND-189 Staining to Detect Phagosome Acidification

1. Dissect adult animals (36 or 48 h post L4 molt) to expose gonads in gonadal buffer with 1 μM Lysosensor Green DND-189 (Invitrogen).

2. Transfer the dissected worms together with 10–20 μL medium to a microscope slide, and cover the samples with a coverslip after adding Polybeads with a diameter of 20 μm (20–50 beads).

3. Examine the fluorescence under an epifluorescent microscope (*see* **Note 17**).

4. Capture DIC and fluorescent images. Calculate the percentage of cell corpses labeled by Lysosensor Green by dividing the total number of raised button-like cell corpses by the number of labeled ones. At least 15 animals should be scored in each strain.

3.7.2 Assessing Phagosome Maturation by Monitoring RAB-5-to-RAB-7 Switch

1. Perform time-lapse recording in strains expressing both mCHERRY::MTM-1(C378S) (enzyme-dead form of the PtdIns3P phosphatase MTM-1) and GFP::RAB-5 (or GFP::RAB-7) as described in Subheading 3.4.2 (*see* **Note 18**) (Table 3).

2. Follow the disappearance of MTM-1(C378S) and the appearance of RAB-5 or RAB-7 on phagosomes ($n \geq 10$). If RAB-5 appears normally, while the appearance of RAB-7 is delayed or abrogated, defects in the phagosomal switch from RAB-5 to RAB-7 should be considered.

3.7.3 Detecting Phagolysosome Formation by Monitoring Phagosomal Recruitment of LAAT-1, NUC-1 or CPL-1

1. Perform time-lapse recording in strains expressing both YFP::2xFYVE and LAAT-1::mCHERRY as described in Subheading 3.4.2 (*see* **Note 19**) (Table 3).

2. Follow the disappearance of YFP::FYVE and the appearance of LAAT-1::mCHERRY on phagosomes ($n \geq 10$). Defective LAAT-1 appearance on phagosomes indicates defects in phagolysosome formation.

3. Perform time-lapse recording in strains expressing both YFP::2xFYVE and NUC-1::nCHERRY (CPL-1::mChOint) as described in Subheading 3.4.2 (Table 3).

4. Follow the attachment, clustering and incorporation of NUC-1- or CPL-1-positive vesicles with phagosomes, and identify which step of phagolysosome formation is affected [28, 32] (*see* **Note 20**).

3.8 Genetic Screens and Analyses (See Note 21)

3.8.1 Forward Genetic Screen Using Ethyl Methanesulfonate (EMS) in Worms Carrying the sem-4(n1378) Mutation

1. Bleach adult animals (~2000) to obtain synchronized eggs, and let them grow to L4 stage.

2. Wash the L4 worms (~5000) off the plates with M9, and treat the worms with 4 mL EMS (47 mM in M9) in a 15 mL tube for 4 h at 20 °C.

3. Wash worms 3 times with M9 and put them on 10 plates for recovery for 2 h at 20 °C. Then pick late L4-stage animals to new plates, 5–10 animals per plate for a total of 50 plates.

4. Pick old adult F1 worms to observe the phenotype of F2 embryos.

5. For genes with maternal effect, put ~10F1 worms on a new plate to produce F2 progeny. Observe F3 embryos in the body of adult F2 animals (*see* **Note 22**).

3.8.2 Reverse Genetic Screen for New Players in Phagocytic Clearance of Germ Cell Corpses

1. Grow bacteria (HT115) expressing dsRNA (generated by the Ahringer laboratory) on ampicillin-resistant plates to get single colonies.

2. Pick a single colony into 3 mL LB medium containing 100 μg/mL ampicillin and 12.5 μg/mL tetracycline; grow overnight at 37 °C.

3. Dilute the overnight cultures at 1/20 into fresh LB medium containing 100 μg/mL ampicillin and 12.5 μg/mL tetracycline; grow bacteria to OD_{600} 0.6–0.8 at 37 °C (*see* **Note 23**).

4. Spot the bacterial culture on NGM agar plates containing 100 μg/mL ampicillin, 12.5 μg/mL tetracycline, and 1 mM IPTG (isopropyl β-d-1-thiogalactopyranoside); leave plates overnight at room temperature to allow dsRNA expression (*see* **Note 24**).

5. Transfer 5–10 synchronized L3-L4 animals (P0) to a plate seeded with RNAi bacteria and culture at 20 °C (*see* **Note 25**).

6. When the F1 progeny reach 36–48 h post L4 molt, analyze them microscopically for cell corpses (*see* **Note 26**).

4 Notes

1. Detailed information can be found in the chapter "Transformation and microinjection" in WormBook (http://www.wormbook.org).

2. Detailed information can be found in the chapter "*C. elegans* gene transformation by microparticle bombardment" in WormBook (http://www.wormbook.org).

3. More information can be found in the "Anatomical Methods" section in WormAtlas (www.wormatlas.org).

4. Worms grow faster at higher temperature. Normally, they are cultured at 20 °C. Higher temperatures may adversely affect worms, while low temperatures may decrease the mating efficiency for genetic crosses. Temperature-sensitive mutants, such as *tbc-2* mutants, may show stronger phenotypes at 25 °C.

5. Both somatic and germ cell corpses are characterized by their raised button-like morphology using a DIC microscope.

6. If cell corpse numbers are significantly higher in a given strain than in wild type at all stages, defective clearance of cell corpses or increased apoptosis should be considered.

7. Avoid too much OP50 when transferring embryos to agar pads. To get enough two-cell stage embryos, adult worms carrying embryos can be split with two needles, and then transferred onto agar pads by mouth pipette.

8. Cell corpse clearance-defective mutants show a normal number of cell death events but have longer cell corpse duration than wild type.

9. Stain the samples in a closed box with NaOH powder inside to absorb CO_2 from the air. This prevents precipitation of lead in the sample. After staining, samples are washed three times with a large volume of ddH_2O (>500 mL).

10. In the germ line, apoptotic cells are quickly recognized and engulfed by sheath cells encasing the germ line. The sheath cell extends pseudo arms between the membranes of a cell corpse and the germ line syncytium until the cell corpse is fully encircled. Under TEM, the internalized cell corpse is separated from healthy cells by four membranes: one cell corpse membrane, two membranes of the engulfing sheath cell, and one membrane of the germ line syncytium (Fig. 3a–c). In mutants defective for engulfment, the sheath cell is unable to extend pseudo arms around the cell corpse. Thus the cell corpse membrane directly contacts the germ line syncytium (Fig. 3d–f).

11. pPD49.78 and pPD49.83 contain *hsp-16.2* and *hsp-16.41* promoters respectively, and the combination of these two vectors is used to induce global expression by heat-shock treatment. All worm expression vectors are available from Addgene (http://www.addgene.org).

12. Transgenic lines typically contain a high copy number of extrachromosomal arrays. Gamma irradiation is used to generate genome-integrated arrays containing relatively low copy

numbers. Integrative transformation can also be achieved by bombardment of worms with gold particles coated with DNA (expression vector) using a bombardment instrument (gene gun). For detailed protocols, please see the chapter "*C. elegans* gene transformation by microparticle bombardment" in WormBook (http://www.wormbook.org).

13. Phagocytosis of germ cell corpses takes much longer than that of embryonic cell corpses. Thus a longer imaging time is required to monitor phagocytosis of germ cell corpses. Gonadal sheath cells, which act as phagocytes of germ cell corpses, are sensitive to heat, so the laser strength and exposure time must be finely controlled.

14. The binding of Annexin V with PtdSer requires Ca^{2+}. The gonadal buffer contains 2 mM $CaCl_2$, so there is no need to add Ca^{2+} again.

15. The PtdIns3P phosphatase MTM-1 is effector of $PtdIns(4,5)P_2$, and MTM-1(C378S) (enzyme-dead form of MTM-1) shows identical dynamic with $PtdIns(4,5)P_2$. The dynamics of mCHERRY::MTM-1(C378S) can also be monitored simultaneously with YFP::2xFYVE to indicate phagosome formation.

16. Though FM4-64 and Annexin V show nice co-localization in unsealed phagosomes, the fluorescence intensity of Annexin V conjugates varies between different apoptotic cells, probably due to differences in PtdSer levels on individual cell corpses. FM4-64 staining can also be performed in strains expressing GFP::PLCδ1-PH.

17. Acidified phagosomes show much stronger fluorescence than non-acidified ones or living cells.

18. MTM-1(C378S) (enzyme-dead form of MTM-1) shows identical dynamic with $PtdIns(4,5)P_2$, and it can be used to indicate forming phagosome.

19. PtdIns3P labeled by 2xFYVE shows very nice co-localization with RAB-5 on phagosomes, so 2xFYVE can be used to indicate early stage phagosomes.

20. The shrinkage of LAAT-1- or NUC-1-positive phagolysosomes represents the final degradation of the enclosing apoptotic cells and the reformation of lysosomes from phagolysosomes.

21. EMS mutagenesis can be performed in wild type or mutants with different genetic backgrounds depending on the purpose of the screen, e.g., enhancer or suppressor screen. F2 progeny with the expected phenotype, e.g., change in embryonic cell corpse numbers, can be screened in F1 worms under a DIC microscope if P0 worms carry a *sem-4* mutation that causes egg-laying defects [39].

22. Mutants might grow more slowly than the wild type, so screens should be performed when most progeny have grown to the adult stage. Live F1 or F2 embryos in dead adult animals (P0 or F1) can be examined for cell corpse phenotypes.

23. In glass tubes, it takes 1.5–2.5 h for bacteria grow to OD 0.6. In 96-well plates, however, it usually takes 6–8 h.

24. Prepare NGM plates 3 days before seeding RNAi bacteria to allow them to dry. 300 µL of bacterial culture are usually seeded in a 35-mm plate.

25. Contamination with OP50 bacteria reduces RNAi efficiency, so worms must be purged of OP50 by transferring them to empty NGM plates before placing them on RNAi plates.

26. If knockdown of a gene causes sterility or lethality, RNAi can be performed by feeding L3-L4 animals and examining the phenotype in the same generation. The essential role of clathrin and AP2 complex subunits in phagocytosis was revealed by this approach [13].

Acknowledgments

We thank Dr. Isabel Hanson for proofreading. Research in the authors' laboratories is supported by the National Natural Science Foundation of China (31325015 to X.W., 31230043 and 31025015 to C.Y.), the National Basic Research Program of China (2010CB835202, 2013CB910101, and 2014CB849700 to X.W., 2013CB910102 and 2011CB910102 to C.Y.), the Chinese Academy of Sciences (KJZD-EW-L08 to C.Y.), and an International Early Career Scientist grant from the Howard Hughes Medical Institute to X.W.

References

1. Lettre G, Hengartner MO (2006) Developmental apoptosis in C. elegans: a complex CEDnario. Nat Rev Mol Cell Biol 7(2):97–108. doi:10.1038/nrm1836

2. Pinto SM, Hengartner MO (2012) Cleaning up the mess: cell corpse clearance in Caenorhabditis elegans. Curr Opin Cell Biol 24(6):881–888. doi:10.1016/j.ceb.2012.11.002

3. Wang X, Li W, Zhao D, Liu B, Shi Y, Chen B, Yang H, Guo P, Geng X, Shang Z, Peden E, Kage-Nakadai E, Mitani S, Xue D (2010) Caenorhabditis elegans transthyretin-like protein TTR-52 mediates recognition of apoptotic cells by the CED-1 phagocyte receptor. Nat Cell Biol 12(7):655–664. doi:10.1038/ncb2068

4. Kang Y, Zhao D, Liang H, Liu B, Zhang Y, Liu Q, Wang X, Liu Y (2012) Structural study of TTR-52 reveals the mechanism by which a bridging molecule mediates apoptotic cell engulfment. Genes Dev 26(12):1339–1350. doi:10.1101/gad.187815.112

5. Zhang Y, Wang H, Kage-Nakadai E, Mitani S, Wang X (2012) C. elegans secreted lipid-binding protein NRF-5 mediates PS appearance on phagocytes for cell corpse engulfment. Curr Biol 22(14):1276–1284. doi:10.1016/j.cub.2012.06.004

6. Mapes J, Chen YZ, Kim A, Mitani S, Kang BH, Xue D (2012) CED-1, CED-7, and TTR-52 regulate surface phosphatidylserine expression

on apoptotic and phagocytic cells. Curr Biol 22(14):1267–1275. doi:10.1016/j.cub.2012. 05.052

7. Wang X, Wu YC, Fadok VA, Lee MC, Gengyo-Ando K, Cheng LC, Ledwich D, Hsu PK, Chen JY, Chou BK, Henson P, Mitani S, Xue D (2003) Cell corpse engulfment mediated by *C. elegans* phosphatidylserine receptor through CED-5 and CED-12. Science 302(5650):1563–1566. doi:10.1126/science.1087641

8. Yang H, Chen YZ, Zhang Y, Wang X, Zhao X, Godfroy JI 3rd, Liang Q, Zhang M, Zhang T, Yuan Q, Ann Royal M, Driscoll M, Xia NS, Yin H, Xue D (2015) A lysine-rich motif in the phosphatidylserine receptor PSR-1 mediates recognition and removal of apoptotic cells. Nat Commun 6:5717. doi:10.1038/ ncomms6717

9. Hsu TY, Wu YC (2010) Engulfment of apoptotic cells in *C. elegans* is mediated by integrin alpha/SRC signaling. Curr Biol 20(6):477–486. doi:10.1016/j.cub.2010.01.062

10. Hsieh HH, Hsu TY, Jiang HS, Wu YC (2012) Integrin alpha PAT-2/CDC-42 signaling is required for muscle-mediated clearance of apoptotic cells in *Caenorhabditis elegans*. PLoS Genet 8(5), e1002663. doi:10.1371/journal. pgen.1002663

11. Cabello J, Neukomm LJ, Gunesdogan U, Burkart K, Charette SJ, Lochnit G, Hengartner MO, Schnabel R (2010) The Wnt pathway controls cell death engulfment, spindle orientation, and migration through CED-10/Rac. PLoS Biol 8(2), e1000297. doi:10.1371/journal.pbio.1000297

12. Conradt B, Xue D (2005) Programmed cell death. WormBook 1–13. doi: 10.1895/ wormbook.1.32.1

13. Chen D, Jian Y, Liu X, Zhang Y, Liang J, Qi X, Du H, Zou W, Chen L, Chai Y, Ou G, Miao L, Wang Y, Yang C (2013) Clathrin and AP2 are required for phagocytic receptor-mediated apoptotic cell clearance in *Caenorhabditis elegans*. PLoS Genet 9(5), e1003517. doi:10.1371/journal.pgen.1003517

14. Shen Q, He B, Lu N, Conradt B, Grant BD, Zhou Z (2013) Phagocytic receptor signaling regulates clathrin and epsin-mediated cytoskeletal remodeling during apoptotic cell engulfment in *C. elegans*. Development 140(15): 3230–3243. doi:10.1242/dev.093732

15. Cheng S, Wang K, Zou W, Miao R, Huang Y, Wang H, Wang X (2015) PtdIns(4,5)P(2) and PtdIns3P coordinate to regulate phagosomal sealing for apoptotic cell clearance. J Cell Biol 210(3):485–502. doi:10.1083/jcb. 201501038

16. Zou W, Lu Q, Zhao D, Li W, Mapes J, Xie Y, Wang X (2009) *Caenorhabditis elegans* myotubularin MTM-1 negatively regulates the engulfment of apoptotic cells. PLoS Genet 5(10), e1000679. doi:10.1371/journal.pgen. 1000679

17. Lu N, Shen Q, Mahoney TR, Neukomm LJ, Wang Y, Zhou Z (2012) Two PI 3-kinases and one PI 3-phosphatase together establish the cyclic waves of phagosomal PtdIns(3)P critical for the degradation of apoptotic cells. PLoS Biol 10(1), e1001245. doi:10.1371/journal. pbio.1001245

18. Cheng S, Wu Y, Lu Q, Yan J, Zhang H, Wang X (2013) Autophagy genes coordinate with the class II PI/PtdIns 3-kinase PIKI-1 to regulate apoptotic cell clearance in *C. elegans*. Autophagy 9(12):2022–2032. doi:10.4161/ auto.26323

19. Sarantis H, Balkin DM, De Camilli P, Isberg RR, Brumell JH, Grinstein S (2012) Yersinia entry into host cells requires Rab5-dependent dephosphorylation of PI(4,5)P(2) and membrane scission. Cell Host Microbe 11(2):117–128. doi:10.1016/j.chom.2012.01.010

20. Kinchen JM, Ravichandran KS (2008) Phagosome maturation: going through the acid test. Nat Rev Mol Cell Biol 9(10):781–795. doi:10.1038/nrm2515

21. Kinchen JM, Doukoumetzidis K, Almendinger J, Stergiou L, Tosello-Trampont A, Sifri CD, Hengartner MO, Ravichandran KS (2008) A pathway for phagosome maturation during engulfment of apoptotic cells. Nat Cell Biol 10(5):556–566. doi:10.1038/ncb1718

22. Chen D, Xiao H, Zhang K, Wang B, Gao Z, Jian Y, Qi X, Sun J, Miao L, Yang C (2010) Retromer is required for apoptotic cell clearance by phagocytic receptor recycling. Science 327(5970):1261–1264. doi:10.1126/science. 1184840

23. Lu N, Shen Q, Mahoney TR, Liu X, Zhou Z (2011) Three sorting nexins drive the degradation of apoptotic cells in response to PtdIns(3) P signaling. Mol Biol Cell 22(3):354–374. doi:10.1091/mbc.E10-09-0756

24. Kinchen JM, Ravichandran KS (2010) Identification of two evolutionarily conserved genes regulating processing of engulfed apoptotic cells. Nature 464(7289):778–782. doi:10.1038/nature08853

25. Li W, Zou W, Zhao D, Yan J, Zhu Z, Lu J, Wang X (2009) *C. elegans* Rab GTPase activating protein TBC-2 promotes cell corpse degradation by regulating the small GTPase RAB-5. Development 136(14):2445–2455. doi:10.1242/dev.035949

26. Lu Q, Zhang Y, Hu T, Guo P, Li W, Wang X (2008) *C. elegans* Rab GTPase 2 is required for the degradation of apoptotic cells. Development 135(6):1069–1080. doi:10.1242/dev.016063

27. Mangahas PM, Yu X, Miller KG, Zhou Z (2008) The small GTPase Rab2 functions in the removal of apoptotic cells in *Caenorhabditis elegans*. J Cell Biol 180(2):357–373. doi:10.1083/jcb.200708130

28. Guo P, Hu T, Zhang J, Jiang S, Wang X (2010) Sequential action of *Caenorhabditis elegans* Rab GTPases regulates phagolysosome formation during apoptotic cell degradation. Proc Natl Acad Sci U S A 107(42):18016–18021. doi:10.1073/pnas.1008946107

29. Xiao H, Chen D, Fang Z, Xu J, Sun X, Song S, Liu J, Yang C (2009) Lysosome biogenesis mediated by vps-18 affects apoptotic cell degradation in *Caenorhabditis elegans*. Mol Biol Cell 20(1):21–32. doi:10.1091/mbc.E08-04-0441

30. Sasaki A, Nakae I, Nagasawa M, Hashimoto K, Abe F, Saito K, Fukuyama M, Gengyo-Ando K, Mitani S, Katada T, Kontani K (2013) Arl8/ARL-8 functions in apoptotic cell removal by mediating phagolysosome formation in *Caenorhabditis elegans*. Mol Biol Cell 24(10):1584–1592. doi:10.1091/mbc.E12-08-0628

31. Liu B, Du H, Rutkowski R, Gartner A, Wang X (2012) LAAT-1 is the lysosomal lysine/arginine transporter that maintains amino acid homeostasis. Science 337(6092):351–354. doi:10.1126/science.1220281

32. Xu M, Liu Y, Zhao L, Gan Q, Wang X, Yang C (2014) The lysosomal cathepsin protease CPL-1 plays a leading role in phagosomal degradation of apoptotic cells in *Caenorhabditis elegans*. Mol Biol Cell 25(13):2071–2083. doi:10.1091/mbc.E14-01-0015

33. Zhou Z, Hartwieg E, Horvitz HR (2001) CED-1 is a transmembrane receptor that mediates cell corpse engulfment in *C. elegans*. Cell 104(1):43–56

34. Kritikou EA, Milstein S, Vidalain PO, Lettre G, Bogan E, Doukoumetzidis K, Gray P, Chappell TG, Vidal M, Hengartner MO (2006) *C. elegans* GLA-3 is a novel component of the MAP kinase MPK-1 signaling pathway required for germ cell survival. Genes Dev 20(16):2279–2292. doi:10.1101/gad.384506

35. Kutscher LM, Shaham S (2014) Forward and reverse mutagenesis in *C. elegans*. WormBook 1–26. doi: 10.1895/wormbook.1.167.1

36. Dickinson DJ, Ward JD, Reiner DJ, Goldstein B (2013) Engineering the *Caenorhabditis elegans* genome using Cas9-triggered homologous recombination. Nat Methods 10(10):1028–1034. doi:10.1038/nmeth.2641

37. Schweinsberg PJ, Grant BD (2013) *C. elegans* gene transformation by microparticle bombardment. WormBook 1–10. doi: 10.1895/wormbook.1.166.1

38. Hall DH (1995) Electron microscopy and three-dimensional image reconstruction. Methods Cell Biol 48:395–436

39. Ellis RE, Jacobson DM, Horvitz HR (1991) Genes required for the engulfment of cell corpses during programmed cell death in *Caenorhabditis elegans*. Genetics 129(1):79–94

Chapter 19

Measurement of *Salmonella enterica* Internalization and Vacuole Lysis in Epithelial Cells

Jessica A. Klein, TuShun R. Powers, and Leigh A. Knodler

Abstract

Establishment of an intracellular niche within mammalian cells is key to the pathogenesis of the gastrointestinal bacterium, *Salmonella enterica* serovar Typhimurium (*S.* Typhimurium). Here we will describe how to study the internalization of *S.* Typhimurium into human epithelial cells using the gentamicin protection assay. The assay takes advantage of the relatively poor penetration of gentamicin into mammalian cells; internalized bacteria are effectively protected from its antibacterial actions. A second assay, the chloroquine (CHQ) resistance assay, can be used to determine the proportion of internalized bacteria that have lysed or damaged their *Salmonella*-containing vacuole and are therefore residing within the cytosol. Its application to the quantification of cytosolic *S.* Typhimurium in epithelial cells will also be presented. Together, these protocols provide an inexpensive, rapid and sensitive quantitative measure of bacterial internalization and vacuole lysis by *S.* Typhimurium.

Key words *Salmonella enterica*, Gentamicin protection assay, Chloroquine resistance assay, *Salmonella*-containing vacuole, Type III secretion system, Vacuole lysis, Epithelial cells, Colony forming units

1 Introduction

Many years ago, it was noted that intracellular bacteria are protected from the bactericidal action of antibiotics [1, 2]. This observation forms the basis for the gentamicin protection assay [3], a widely used technique to study the internalization of bacteria into mammalian cells. Gentamicin is an aminoglycoside antibiotic that has proven efficacy against many Gram-negative and Gram-positive bacteria. It is widely believed to be ineffective against intracellular bacteria because it does not penetrate mammalian cells. However, gentamicin can penetrate eukaryotic cells via endocytic and non-endocytic routes [4–6], albeit poorly, and accumulates in lysosomes, as well as the Golgi complex, cytoplasm, and nucleus [7, 8]. Rather than being membrane-impermeant, the lack of bactericidal activity of gentamicin against internalized bacteria is more likely due its relatively low cellular accumulation. The gentamicin protection

Roberto Botelho (ed.), *Phagocytosis and Phagosomes: Methods and Protocols*, Methods in Molecular Biology, vol. 1519,
DOI 10.1007/978-1-4939-6581-6_19, © Springer Science+Business Media New York 2017

assay requires no expensive equipment and generates quantitative, reproducible data in a relatively short time. Whilst generally conducted in 24-well tissue culture plates, it can easily be adapted to 48-well or 96-well plates for the screening of bacterial mutant libraries, or testing the effects of either chemical compounds or siRNA knockdown of mammalian genes on bacterial invasion, for example.

Salmonella enterica serovar Typhimurium (*S.* Typhimurium) is a facultative, intracellular pathogen that causes self-limiting gastroenteritis in many animal species, including humans, after the ingestion of contaminated food and water. *S.* Typhimurium colonizes epithelial cells (enterocytes and goblet cells) and macrophages in the gastrointestinal tract of infected hosts. Bacterial entry into non-phagocytic epithelial cells is dependent upon a type III secretion system (T3SS1) encoded on *Salmonella* pathogenicity island-1 (SPI-1) [9]. T3SS1 translocates numerous type III effector proteins into host cells that alter host cytoskeletal architecture to generate large actin-rich plasma membrane protrusions, known as "ruffles", that engulf the bacteria into a membrane-bound vacuole, the *Salmonella*-containing vacuole (SCV) (reviewed in [10]). Flagella-based motility is also required for *S.* Typhimurium entry into non-phagocytic cells [11–13] and SPI-1 and flagella are co-regulated under a variety of in vitro conditions [14–16]. Bacterial internalization is independent of T3SS2, a second T3SS encoded on SPI-2 [17, 18], which functions intracellularly to translocate type III effectors required for SCV maturation and positioning (reviewed in [19]). Here, we will describe the use of the gentamicin protection assay to quantify *S.* Typhimurium invasion into HeLa epithelial cells, a non-phagocytic human cell line that has been used extensively to decipher many aspects of the type III effector-mediated internalization event.

Whilst the majority of internalized *S.* Typhimurium remain confined within a membrane-bound vacuole in epithelial cells, a small but significant proportion lyse or damage their nascent phagosome to enter the host cell cytosol [20, 21]. Some of these cytosolic bacteria are eliminated by xenophagy [20, 22, 23], which is the autophagic degradation of bacteria, but some can escape recognition by selective autophagy receptors [24, 25] and ultimately hyper-replicate in the cytosol of epithelial cells [21, 25–27]. This eventually leads to epithelial cell death by pyroptosis and lytic release of bacteria [26]. Chloroquine (CHQ) is a lysosomotropic agent that accumulates to high concentrations within endosomes, but does not access the cytosol [28]. The differential intracellular distribution of CHQ imparts its specificity against vacuolar but not cytosolic bacteria [29], although the mechanism of action of CHQ-dependent killing of vacuolar bacteria remains unknown. The CHQ resistance assay was initially used to identify *Shigella flexneri* mutants that did not lyse their internalization vacuole [30].

Recently, we adapted this assay to quantify cytosolic *S.* Typhimurium at different times post-infection (p.i.) in epithelial cells [21] and provide a detailed description of this protocol here. Like the gentamicin protection assay, the CHQ resistance assay is relatively inexpensive, readily scalable and generates quantitative data in a short timeframe.

2   Materials

1. Use double-distilled water to make all reagents.

2. HeLa cervical epithelial cells (American Type Culture Collection (ATCC), CCL-2), low passage number (*see* **Notes 1** and **2**).

3. Growth medium: Eagle's minimum essential medium with Earle's balanced salts (EMEM), 2 mM L-glutamine, 1 mM sodium pyruvate, and 10 % heat-inactivated fetal bovine serum (FBS) (*see* **Notes 3–5**).

4. Trypsin EDTA: 0.25 % trypsin, 2.21 mM EDTA in HBSS without sodium bicarbonate, calcium and magnesium (*see* **Note 3**).

5. Phosphate-buffered saline without divalent cations (PBS⁻⁻): 0.14 g/L KH_2PO_4, 9 g/L NaCl, 0.80 g/L Na_2HPO_4 (anhydrous), pH 7.4 (*see* **Note 3**).

6. Dulbecco's PBS with divalent cations (DPBS⁺⁺): 0.2 g/L KCl, 0.2 g/L KH_2PO_4, 8 g/L NaCl, 2.17 g/L $Na_2HPO_4 \cdot 7H_2O$, 0.1 g/L $CaCl_2$ (anhydrous), 0.1 g/L $MgCl_2 \cdot 6H_2O$, pH 7.4 (*see* **Note 3**).

7. Hanks' balanced salt solution (HBSS): 0.14 g/L $CaCl_2$ (anhydrous), 0.40 g/L KCl, 0.06 g/L KH_2PO_4, 0.098 g/L $MgSO_4$ (anhydrous), 8 g/L NaCl, 0.048 g/L Na_2HPO_4 (anhydrous), 0.35 g/L $NaHCO_3$, 1 g/L glucose, pH 7.25 (*see* **Note 3**).

8. 75 cm^2 tissue culture flasks and 24-well tissue-culture treated plates (*see* **Note 3**).

9. LB-Miller broth: 10 g/L tryptone, 5 g/L yeast extract, 10 g/L NaCl.

10. LB-Miller agar plates: 10 g/L tryptone, 5 g/L yeast extract, 10 g/L NaCl, 15 g/L agar.

11. 100 mg/mL streptomycin in water (filter sterilize and store aliquots at –20 °C).

12. 50 mg/mL gentamicin in water (filter sterilize and store aliquots at –20 °C).

13. 52 mg/mL CHQ diphosphate salt (100 mM) in water (*see* **Note 6**).

14. Lysis buffer: 0.2 g sodium deoxycholate per 100 mL water (filter sterilize and store at room temperature).

15. Glycerol stock of *Salmonella enterica* serovar Typhimurium SL1344. Dilute 0.75 mL of an overnight culture of *S.* Typhimurium (LB-Miller broth + 100 μg/mL streptomycin) with an equal volume of 30 % (v/v) sterile glycerol in a 2 mL screw cap cryovial. Gently mix and freeze the tube at −80 °C (*see* **Note 7**).

16. Freshly streaked plate of *S.* Typhimurium. Using a sterile loop, pipette tip or toothpick, spread some bacteria from the top of the glycerol stock onto a LB-Miller agar plate containing 100 μg/mL streptomycin and incubate at 37 °C overnight. Keep plate at 4 °C for <1 week.

3   Methods

3.1   Gentamicin Protection Assay

1. Culture HeLa cells to confluency in growth medium in a 75 cm^2 tissue culture flask at 37 °C in a 5 % CO_2 incubator (*see* **Note 7**).

2. Remove and discard the culture medium. Rinse the cell monolayer with 5 mL PBS^{--} and discard.

3. Add 1 mL trypsin-EDTA and gently swirl the flask to cover the monolayer completely. Allow the cells to incubate for 2–5 min and dislodge by gently tapping the flask.

4. Add 4 mL growth medium and resuspend the cells by gentle pipetting.

5. Count the HeLa cells on a hemocytometer or cell counter.

6. Dilute the HeLa cells to 5×10^4/mL in growth medium and add 1 mL per well in a 24-well plate. Incubate at 37 °C in a CO_2 incubator for 18–24 h prior to infection.

7. Inoculate 2 mL of LB-Miller broth containing 100 μg/mL streptomycin with a single colony of *S.* Typhimurium (from a freshly streaked agar plate) in a 10 mL snap cap polypropylene tube. Shake at 220 rpm for 16–18 h at 37 °C (*see* **Note 8**).

8. Inoculate 10 mL LB-Miller broth (no antibiotics) with 0.3 mL overnight culture in a 125 mL Erlenmeyer flask. Shake at 220 rpm for 3.5 h at 37 °C. The optical density (OD_{600}) of the culture should be ~3.0, which corresponds to 3×10^9 CFU/mL (Fig. 1); (*see* **Notes 8** and **9**).

9. Centrifuge 1 mL of bacterial subculture in a microcentrifuge tube at $6000 \times g$ for 90 s. Remove the supernatant and resuspend the bacterial pellet in 1 mL HBSS by gently pipetting up and down. Proceed directly to infect monolayers (*see* **Note 10**).

10. Add washed bacteria to growth media in each well at a multiplicity of infection (MOI, the number of bacteria added

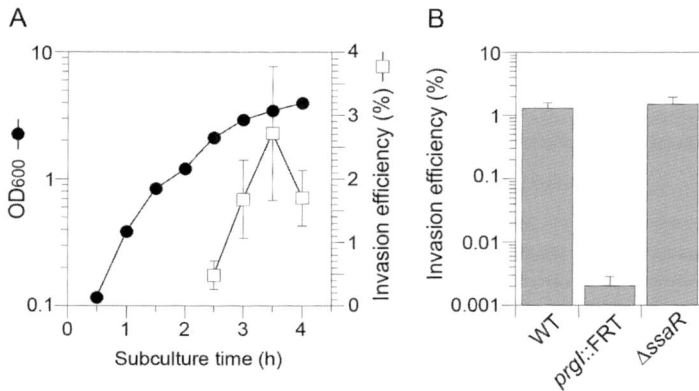

Fig. 1 *S*. Typhimurium SL1344 invasion efficiency peaks at 3.5 h subculture growth and is dependent upon SPI-1. (**a**) An aliquot of overnight culture (0.3 mL) of SL1344 wild-type bacteria was used to inoculate 10 mL LB-Miller broth and growth continued at 37 °C with shaking at 220 rpm. Bacterial growth (OD_{600}; *y*-axis on left side; *black circles*) and invasion efficiency into HeLa epithelial cells at 1 h p.i. (*y*-axis on right side; *open squares*) was monitored over subculture time. The peak of invasion efficiency (% bacterial inoculum that has been internalized) occurs at 3.5 h of subculture growth. Data is shown as mean ± SD from three independent experiments. (**b**) Bacterial subcultures were grown for 3.5 h as described in (**a**). The efficiency of SL1344 wild type, a SPI-1 deficient mutant (*prgI*::FRT) and a SPI-2 deficient mutant (Δ*ssaR*) internalization into HeLa epithelial cells was determined at 1 h p.i. Note that the invasion efficiency of the *prgI*::FRT mutant is reduced by >500-fold

per HeLa cell) of ~100. Start the lab timer (t_0). Two to three technical replicates per bacterial culture are recommended. Allow infection to proceed for 10 min at 37 °C in a 5 % CO_2 incubator (*see* **Note 11**).

11. Whilst the infection is underway, make serial dilutions of bacterial subcultures in DPBS^{++} and spot quintuple 10 µL aliquots on dry LB-Miller agar plates. Invert plates and incubate overnight at 37 °C. Count colony forming units (CFU) the next day to calculate the "bacterial inoculum".

12. Aspirate the media and wash monolayers thrice with 1 mL HBSS per well to remove most of the extracellular bacteria. Add fresh growth medium and continue incubation for 20 min at 37 °C in a CO_2 incubator (*see* **Note 12**).

13. At 30 min post-infection (p.i.), replace media with growth media containing 100 µg/mL gentamicin to kill any remaining non-internalized bacteria. Continue incubation for an additional 30 min (*see* **Note 13**).

14. At 1 h p.i., wash monolayers once with DPBS^{++} and solubilize by pipetting up and down in 1 mL lysis buffer (*see* **Note 14**).

15. Collect lysis buffer and transfer to 1.5 mL Eppendorf tubes. Immediately make serial dilutions in DPBS^{++} and spot quintuple 10 µL aliquots of each dilution on dry LB-Miller agar plates.

16. Invert plates, incubate at 37 °C overnight and count CFU for number of "internalized bacteria."

17. Invasion efficiency is calculated as internalized bacteria/bacterial inoculum $\times 100\%$ (Fig. 1); (see **Note 11**).

3.2 Chloroquine Resistance Assay

1. Culture HeLa cells to confluency in growth medium in a 75 cm^2 tissue culture flask at 37 °C in a 5% CO_2 incubator.

2. Remove and discard the culture medium. Rinse the cell monolayer with 5 mL PBS^{--} and discard.

3. Add 1 mL trypsin-EDTA and gently swirl the flask to cover the monolayer completely. Allow the cells to incubate for 2–5 min and dislodge by gently tapping the flask.

4. Add 4 mL growth medium and resuspend the cells by gentle pipetting.

5. Count the HeLa cells on a hemocytometer or cell counter.

6. Dilute the HeLa cells to 5×10^4/mL in growth medium and add 1 mL per well in a 24-well plate. For this assay, you will need twice as many wells for each bacterial strain or time point compared to the gentamicin protection assay (Fig. 2). Incubate at 37 °C in a CO_2 incubator for 18–24 h prior to infection.

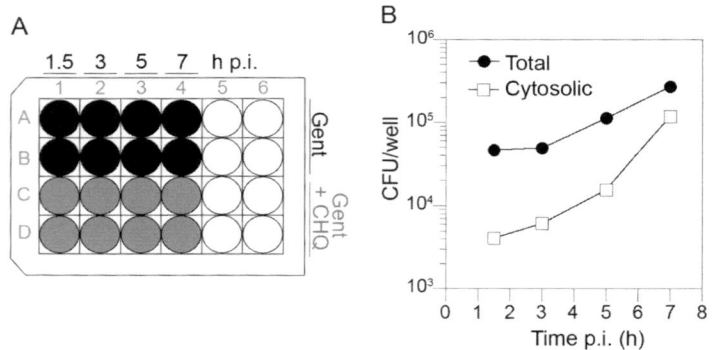

Fig. 2 CHQ resistance assay for determining cytosolic *S.* Typhimurium. (**a**) A typical 24-well tissue culture plate set-up for the CHQ resistance assay. For each time point (1.5, 3, 5, and 7 h p.i.), set aside duplicate wells for total bacteria (gentamicin, *black* wells) and duplicate wells for cytosolic bacteria (gentamicin plus CHQ, *grey* wells). (**b**) HeLa cells were seeded in a 24-well plate as shown in (**a**) and infected with *S.* Typhimurium SL1344 wild type. One hour prior to each time point, two wells were treated with CHQ. At the indicated times, monolayers were solubilized and CFU enumerated by plating on LB agar. *Black* circles = total bacteria, *open squares* = cytosolic bacteria. At 1.5 and 7 h p.i., approximately 15 and 50% of bacteria are present in the cytosol, respectively [21]

7. Inoculate 2 mL of LB-Miller broth containing 100 µg/mL streptomycin with a single colony of *S.* Typhimurium (from a freshly streaked agar plate) in a 10 mL snap cap polypropylene tube. Shake at 220 rpm for 16–18 h at 37 °C (*see* **Note 8**).

8. Inoculate 10 mL LB-Miller broth (no antibiotics) with 0.3 mL overnight culture in a 125 mL Erlenmeyer flask. Shake at 220 rpm for 3.5 h at 37 °C. The optical density (OD_{600}) of the culture should be ~3.0, which corresponds to 3×10^9 CFU/mL (Fig. 1); (*see* **Note 8** and **9**).

9. Centrifuge 1 mL of bacterial subculture in a microcentrifuge tube at $6000 \times g$ for 90 s. Remove the supernatant and resuspend the bacterial pellet in 1 mL HBSS by gently pipetting up and down. Proceed directly to infect monolayers (*see* **Note 10**).

10. Add washed bacteria to growth media in each well at an MOI of ~100. Start the lab timer (t_0). Two to three technical replicates per bacterial culture are recommended. Allow infection to proceed for 10 min at 37 °C in a 5 % CO_2 incubator (*see* **Note 11**).

11. Aspirate the media and wash monolayers thrice with 1 mL HBSS per well to remove most of the extracellular bacteria. Add fresh growth medium and continue incubation for 20 min at 37 °C in a CO_2 incubator.

12. At 30 min p.i., replace media with growth media containing 100 µg/mL gentamicin to eliminate any remaining non-internalized bacteria. Continue incubation for an additional 1 h (*see* **Notes 6** and **15**).

13. Reduce gentamicin concentration to 10 µg/mL thereafter.

14. To determine the extent of SCV lysis, infected cells are incubated in the presence of gentamicin with or without CHQ for 1 h (Fig. 2). For example, to quantify the percentage of bacteria in the cytosol at 90 min p.i., incubate half the designated wells with 100 µg/mL gentamicin and half with 100 µg/mL gentamicin plus 400 µM CHQ from 30 to 90 min p.i. For 5 h p.i., incubate half the designated wells with 10 µg/mL gentamicin and half with 10 µg/mL gentamicin plus 400 µM CHQ from 4 to 5 h p.i.

15. Wash monolayers once with DPBS[++] and solubilize by pipetting up and down in 1 mL lysis buffer (*see* **Note 14**).

16. Collect lysis buffer and transfer to 1.5 mL Eppendorf tubes. Immediately make serial dilutions in DPBS[++] and spot quintuple 10 µL aliquots of each dilution on dry LB-Miller agar plates.

17. Invert and incubate overnight at 37 °C.

18. Cytosolic bacteria (CHQ-resistant bacteria)=CFU from HeLa cells treated with gentamicin and CHQ; total internalized bacteria=CFU from HeLa cells treated with gentamicin alone (Fig. 2).

The percentage of cytosolic *S.* Typhimurium at any time point is calculated as CHQ-resistant bacteria/total internalized bacteria×100% (Fig. 2).

4 Notes

1. Many journals are adopting requirements for cell line validation for publication. Obtain cell lines from a reputable source, such as American Type Culture Collection (ATCC).

2. Passage number affects numerous characteristics of cell lines, including morphology, proliferation, responses to stimuli and genotype. Record cell line passage number and use low passage number cells (<passage 15 upon receipt from ATCC) to maximize experimental consistency.

3. Prepared tissue culture media and tissue culture consumables are available from numerous commercial suppliers.

4. L-glutamine is an unstable essential amino acid required by mammalian cells in tissue culture. Its stability is dependent upon temperature and pH. To minimize L-glutamine degradation to ammonia, L-glutamine can be stored in aliquots at −20 °C and added to media just before use. Alternatively, a stabilized form of L-glutamine, GlutaMAX, is available from Invitrogen.

5. There is considerable lot-to-lot variation in the composition of fetal bovine serum. To reduce experimental variation, it is recommended that users evaluate different serum samples (from different lots and/or suppliers) for mammalian cell growth and attachment, then purchase large quantities of the same serum lot. Serum with low endotoxin contamination is recommended. We have found that serum can be stored frozen at −20 °C for many years without performance loss.

6. CHQ is sensitive to light. We have found considerably less experimental variability if working solutions are prepared extemporaneously in growth media from a CHQ stock solution prepared on the day.

7. Here we describe the use of the gentamicin protection assay in HeLa epithelial cells, perhaps the most widely used cell line to study type III effector-mediated invasion and subsequent trafficking of the SCV. Theoretically, this assay is adaptable for the determination of *Salmonella* internalization into any adherent tissue culture cell line or primary cells. Parameters that may need to be modified include the cell seeding density, MOI, infection time and lysis buffer.

8. This protocol is tailored for *S.* Typhimurium SL1344 (*hisG rpsL*) [31], which is naturally resistant to streptomycin.

This strain is available from the *Salmonella* Genetic Stock Centre (SGSC). The gentamicin protection assay is applicable to other *S.* Typhimurium strains, such as ATCC14028 [32] and ST4/74 [33], or *S. enterica* serovars, but will require optimization of subculture conditions to determine the peak of invasion efficiency. *S.* Typhimurium 14028 is known to be less invasive than SL1344 due to an overall decreased expression of SPI-1 [34].

9. SPI-1 gene expression is induced under high salt conditions [35]. For this reason, we recommend growing bacterial cultures in LB-Miller broth (10 g NaCl/L, which is equivalent to 1% (w/v) NaCl) rather than Luria–Bertani (LB) broth, which contains 0.5% (w/v) NaCl by comparison.

10. To achieve an MOI of ~100, centrifuge 1 mL bacterial subculture at $6000 \times g$ for 90 s, discard the supernatant and gently resuspend the bacterial pellet in an equal volume of HBSS. Dilute tenfold (0.1 mL plus 0.9 mL HBSS) and add 10 μL of this diluted culture to each well (equivalent to 1 μL of subculture).

11. SPI-1 and flagella are co-regulated under a variety of conditions [14–16], and both are required for the efficient internalization of *S.* Typhimurium into non-phagocytic cells [9, 11, 12]. The invasion deficiency of flagella mutants, but not SPI-1 mutants, can be overcome by centrifuging bacteria ($700 \times g$ for 5 min) onto epithelial cells to initiate close contact [36], followed by the 10 min incubation step.

12. Bacteria that have recently been internalized into mammalian cells, i.e., those in nascent vacuoles, are likely to be vulnerable to the antibacterial actions of gentamicin. By including a 20 min chase period in the absence of antibiotic, this minimizes the susceptibility of bacteria in nascent vacuoles to gentamicin by allowing sufficient time for *Salmonella*-containing vacuole trafficking to occur.

13. The antibacterial activity of gentamicin is drastically affected by pH; the minimum inhibitory concentration (MIC) for gentamicin against *S.* Typhimurium is 8 μg/mL at pH 7.0 and 125 μg/mL at pH 5.7 [37]. Under certain conditions, internalized bacteria may be susceptible to the actions of gentamicin. For example, if infection or experimental treatment leads to a compromised plasma membrane, gentamicin can freely penetrate the host cell and cytosolic *S.* Typhimurium residing at neutral pH within the cytosol will be especially susceptible to gentamicin [27]. Alternatively, drug treatments that lead to neutralization of the SCV, e.g., bafilomycin, could increase the bioactivity of gentamicin against vacuolar *Salmonella*.

14. Sodium deoxycholate is an ionic detergent that lyses eukaryotic cell membranes, thereby releasing intracellular *Salmonella*. We have found that a working concentration of 0.2 % (w/v) has no effect on the viability of *S*. Typhimurium SL1344. Other detergents, e.g., Triton X-100, saponin, or even water, can be used to lyse eukaryotic cells, but care should be taken to ensure that these treatments do not kill intracellular bacteria or inhibit their growth on LB-Miller agar.

15. 400 μM CHQ is sufficient to kill vacuolar *Salmonella* in HeLa epithelial cells, without affecting cell viability. This concentration of CHQ is also suitable for determining cytosolic *S*. Typhimurium in HCT 116 colonic epithelial cells [21], HCT-8 colonic epithelial cells and J774A.1 macrophage-like cells. For adapting the CHQ assay to other tissue culture cells, the optimal CHQ concentration should be determined by the end user [21]. We typically test a concentration range of CHQ (200–800 μM) for its effect on vacuolar CFU and epithelial cell viability.

Acknowledgment

This work was supported by start-up funds from the Paul G. Allen School for Global Animal Health and the Stanley L. Adler Research Fund (LAK). TRP is supported by an NIH/NIAID Training Grant 2T32AI007025. JAK is the recipient of a Poncin Scholarship.

References

1. Magoffin RL, Spink WW (1950) The protection of intracellular Brucella against streptomycin alone and in combination with other antibiotics. J Lab Clin Med 36(6):959–960

2. Holmes B, Quie PG, Windhorst DB, Pollara B, Good RA (1966) Protection of phagocytized bacteria from the killing action of antibiotics. Nature 210(5041):1131–1132

3. Elsinghorst EA (1994) Measurement of invasion by gentamicin resistance. Meth Enzymol 236:405–420

4. Takano M, Ohishi Y, Okuda M, Yasuhara M, Hori R (1994) Transport of gentamicin and fluid-phase endocytosis markers in the LLC-PK1 kidney epithelial cell line. J Pharmacol Exp Ther 268(2):669–674

5. Myrdal SE, Steyger PS (2005) TRPV1 regulators mediate gentamicin penetration of cultured kidney cells. Hear Res 204(1–2):170–182

6. Seral C, Van Bambeke F, Tulkens PM (2003) Quantitative analysis of gentamicin, azithromycin, telithromycin, ciprofloxacin, moxifloxacin, and oritavancin (LY333328) activities against intracellular Staphylococcus aureus in mouse J774 macrophages. Antimicrob Agents Chemother 47(7):2283–2292

7. Sandoval RM, Dunn KW, Molitoris BA (2000) Gentamicin traffics rapidly and directly to the Golgi complex in LLC-PK(1) cells. Am J Physiol Renal Physiol 279(5):F884–F890

8. Myrdal SE, Johnson KC, Steyger PS (2005) Cytoplasmic and intra-nuclear binding of gentamicin does not require endocytosis. Hear Res 204(1–2):156–169

9. Galán JE, Curtiss R (1989) Cloning and molecular characterization of genes whose products allow Salmonella typhimurium to penetrate tissue culture cells. Proc Natl Acad Sci U S A 86(16):6383–6387

10. Haraga A, Ohlson MB, Miller SI (2008) Salmonellae interplay with host cells. Nat Rev Microbiol 6(1):53–66

11. Khoramian-Falsafi T, Harayama S, Kutsukake K, Pechère JC (1990) Effect of motility and

chemotaxis on the invasion of Salmonella typhimurium into HeLa cells. Microb Pathog 9(1):47–53

12. Jones BD, Lee CA, Falkow S (1992) Invasion by Salmonella typhimurium is affected by the direction of flagellar rotation. Infect Immun 60(6):2475–2480

13. van Asten FJAM, Hendriks HGCJM, Koninkx JFJG, van Dijk JE (2004) Flagella-mediated bacterial motility accelerates but is not required for Salmonella serotype Enteritidis invasion of differentiated Caco-2 cells. Int J Med Microbiol 294(6):395–399

14. Chubiz JEC, Golubeva YA, Lin D, Miller LD, Slauch JM (2010) FliZ regulates expression of the Salmonella pathogenicity island 1 invasion locus by controlling HilD protein activity in Salmonella enterica serovar typhimurium. J Bacteriol 192(23):6261–6270

15. Golubeva YA, Sadik AY, Ellermeier JR, Slauch JM (2012) Integrating global regulatory input into the Salmonella pathogenicity island 1 type III secretion system. Genetics 190(1):79–90

16. Singer HM, Kühne C, Deditius JA, Hughes KT, Erhardt M (2014) The Salmonella Spi1 virulence regulatory protein HilD directly activates transcription of the flagellar master operon flhDC. J Bacteriol 196(7):1448–1457

17. Ochman H, Soncini FC, Solomon F, Groisman EA (1996) Identification of a pathogenicity island required for Salmonella survival in host cells. Proc Natl Acad Sci U S A 93(15):7800–7804

18. Shea JE, Hensel M, Gleeson C, Holden DW (1996) Identification of a virulence locus encoding a second type III secretion system in Salmonella typhimurium. Proc Natl Acad Sci U S A 93(6):2593–2597

19. Figueira R, Holden DW (2012) Functions of the Salmonella pathogenicity island 2 (SPI-2) type III secretion system effectors. Microbiology 158(Pt 5):1147–1161

20. Birmingham CL, Brumell JH (2006) Autophagy recognizes intracellular Salmonella enterica serovar Typhimurium in damaged vacuoles. Autophagy 2(3):156–158

21. Knodler LA, Nair V, Steele-Mortimer O (2014) Quantitative assessment of cytosolic Salmonella in epithelial cells. PLoS One 9(1), e84681

22. Thurston TLM, Ryzhakov G, Bloor S, von Muhlinen N, Randow F (2009) The TBK1 adaptor and autophagy receptor NDP52 restricts the proliferation of ubiquitin-coated bacteria. Nat Immunol 10(11):1215–1221

23. Wild P et al (2011) Phosphorylation of the autophagy receptor optineurin restricts Salmonella growth. Science 333(6039):228–233

24. Tattoli I et al (2012) Amino acid starvation induced by invasive bacterial pathogens triggers an innate host defense program. Cell Host Microbe 11(6):563–575

25. Yu HB et al (2014) Autophagy facilitates salmonella replication in HeLa cells. MBio 5(2):e00865–14

26. Knodler LA et al (2010) Dissemination of invasive Salmonella via bacterial-induced extrusion of mucosal epithelia. Proc Natl Acad Sci U S A 107(41):17733–17738

27. Malik-Kale P, Winfree S, Steele-Mortimer O (2012) The bimodal lifestyle of intracellular Salmonella in epithelial cells: replication in the cytosol obscures defects in vacuolar replication. PLoS One 7(6), e38732

28. Steinberg TH (1994) Cellular transport of drugs. Clin Infect Dis 19(5):916–921

29. Finlay BB, Falkow S (1988) Comparison of the invasion strategies used by Salmonella choleraesuis, Shigella flexneri and Yersinia enterocolitica to enter cultured animal cells: endosome acidification is not required for bacterial invasion or intracellular replication. Biochimie 70(8):1089–1099

30. Zychlinsky A et al (1994) IpaB mediates macrophage apoptosis induced by Shigella flexneri. Mol Microbiol 11(4):619–627

31. Hoiseth SK, Stocker BA (1981) Aromatic-dependent Salmonella typhimurium are non-virulent and effective as live vaccines. Nature 291(5812):238–239

32. Jarvik T, Smillie C, Groisman EA, Ochman H (2010) Short-term signatures of evolutionary change in the Salmonella enterica serovar typhimurium 14028 genome. J Bacteriol 192(2):560–567

33. Jones PW, Collins P, Aitken MM (1988) Passive protection of calves against experimental infection with Salmonella typhimurium. Vet Rec 123(21):536–541

34. Clark L et al (2011) Differences in Salmonella enterica serovar Typhimurium strain invasiveness are associated with heterogeneity in SPI-1 gene expression. Microbiology 157(Pt 7):2072–2083

35. Bajaj V, Lucas RL, Hwang C, Lee CA (1996) Co-ordinate regulation of Salmonella typhimurium invasion genes by environmental and regulatory factors is mediated by control of hilA expression. Mol Microbiol 22(4):703–714

36. Ibarra JA et al (2010) Induction of Salmonella pathogenicity island 1 under different growth conditions can affect Salmonella-host cell interactions in vitro. Microbiology 156(Pt 4):1120–1133

37. Menashe O, Kaganskaya E, Baasov T, Yaron S (2008) Aminoglycosides affect intracellular Salmonella enterica serovars typhimurium and virchow. Antimicrob Agents Chemother 52(3):920–926

Chapter 20

Bacterial Binding, Phagocytosis, and Killing: Measurements Using Colony Forming Units

Kyle E. Novakowski, Dessi Loukov, Vikash Chawla, and Dawn M.E. Bowdish

Abstract

Herein we provide a colony forming unit (CFU)-based counting method for quantitating the bacterial binding, phagocytosis, and killing capacity of phagocytes. Although these functions can be measured by immuno-fluorescence and dye-based assays, quantitating CFUs is comparatively inexpensive and easy to perform. The protocol described below is easily modified for use with different phagocytes (e.g., macrophages, neutrophils, cell lines), types of bacteria or opsonic conditions.

Key words Phagocytosis, Bacterial killing, Particle binding, Macrophage, Gentamicin protection, Phagocyte

1 Introduction

Essential functions of the innate immune system include the detection, engulfment, and destruction of bacteria, which is primarily performed by neutrophils and macrophages. Profiling the functions of professional phagocytes provides valuable insight into innate immune function and can be used in a wide range of biological applications. These include testing the effect of a chemical compound, or characterizing the phenotype of a genetic knockout. Methods to quantify cellular association, phagocytosis, and killing by enumeration of viable microbial units were first described in the 1930s and 1940s [1]. Since then assays have diversified to include quantification of bacterial binding, uptake, and killing of fluorescently labeled bacteria by flow cytometry or fluorescence microscopy. However viability-based assays remain the most cost-effective and versatile.

To better understand the principles of this assay, a distinction must be made between binding and total cell association. Total cell association refers to bacteria that have been internalized by a phagocyte, *in addition to* bacteria that are bound, but remain

Roberto Botelho (ed.), *Phagocytosis and Phagosomes: Methods and Protocols*, Methods in Molecular Biology, vol. 1519, DOI 10.1007/978-1-4939-6581-6_20, © Springer Science+Business Media New York 2017

outside the cell. This distinction is important, because not all bound bacteria are phagocytosed. If, for example, the bacteria produce anti-phagocytic factors, or the presence of specific chemical inhibitors or gene deletions interfere with receptor expression, bacterial binding may occur without internalization [2–5]. One method to distinguish between bound and internalized bacteria is to differentially label internal and external bacteria using two different fluorescent dyes, or to use a pH-sensitive dye which only fluoresces in the low pH-environment of phagosome [6]. Unfortunately these methods require expensive reagents and analysis equipment (e.g., flow cytometers, fluorescence microscopes). Furthermore, examination of samples by microscopy can be time consuming and is relatively low-throughput.

An alternative approach is the use of viability-based assays. These assays are simpler and more affordable alternatives that can be used with a wide range of bacteria, some of which may not be amenable to fluorescent labeling by dyes or fluorescent proteins. Bacterial binding can be quantified by incubating phagocytes and bacteria at cold temperatures, which allows for phagocyte–bacteria interaction, but prevents internalization because processes such as actin polymerization and membrane trafficking do not occur at sub-physiological temperatures [7]. Total cell association can be quantified by incubating phagocytes and bacteria at physiologic temperatures (i.e. 37 °C), which allows for both bacterial binding and internalization. In order to distinguish between bound and internalized bacteria, extracellular bacteria can be killed through the addition of antibiotics. In order to determine the rate at which phagocytosed bacteria are killed, one can measure bacterial numbers over time. We have successfully used this assay to characterize the role of a macrophage receptor in anti-pneumococcal immunity, to study the effects of age on macrophage killing capacity using human and mouse macrophages and to discover drugs that increased macrophage phagocytosis [8–11]. Herein, we describe methods to characterize three distinct phagocyte functions: binding of particles/bacteria, phagocytosis/internalization of bacteria, and killing of bacteria.

2 Materials

Reagents used in this assay are specific to assays performed using *Streptococcus pneumoniae*. When performing this assay with other bacteria, reagents such as liquid culture media and agar growth plates can be substituted for appropriate growth media. This protocol can also be adapted for use with inert particles (*see* Subheading 3.1, **step 1**).

2.1 Bacteria and Microbiology Components

1. Early-to-mid-log phase bacteria (*see* **Notes 1** and **2**).

2. Appropriate liquid culture media (e.g., Luria–Bertani broth or tryptic soy broth). Follow manufacturer's instructions for broth preparation. Autoclave to sterilize. Store at 4 °C. Only open under sterile conditions (*see* **Note 3**).

3. Sterile water.

4. Hank's Balanced Salt Solution (HBSS).

5. Gentamicin solution: Dissolve gentamicin sulfate salt to a final concentration of 50 mg/mL in sterile water. Store at –20 °C in aliquots (*see* **Note 2**).

6. Tryptic soy agar (TSA)/Sheep's blood plates: dissolve 15 g of TSA in 500 mL deionized water. Autoclave to sterilize. Cool for 1.5 h in a 60 °C water bath. Under sterile conditions, add 25 mL sterile, defibrinated sheep's blood, and swirl gently to mix (*see* **Note 3**).

7. Spectrophotometer.

8. 15 mL polystyrene culture centrifuge tubes.

9. Incubator. The incubation conditions of plates for quantification of bacteria will vary depending on the bacteria being used.

2.2 Macrophage Components

1. Macrophages: any source of macrophages can be used, including bone marrow-derived, alveolar, resident or recruited peritoneal, blood monocyte-derived (*see* **Note 4**).

2. Phosphate buffered saline (PBS): Warm PBS to 37 °C prior to use.

3. Accutase cell detachment solution: Store at –20 °C in 10 mL aliquots. Warm to 37 °C prior to use (*see* **Note 5**).

4. Cell lifters (*see* **Note 6**).

5. Trypan blue, 0.4 %.

6. Hemocytometer.

7. Light microscope with a minimum objective magnification of 20×.

8. 50 mL conical tubes.

2.3 General Assay Components

1. 2 mL microcentrifuge tubes.

2. Vortex.

3. Microcentrifuge.

4. Serological pipettes (5, 10, and 25 mL).

5. Micropipettes and micropipette tips.

6. Appropriate disposal reagents and equipment for liquid biological waste. Please follow institutional requirements.

7. Nutating Mixer.

8. Black 96-well plate.

9. Spectrophotometer capable of reading fluorescence.

10. *If performing assays for particle uptake:* Fluorescent polystyrene microspheres, 0.5–2 µM (Polysciences, Warrington, PA, USA).

---

3   Methods

3.1 Bacterial Binding Assay

Carry out all procedures under sterile conditions in a biological safety cabinet. For the experiments using *S. pneumoniae* in Figs. 1, 2, and 3, TSB/Sheep's blood plates were used.

1. Grow liquid cultures to early or mid-log phase (*see* **Notes 1** and **7**). While waiting, prepare and label tubes containing 900 µL sterile water and appropriate media plates for the experiments.

2. When the liquid bacterial culture reaches early to mid-log growth (measured using OD_{600}, *see* **Note 1**) remove 1 mL into a 2 mL microcentrifuge tube and centrifuge at $12,000 \times g$ for 1 min.

3. Carefully remove the supernatant and resuspend the pellet in 1 mL HBSS. Keep on ice to prevent additional bacterial growth.

4. Visually inspect macrophage culture under a light microscope to ensure high cell viability and that the culture is free of contamination. Viable macrophages typically have a dendritic-like

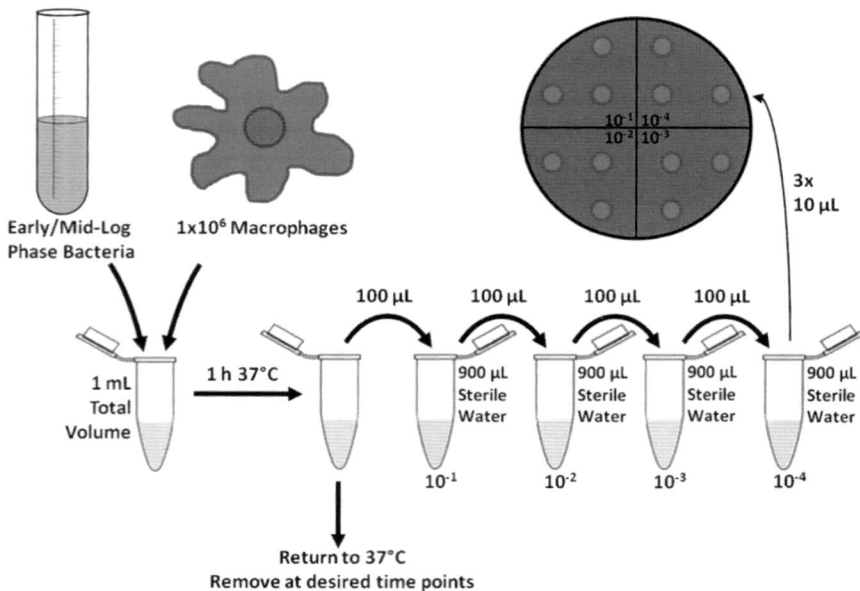

Fig. 1 A schematic outlining the major steps to perform a CFU-based bacterial killing assay

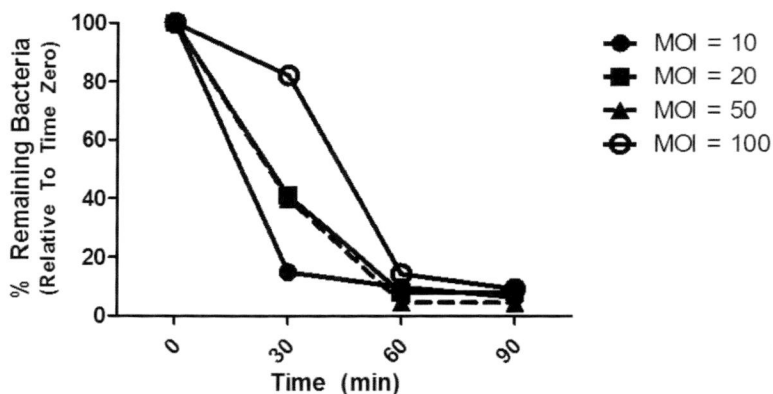

Fig. 2 A representative killing assay highlighting differences in kill curves with varying multiplicities of infection (MOI) using bone marrow-derived macrophages

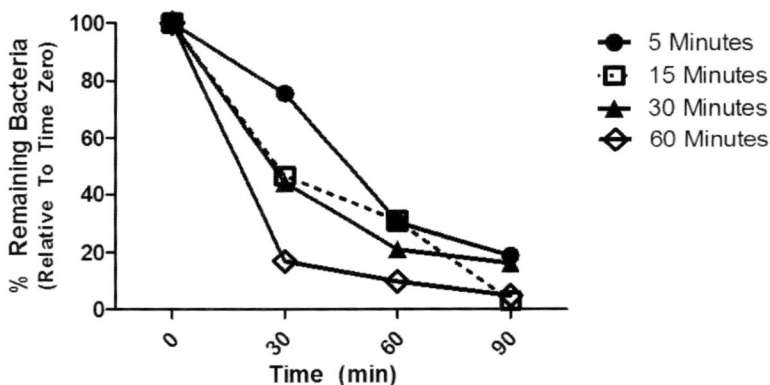

Fig. 3 A representative killing assay highlighting differences in kill curves with varying pre-incubation times before assaying bacterial killing using bone marrow-derived macrophages

morphology and should be strongly adherent. Floating or rounded cells generally indicate a loss of viability, which can be confirmed by trypan blue staining.

5. Remove macrophage culture media and gently wash with warm PBS.

6. Remove PBS and add 10 mL of warm Accutase cell detachment solution. Cover the plate and place in a 37 °C incubator for 10 min. Visually inspect the plate under a light microscope to determine cell detachment. Some cells should be completely detached and remaining adherent cells should be round in morphology. If cells still remain adherent, incubate for an additional 5–10 min.

7. Gently scrape the plate with a cell lifter, to remove any remaining attached cells (*see* **Note 6**). Pipette Accutase solution containing macrophages into a 50 mL conical tube. To maximize

macrophage recovery, wash the plate with 10 mL warm PBS and add to the 50 mL conical tube.

8. Centrifuge for 10 min at $500 \times g$.

9. Carefully remove all supernatant and gently resuspend the pellet in 1 mL warm PBS. Do not vortex.

10. Determine the cells/mL by diluting cells 1:1 with trypan blue and pipette 10 μL into a hemocytometer. Count viable cells following standard usage of a hemocytometer.

11. Add up to 1×10^6 viable macrophages into a 2 mL microcentrifuge tube. Bring the volume of solution up to 900 μL by adding PBS. Repeat for each desired time point. For example, if performing time points of 0, 30, 60, and 90 min, you will require 4 tubes containing equal numbers of macrophages (*see* **Note 8**).

12. Briefly vortex bacteria and add the desired multiplicity of infection (MOI) to each tube. The MOI will vary depending on the bacteria used in the assay and number/type of macrophages, but generally ranges from 10 to 100 (*see* **Note 9**, Fig. 2). In the case of *Streptococcus pneumoniae*, an MOI of 20 is optimal. The final volume of the macrophage/bacteria solution should be brought to 1 mL with PBS.

13. Incubate with gentle mixing at 4 °C for 1 h. This incubation allows for cell to cell association. At time points 0, 20, 40, and 60 min, remove the tube and centrifuge at $500 \times g$ for 5 min. At low speeds, macrophages (and bound bacteria) will form a pellet whereas unbound bacteria will remain in the supernatant.

14. Carefully remove the supernatant to remove unbound bacteria and gently resuspend in 1 mL cold PBS. Centrifuge at $500 \times g$, 4 °C for 5 min. Repeat wash two times.

15. Resuspend in 1 mL of PBS and lyse macrophages by pipetting 100 μL of the macrophage solution into 900 μL sterile water (prepared in **step 1**) (*see* **Note 10**). Briefly vortex. This tube is now a 10^{-1} dilution.

16. Perform a serial dilution by repeating **step 14** using 100 μL of the 10^{-1} dilution into 900 μL sterile water. Briefly vortex. This tube is now a 10^{-2} dilution. Repeat for 10^{-3} and 10^{-4} dilutions.

17. Working from the most dilute to the most concentrated dilution, briefly vortex and pipette three separate 10 μL drops onto the agar plate. Repeat for all other dilutions (*see* Fig. 1).

18. Allow the plate to dry at room temperature for approximately 20 min or until the droplets are no longer visible on the agar plate.

19. Incubate plates inverted in a 37 °C incubator overnight.

20. Count colonies for each dilution. This will quantitate all cell-associated bacteria. To determine CFU/mL or total number of internalized bacteria, multiply the number of colonies by the inverse of the dilution factor and by 10^2, to account for the 10 μL plating volume. For example, 17 colonies counted on a 10^{-3} droplet would be calculated as:

$$17 \times 10^3 \text{ (dilution factor)} \times 10^2 \text{ (plating volume)} = 1.7 \times 10^6 \text{ CFU/mL}$$
(*see* Fig. 1).

3.2 Gentamicin Protection Assay for Bacterial Phagocytosis

Carry out all procedures under sterile conditions in a biological safety cabinet. This protocol has been optimized for use with *Streptococcus pneumoniae*, but can be performed with other bacteria. Ensure that the bacteria are susceptible to this concentration of gentamicin prior to beginning the experiment (*see* **Note 2**).

1. Grow bacteria to early or mid-log phase in liquid culture (*see* **Note 1**). While waiting, prepare and label tubes containing 900 μL sterile water and TSB/Sheep's blood plates to be used in **steps 16** and **18**.

2. When the liquid bacterial culture reaches early to mid-log growth (as measured by OD_{600}) remove 1 mL into a 2 mL microcentrifuge tube and centrifuge at $12,000 \times g$ for 1 min. For *S. pneumoniae*, early log phase ($OD_{600} = 0.5$) is generally obtained from 1 to 4 h when prepared from a 1 mL freezer stock. Carefully remove the supernatant and resuspend the pellet in 1 mL HBSS. Keep on ice until use.

3. Visually inspect macrophage culture under a light microscope to ensure high cell viability and the culture is free of contamination. Macrophages typically have a dendritic-like morphology and should be strongly adherent.

4. Remove macrophage culture media and gently wash with warm PBS.

5. Remove PBS and add 10 mL of warm Accutase cell detachment solution. Cover the plate and place in a 37 °C incubator for 10 min. Visually inspect the plate under a light microscope to determine whether cells have detached. If cells still remain adherent, incubate for an additional 5–10 min.

6. Gently scrape the plate with a cell lifter to remove any loosely adherent cells (*see* **Note 6**). Pipette Accutase solution containing macrophages into a 50 mL conical tube. To maximize macrophage recovery, wash the plate with = 10 mL warm PBS and add to the 50 mL conical tube.

7. Centrifuge for 10 min at $500 \times g$.

8. Carefully remove all supernatant and gently resuspend the pellet in 1 mL of warm PBS. Do not vortex.

9. Determine the CFU/mL by diluting cells 1:1 with trypan blue and pipette 10 μL into a hemocytometer. Count viable cells following standard usage of a hemocytometer.

10. Add up to 1×10^6 viable macrophages into a 2 mL microcentrifuge tube. Bring the volume of solution up to 900 μL by adding PBS. Repeat for each desired time point. For example, if performing time points of 0, 30, 60, and 90 min, you will require four tubes containing equal numbers of macrophages (*see* **Note 8**).

11. Briefly vortex bacteria and add the desired multiplicity of infection (MOI) to each tube (*see* Fig. 2). The MOI will vary depending on the bacteria used in the assay and number/type of macrophages. In the examples in Figs. 2 and 3, *Streptococcus pneumoniae* was added at an MOI of 20. The final volume of the macrophage/bacteria solution should be brought to 1 mL with PBS.

12. Incubate with gentle mixing at 37 °C for 30 min to allow for bacterial binding and internalization (*see* Fig. 3). After this preincubation, at time points +0, +20, +40, and +60 min, remove the tube and centrifuge at $500 \times g$ for 5 min.

13. Carefully remove the supernatant to remove unbound bacteria and gently resuspend in 1 mL HBSS + 50 μg/mL gentamicin. Incubate with gentle mixing at 37 °C for 40 min (*see* **Note 2**).

14. Add 1 mL HBSS and centrifuge at $500 \times g$ for 5 min. Wash cells two additional times with 1 mL HBSS to completely remove gentamicin.

15. Following the final wash, resuspend in 1 mL HBSS and lyse macrophages by pipetting 100 μL of the macrophage solution into 900 μL sterile water (prepared in **step 1**) (*see* **Note 9**). Briefly vortex. This tube is now a 10^{-1} dilution.

16. Perform a serial dilution by repeating **step 16** using 100 μL of the 10^{-1} dilution into 900 μL sterile water. Briefly vortex. This tube is now a 10^{-2} dilution. Repeat for 10^{-3} and 10^{-4} dilutions.

17. Working from the most dilute to the most concentrated dilution, briefly vortex and pipette three separate 10 μL drops onto the agar plate. Repeat for all other dilutions.

18. Allow the plate to dry at room temperature for approximately 20 min or until the droplets are no longer visible on the agar plate.

19. Incubate plates inverted in a 37 °C incubator overnight.

20. Count colonies for each dilution. To determine CFU/mL or total number of internalized bacteria, multiply the number of colonies by the inverse of the dilution factor and by 10^2, to

account for plating volume. For example, 17 colonies counted on a 10^{-3} droplet would be calculated as:

$$17 \times 10^3 \,(\text{dilution factor}) \times 10^2 \,(\text{plating volume}) = 1.7 \times 10^6 \,\text{CFU/} \\ \text{mL} \,(see \text{ Fig. 1}).$$

3.3 Particle Binding Assay 12

This adapted protocol uses inert fluorescent particles such as polystyrene microspheres in place of bacteria. When using inert particles, attention should be given to particle size since some cell lines and phagocytes cannot internalize particles larger than 1–2 μm. In general, particles greater than 0.5 μm are internalized by phagocytosis and those smaller than 0.5 μm are internalized by endocytosis [13]. Particles may be coated with ligands, such as bacterial cell wall components or antibodies either by passive adsorption or covalent coupling. Finally, it is advisable to perform a dose-response experiment to determine optimal particle–phagocyte ratios.

1. Visually inspect macrophage culture under a light microscope to ensure high cell viability and the culture is free of contamination. Macrophages typically have a dendritic-like morphology and should be strongly adherent.

2. Remove macrophage culture media and gently wash with warm PBS.

3. Remove PBS and add 10 mL of warm Accutase cell detachment solution. Cover the plate and place in a 37 °C incubator for 10 min. Visually inspect the plate under a light microscope to determine whether cells have detached. If cells still remain adherent, incubate for an additional 5–10 min.

4. Gently scrape the plate with a cell lifter to remove any loosely adherent cells (*see* **Note 6**). Pipette Accutase solution containing macrophages into a 50 mL conical tube. To maximize macrophage recovery, wash the plate with 10 mL warm PBS and add to the 50 mL conical tube.

5. Centrifuge for 10 min at $500 \times g$.

6. Carefully remove all supernatant and gently resuspend the pellet in 1 mL of warm PBS. Do not vortex.

7. Determine the CFU/mL by diluting cells 1:1 with trypan blue and pipette 10 μL into a hemocytometer. Count viable cells following standard usage of a hemocytometer.

8. Add up to 1×10^6 viable macrophages into a 2 mL microcentrifuge tube. Bring the volume of solution up to 900 μL by adding PBS. Repeat for each desired time point. For example, if performing time points of 0, 30, 60 and 90 min, you will require four tubes containing equal numbers of macrophages (*see* **Note 8**).

9. Vortex particles and add at the desired particle–macrophage ratio and bring the total volume up to 1 mL with either PBS or cell culture media

10. Incubate on with gentle agitation at 4 °C for 1 h. At time points 0, 20, 40, and 60 min, remove the tube and centrifuge at 500 ×g for 5 min.

11. Carefully remove the supernatant to remove unbound particles and gently resuspend in 1 mL PBS or media. Centrifuge at 500 ×g for 5 min. Repeat wash twice more.

12. Resuspend in 200 μL PBS or media.

13. Add suspension to one well of a black 96-well plate. Read at the excitation and emission recommended by the particle manufacturer.

3.4 Bacterial Killing Assay

Carry out all procedures under sterile conditions in a biological safety cabinet. This protocol has been optimized for use with *Streptococcus pneumoniae*, but can be performed with other bacteria. Please see additional notes for assay modifications.

1. Grow bacteria to early or mid-log phase in liquid culture (*see* **Note 1**). While waiting, prepare and label tubes containing 900 μL sterile water and TSB/Sheep's blood plates to be used in **steps 16** and **18** (*see* Fig. 1 for experimental schematic).

2. When the liquid bacterial culture reaches the desired OD_{600}, remove 1 mL into a 2 mL microcentrifuge tube and centrifuge at 12,000 ×g for 1 min.

3. Carefully remove the supernatant and resuspend the pellet in 1 mL HBSS. Keep on ice to prevent additional bacterial growth.

4. Visually inspect the macrophage culture under a light microscope to ensure high cell viability and that the culture is free of contamination. Macrophages typically have a dendritic-like morphology and should be strongly adherent.

5. Remove macrophage culture media and gently wash with warm PBS.

6. Remove PBS and add 10 mL of warm Accutase cell detachment solution. Cover the plate and place in a 37 °C incubator for 10 min. Visually inspect the plate under a light microscope to determine cell detachment. Some cells should be completely detached and remaining adherent cells should be round in morphology. If cells still remain adherent, incubate for an additional 5–10 min.

7. Gently scrape the plate with a cell lifter, covering all surface area (*see* **Note 6**). Pipette Accutase solution containing macrophages into a 50 mL conical tube. To maximize macrophage recovery, wash the plate with 10 mL warm PBS and add to the 50 mL conical tube.

8. Centrifuge for 10 min at 500 ×g.

9. Carefully remove all supernatant and gently resuspend the pellet in 1 mL warm PBS. Do not vortex.

10. Determine the CFU/mL by diluting cells 1:1 with trypan blue and pipette 10 μL into a hemocytometer. Count viable cells following standard usage of a hemocytometer.

11. Add up to 1×10^6 viable macrophages into a 2 mL microcentrifuge tube. Bring the volume of solution to 900 μL by adding PBS.

12. Briefly vortex bacteria and add the desired multiplicity of infection (MOI) to each tube (*see* Fig. 2). The MOI will vary depending on the bacteria used in the assay and number/type of macrophages. In the examples in Figs. 2 and 3, *Streptococcus pneumoniae* was added at an MOI of 20. The final volume of the macrophage/bacteria solution should be brought to 1 mL with PBS.

13. Incubate with gentle mixing at 37 °C for 30 min to allow for bacterial binding and internalization (*see* Fig. 3). After this pre-incubation, remove the tube and centrifuge at $500 \times g$ for 5 min.

14. Carefully remove the supernatant to remove unbound bacteria and gently resuspend in 1 mL HBSS.

15. Wash cells 1 additional time with 1 mL HBSS to completely remove unbound bacteria.

16. Following the final wash, resuspend in 1 mL HBSS.

17. Lyse macrophages by pipetting 100 μL of the macrophage solution into 900 μL sterile water (prepared in **step 1**) (*see* **Note 9**). Briefly vortex. This tube is now a 10^{-1} dilution at time point +0 min. Return the original tube containing macrophages and bound/internalized bacteria to the mixer at 37 °C for time points of +30, +60, and +90 min, +120 and +180 min.

18. Perform a serial dilution using 100 μL of the 10^{-1} dilution into 900 μL sterile water for each time point. Briefly vortex. This tube is now a 10^{-2} dilution. Repeat for 10^{-3} and 10^{-4} dilutions.

19. Working from the most dilute to the most concentrated dilution, briefly vortex and pipette three separate 10 μL drops onto the agar plate. Repeat for all other dilutions.

20. Allow the plate to dry at room temperature for approximately 20 min or until the droplets are no longer visible on the agar plate.

21. Repeat **steps 17–20** for time points of +30, +60, and +90, +120, and +180 min.

22. Incubate plates inverted in a 37 °C incubator overnight.

23. Count colonies for each dilution. To determine CFU/mL or total number of internalized bacteria, multiply the number of colonies by the inverse of the dilution factor and by 10^2, to account for plating volume. For example, 17 colonies counted on a 10^{-3} droplet would be calculated as:

17×10^3 (dilution factor) $\times 10^2$ (plating volume) $= 1.7 \times 10^6$ CFU/mL (*see* Fig. 1).

4   Notes

1. Accurate quantitation of the MOI is essential to ensure reproducibility. Therefore, it is critical to perform a growth curve for the strain of bacteria to be used in the assay in order to prepare the assay with an accurate MOI. Determine the CFU/mL when the culture is analyzed via a spectrophotometer at OD_{600}.

2. The sensitivity of a given strain of bacteria to gentamicin should be titrated by performing a standard kill curve procedure in the presence of varying doses of gentamicin at 10 min time points for approximately 1 h.

3. The type of liquid culture media, agar plate and incubation conditions can be modified depending on the strain of bacteria being used for the assay.

4. Permeability of the phagocytic cells should be given consideration when performing a gentamicin protection assay, as gentamicin can penetrate permeabilized membranes and kill internalized bacteria. This can occur, for example, when cells are forcefully washed or stored at incorrect temperatures.

5. Accutase is preferred over trypsin, as trypsin has been shown to cleave some phagocytic receptors. If assaying the relevance of specific receptors, it is advisable to treat cells with Accutase and check receptor expression via FACS or IF microscopy. Include a saline or untreated control.

6. Cell viability can be significantly enhanced by using cell lifters in place of cell scrapers.

7. Before using this assay to quantify bacterial binding, ensure that the bacteria will survive at this temperature for the desired time course. To do so, grow bacteria to early or mid-log phase, perform a serial dilution and plate droplets to determine CFU/mL. Then incubate bacteria at 4 °C. At desired the time point(s), remove bacteria, perform serial dilutions and plate. Following overnight incubation at optimal growth conditions for the desired bacteria, compare initial CFU/mL to CFU/mL at each time point.

8. A minimum of 1×10^5 macrophages are required to perform this assay. In our experience 1×10^6 macrophages per condition increases reproducibility.

9. Bacteria should be titrated using serial dilutions in PBS, plated on appropriate solid agar medium to count CFUs.

10. Lysis of macrophages ensures that bacterial killing is immediately halted; however plating should be performed immediately to ensure that bacterial viability doesn't decrease.

References

1. Cohn ZA, Morse SI (1959) Interactions between rabbit polymorphonuclear leucocytes and staphylococci. J Exp Med 110:419–443
2. Fox EN (1974) M proteins of group A Streptococci. Bacteriol Rev 38:57–86
3. Cho K, Arimoto T, Igarashi T, Yamamoto M (2013) Involvement of lipoprotein PpiA of Streptococcus gordonii in evasion of phagocytosis by macrophages. Mol Oral Microbiol 28:379–391. doi:10.1111/omi.12031
4. Mukouhara T, Arimoto T, Cho K et al (2011) Surface lipoprotein PpiA of Streptococcus mutans suppresses scavenger receptor MARCO-dependent phagocytosis by macrophages. Infect Immun 79:4933–4940. doi:10.1128/IAI.05693-11
5. Dilworth JA, Hendley JO, Mandell GL (1975) Attachment and ingestion of gonococci by human neutrophils. Infect Immun 11:512–516
6. Campbell PA, Canono BP, Drevets DA (2001) Measurement of bacterial ingestion and killing by macrophages. Curr Protoc Immunol Chapter 14: Unit 14.6. doi: 10.1002/0471142735.im1406s12
7. Underhill DM, Ozinsky A (2002) Phagocytosis of microbes: complexity in action. Annu Rev Immunol 20:825–852. doi:10.1146/annurev.immunol.20.103001.114744
8. Dorrington MG, Roche AM, Chauvin SE et al (2013) MARCO Is required for TLR2- and Nod2-mediated responses to Streptococcus pneumoniae and clearance of pneumococcal colonization in the murine nasopharynx. J Immunol 190:250–258. doi:10.4049/jimmunol.1202113
9. Puchta A, Naidoo A, Verschoor CP et al (2016) TNF drives monocyte dysfunction with age and results in impaired anti-pneumococcal immunity. PLoS Pathog 12, e1005368. doi:10.1371/journal.ppat.1005368
10. Verschoor CP, Johnstone J, Loeb M et al (2014) Anti-pneumococcal deficits of monocyte-derived macrophages from the advanced-age, frail elderly and related impairments in PI3K-AKT signaling. Hum Immunol 75:1192–1196. doi:10.1016/j.humimm.2014.10.004
11. Perry JA, Koteva K, Verschoor CP et al (2015) A macrophage-stimulating compound from a screen of microbial natural products. J Antibiot 68:40–46. doi:10.1038/ja.2014.83
12. Novakowski KE, Huynh A, Han S, Dorrington MG, Yin C, Tu Z, Pelka P, Whyte P, Guarné A, Sakamoto K, Bowdish DM (2016) A naturally occurring transcript variant of MARCO reveals the SRCR domain is critical for function. Immunol Cell Biol 94(7):646–55. doi:10.1038/icb.2016.20
13. Mellman I (1996) Endocytosis and molecular sorting. Annu Rev Cell Dev Biol 12:575–625. doi:10.1146/annurev.cellbio.12.1.575

Chapter 21

Filamentous Bacteria as Targets to Study Phagocytosis

Akriti Prashar, Sana I. S. Khan, and Mauricio R. Terebiznik

Abstract

Filamentous targets are internalized via phagocytic cups that last for several minutes before closing to form a phagosome. This characteristic offers the possibility to study key events in phagocytosis with greater spatial and temporal resolution than is possible to achieve using spherical particles, for which the transition from a phagocytic cup to an enclosed phagosome occurs within a few seconds after particle attachment. In this chapter, we provide methodologies to prepare filamentous bacteria and describe how they can be used as targets to study different aspects of phagocytosis.

Key words Phagocytosis, Target morphology, Filamentous bacteria, Phagosome maturation, Phagocytic cup formation

1 Introduction

The canonical model of phagocytosis has been delineated using spheroidal targets [1]. However, the uptake of these targets occurs within seconds making the study of the initial events of phagocytosis such as the formation of a phagocytic cup, its remodeling and closing, and the scission of the nascent phagosome from the plasma membrane technically challenging. This limitation can be overcome by different techniques that hinder the ability of macrophages to enclose targets including: parachuting phagocytes onto a defined planar surface coated with IgG (also known as frustrated phagocytosis), using pharmacological treatments, dominant negative mutants or by gene silencing approaches [2, 3]. Nonetheless, these methodologies arrest phagocytosis, presenting limitations for the study of phagocytic cup remodelling, phagosome formation and maturation.

In contrast to spheroidal targets, filamentous bacteria are internalized via long-lasting phagocytic cups that extend along the length of the targets, which can be longer than 50 µm. Consequently, the resulting phagocytic cups can easily persist for several minutes, depending on the length of the filament. This allows for the study

Roberto Botelho (ed.), *Phagocytosis and Phagosomes: Methods and Protocols*, Methods in Molecular Biology, vol. 1519, DOI 10.1007/978-1-4939-6581-6_21, © Springer Science+Business Media New York 2017

of phagocytic cup formation and remodelling, and phagosome formation and maturation in unprecedented spatiotemporal detail. Also, since these phagocytic cups fuse with endosomal compartments, as we described elsewhere in [4], these targets could also be utilized for studying events that correspond to phagosome maturation for spheroidal targets. Furthermore, by using biological targets that phagocytes may potentially encounter in vivo, additional parameters such as target surfaces and opsonins can be modified to examine different signalling pathways and their roles in the process of phagocytosis. These filamentous bacterial targets can be easily modified by cross-linking to fluorescent probes and can be used in real-time microscopy-based assays. Using this model we recently demonstrated that filamentous target morphology affects phagosome maturation. This has important consequences for the outcome of phagocytosis as longer filaments of *Legionella pneumophila* survived and replicated more frequently than the bacillary forms in macrophages [4]. The methods described here can be used to study other filamentous pathogens like fungi or specialized host cells like neutrophils to assess the role of target morphology in host–pathogen interactions in addition to studying phagosome morphogenesis.

This chapter provides details regarding the use of filamentous bacteria as phagocytic targets to study the progression of the various stages involved in phagocytosis [4].

2 Materials

1. Luria broth (LB).
2. LB with 25 % glycerol (v/v).
3. 1× phosphate-buffered saline (PBS): 0.144 g/L KH_2PO_4, 9.00 g/L NaCl, 0.795 g/L Na_2HPO_4.
4. Anti-*Legionella pneumophila* IgG monoclonal antibody (EMD Millipore, Etobicoke, ON, Canada).
5. 1× PBS with 5 % (w/v) skim milk.
6. Fluorescently labeled secondary antibodies.
7. 16 % paraformaldehyde (PFA).
8. Dulbecco's Modification of Eagle's Medium (DMEM) supplemented with 10 % heat-inactivated fetal bovine serum (FBS).
9. 1× PBS with 0.1 % (v/v) Triton X-100.
10. Liquid buffered yeast extract (BYE): 54.9 mM ACES buffer ((*N*-(2-Acetamido)-2-aminoethanesulfonic acid)), 10 g/L yeast extract, 6.84 mM α-ketoglutarate, 2.28 mM l-cysteine HCl, and 33.6 µM ferric pyrophosphate.

11. Buffered charcoal yeast extract agar (BCYE): 54.9 mM ACES, 10 g/L yeast extract, 6.84 mM α-ketoglutarate, 2.28 mM l-cysteine HCl, 33.6 µM ferric pyrophosphate, 11.4 g/L bacteriological agar, 2 g/L acid-washed activated charcoal with HCl, and 100 µg/mL thymidine.

12. DQ-BSA resuspended in 1× PBS to stock concentration of 1 mg/mL and stored at 4 °C for 2–4 weeks (Thermo Fisher Scientific, Burlington, ON, Canada).

13. Carbodiimide (or EDAC) resuspended in water, prepared fresh before use.

14. 25 mg/mL Ciprofloxacin in water; stored at –20 °C (Sigma-Aldrich, Oakville, ON, Canada).

15. Bacteria for generating targets: *Escherichia coli*, *Salmonella typhimurium* or *Legionella pneumophila*.

16. RAW 264.7 macrophages.

17. 1.0 N potassium hydroxide.

18. 0.1 M sodium borate, pH 8.0 prepared the day of the experiment.

19. 1.0 M glycine in water prepared fresh before being diluted for use in experiments.

20. Fluorescent dextrans of different molecular weights (Thermo Fisher Scientific, Burlington, ON, Canada).

21. Confocal microscope for fixed and/or live-cell imaging.

3   Methods

All reagents and media were prepared using distilled water. All bacterial culturing procedures were performed using standard aseptic techniques and according to biosafety regulations.

3.1 Obtaining Filamentous Bacterial Targets

Elongated polystyrene particles have been utilized as targets for phagocytosis [5]. However, filamentous bacteria offer an advantage over these particles, as they are more flexible and therefore easily accommodated inside the macrophages. This property allows macrophages to internalize filaments >100 µm. Filamentous bacterial targets can be opsonized, or utilized without opsonization, and can be generated from bacteria expressing fluorescent proteins and conjugated to bioactive, fluorogenic and fluorescent probes for microscopy-based, real-time functional assays [6, 7].

3.1.1 Generating Filamentous Bacterial Targets of Escherichia coli or Salmonella typhimurium

Filamentous growth in bacteria can be induced by mutations in genes controlling cell division [8, 9] or by using antibiotics that inhibit bacterial cell division as described below.

1. Prepare Luria broth (LB) media, autoclave to sterilize and store as per the manufacturer's directions. Generally, sterile LB media can be stored at 4 °C for several months.

2. Take room temperature LB media and transfer 3 mL to a sterile 10 mL bacterial culture tube.

3. Using a sterile inoculating loop, transfer bacteria to this culture tube from stocks maintained at −80 °C in sterile LB media containing 25 % glycerol (v/v).

4. Place the culture tubes in a shaking incubator and shake at 200 rpm at 37 °C overnight.

5. Make a 1:10 dilution of the overnight culture in LB and measure the optical density at 600 nm (OD_{600}). Prepare a subculture in a bacterial culture tube using LB to obtain a final volume of 3 mL and a starting OD_{600} of 0.05. Add 0.5 µg/mL of ciprofloxacin (see **Note 1**) to the culture and place in a shaking incubator to shake at 200 rpm at 37 °C for 4 h.

6. Pellet the bacteria by centrifugation at $1000 \times g$ for 10 min at 4 °C.

7. Wash the pellet twice with 1× PBS and resuspend in 4% PFA (v/v in 1× PBS) to kill and fix bacteria. These PFA-treated bacteria can be stored at 4 °C for several months (see **Note 2**).

3.1.2 Generating Filamentous Bacterial Targets of L. pneumophila in BCYE Agar

Filamentous *L. pneumophila* targets can be obtained by growing bacteria in broth cultures or by resuspending colonies from cultures on agar as described below. Bacteria grown using these procedures yield cultures that are heterogeneous and contain bacteria of different lengths [4]. However, they are enriched in long bacterial filaments [10]. For studies requiring homogeneous populations of filamentous bacterial targets, the cultures can be filtered as described in [11].

1. Prepare BCYE media and autoclave to sterilize. Cool the media to 56 °C and add l-cysteine, ferric pyrophosphate and thymidine supplements that have been filter-sterilized (see **Note 3**). Adjust pH to 6.9 using 1 N KOH that has been filter-sterilized. Keep stirring the media to ensure that the charcoal remains well suspended. Pour 25 mL of media into sterile bacteriological plates and allow them to cool overnight at room temperature. Store the plates at 4 °C (see **Note 4**).

2. Using a sterile inoculating loop, streak *L. pneumophila* from frozen glycerol stocks maintained at −80 °C in sterile LB media containing 25 % glycerol (v/v) over one-third of the plate. Drag the sterile loop over the previously streaked bacteria once and streak over another one-third of the plate. Repeat this step once more over the remaining one-third of the plate. Invert the plate and incubate it at 37 °C and 5 % CO_2 for 3–4 days to obtain bacterial colonies.

3. Bacteria growing on these plates are used as an inoculum to obtain filamentous targets to study phagocytosis (see **Note 5**).

3.1.3 Generating Filamentous Bacterial Targets of L. pneumophila in BYE Media

1. Prepare BYE broth media and sterilize by autoclaving. Cool the media to 56 °C and add l-cysteine and ferric pyrophosphate that have been filter-sterilized (*see* **Note 3**). Adjust pH to 6.9 using 1 N KOH that has been filter-sterilized. Store media at 4 °C (*see* **Note 6**).

2. Transfer 25 mL of media to a sterile 125 mL Erlenmeyer flask. Remove a single bacterial colony from a 3–4 day old BCYE agar plate using a sterile inoculating loop and resuspend the bacteria well in the media.

3. Place the flask containing the bacteria in a shaking incubator set to 37 °C. Shake the cultures at 100 rpm until they reach an OD_{600} of 3.5–4.0, which usually takes 24 h. By using these slow agitation conditions, the cultures obtained will be enriched in filamentous bacteria.

3.1.4 Preparing Fixed Filamentous Bacterial Targets

1. If using BYE broth cultures, transfer 25 mL of the OD_{600} 3.5–4.0 *L. pneumophila* culture to a 50 mL conical tube. Centrifuge at $1000 \times g$ for 10 min to pellet the bacteria.

2. If using bacteria cultured on BCYE agar, use a sterile inoculating loop and resuspend colonies in 1× PBS. Generally, the colonies from an entire plate can be resuspendend in 25 mL of 1× PBS.

3. Wash the pellet twice with 1× PBS and resuspend the pellet in 4% PFA (v/v in 1× PBS) (*see* **Note 7**).

4. The fixed bacteria can be stored in PFA at 4 °C for several months (*see* **Note 2**).

3.2 Phagocytosis Assay Using Filamentous Bacterial Targets

Although the following methodology mentions RAW 264.7 macrophages, it can be used to perform phagocytosis on any cultured macrophage cell line, primary macrophages or monocyte-derived macrophages.

1. Grow RAW 264.7 cells in DMEM supplemented with 10% FBS (hereafter referred to as tissue culture medium) at 37 °C and 5% CO_2. Use 25 cm^3 cell culture-treated flasks to provide cells with a consistent growth surface.

2. Once cells are ~80–85% confluent, remove the medium from the flask containing macrophages and replace with 5 mL of fresh tissue culture medium. Gently scrape cells using a cell scraper and dislodge cells by gently pipetting up and down.

3. Count cells using a haemocytometer and plate macrophages to achieve a final density of 5×10^5 cells per well (18 mm diameter) at the time of performing the assay.

4. Wash glass coverslips in sterile ultrapure water (*see* **Note 8**). Add 1 coverslip to each well of a 12-well tissue culture-treated plate. To each well, add the calculated volume of cell suspension from **step 3** and distribute contents by swirling.

5. Incubate cells at 37 °C with 5 % CO_2 to allow them to adhere.

6. Prepare the fixed bacterial targets by diluting the stock to yield 1 mL of bacteria at OD 2.0 in 1× PBS. Wash targets three times by centrifugation for 2 min at $10,000 \times g$, removing the supernatant and resuspending in 1× PBS.

7. Opsonize targets by incubating with 0.1 mg/mL of anti-Lp1 (*see* **Notes 9** and **10**) for 1 h at room temperature. Wash targets three times as described above and resuspend pellet in 1 mL of 1× PBS.

8. Place the macrophages plated on tissue culture plates in a pre-cooled centrifuge to cool cells at 15 °C for 5 min (*see* **Note 11**). Add 50 µL of the opsonized fixed bacterial targets to each well, yielding a 150:1 target-to-cell ratio (*see* **Note 12**).

9. Centrifuge the plate at 15 °C and $300 \times g$ for 5 min to synchronize target binding. Incubate the cells at 37 °C and 5 % CO_2 for 5 min to enhance the attachment of bacterial targets to the macrophages (*see* **Note 13**).

10. Wash the cells as follows: remove media from the wells and replace with 1 mL of 1× PBS warmed to 37 °C. Swirl contents and remove PBS. Repeat for a total of three washes. Add 1 mL of fresh tissue culture medium equilibrated at 37 °C and 5 % CO_2 to each well and incubate cells for desired periods.

11. The cells can be visualized by live-cell microscopy or fixed for immunofluorescence imaging. To fix cells remove the medium and wash cells three times with 1× PBS as described. Remove PBS and add 1 mL of 4 % PFA (v/v in 1× PBS) to each well. Incubate at room temperature for 20 min and wash three times with 1× PBS.

12. Store coverslips in 1× PBS until immunolabeling of the desired targets is performed.

3.3 Measuring Phagocytic Internalization Using Filamentous Bacteria

A phagocytic index assessing the uptake of spheroidal targets can be obtained by distinguishing between external and intracellular targets by immunofluorescence (also referred to as differential immunofluorescence) [12]. However, the prolonged phagocytic uptake stage observed with filamentous bacteria provides an interval where targets are partially internalized in a phagocytic cup. Therefore, by using differential immunolabeling as outlined below, it is not only possible to determine phagocytic index but also to measure the length of segments of the targets inside the phagocytic cup over time and determine the kinetic parameters of target internalization. This is illustrated in Fig. 1a where PFA fixed RFP-expressing *L. pneumophila* is being internalized through a long-lasting phagocytic cup formed in a RAW 264.7 macrophage 10 min after the binding was synchronized. The external segment of the filament that was differentially immunolabeled as described in

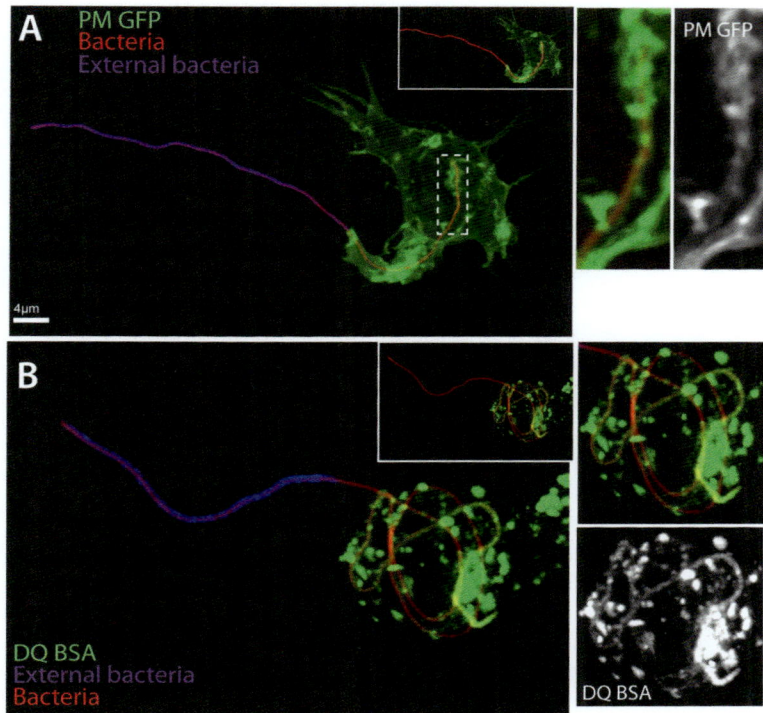

Fig. 1 Internalization of filamentous bacterial targets through a long-lasting phagocytic cup. (**a**) RAW 264.7 macrophage expressing PM-GFP internalizing a PFA fixed *L. pneumophila* filament. Cells were transiently transfected with PM-GFP overnight using Fugene HD following the manufacturer's protocol. PFA fixed RFP-expressing *L. pneumophila* filaments were generated as outlined in Subheading 3.1.2 and used to perform a phagocytosis assay as described above in Subheading 3.2. Cells were washed and fixed 10 min after the binding was synchronized. External segments of the filament were differentially immunolabeled as described in Subheading 3.3, **step 1**. Images to the *right* are higher magnifications of the framed region from the *main panel*. Phagocytic containing a filamentous bacterial target undergoing maturation by fusion with lysosomes. (**b**) RAW 264.7 cells were incubated with DQ-BSA for 1 h to label the late endosomes and lysosomes as outlined in Subheading 3.4, **step 3**. Phagocytosis was performed by adding PFA fixed, RFP-expressing *L. pneumophila* filaments as described in Subheading 3.2. Cells were fixed 20 min after the target attachment was synchronized. External segments of the filament were differentially immunolabeled as outlined in Subheading 3.3, **step 1**. Images to the *right* show a higher magnification of the phagocytic cup labeled with DQ-BSA delivered to this compartment via endosomal fusion

Subheading 3.3, **step 1** is shown in blue. Images to the right are higher magnifications of the framed region from the main panel.

All steps outlined can be carried out at room temperature for 1 h unless otherwise noted.

1. After performing the phagocytosis assay as described in Subheading 3.2, incubate PFA-fixed cells with 1 mL of 5 % skim milk (v/v in 1× PBS) for blocking.

2. Wash cells three times with 1× PBS and incubate with a primary antibody against your bacteria of interest diluted in blocking solution to label the filamentous bacterial targets.

3. Wash cells three times with 1× PBS and incubate with the desired fluorescent dye-conjugated secondary antibody diluted in the blocking solution. If using fluorescent bacterial targets, *see* **Note 14**. This procedure will immunolabel the segments of the filamentous bacteria not yet internalized by the macrophage.

4. Wash cells three times with 1× PBS.

5. Incubate cells with 1 mL of 0.1% Triton X-100 (v/v in 1× PBS) for 20 min to permeabilize cell membranes.

6. Block and incubate again with the primary antibody as described in **steps 2** and **3**.

7. Following three washes with 1× PBS, incubate cells with a different fluorescent dye-conjugated secondary antibody to allow for the distinction between external and internalized bacterial targets. This procedure will immunolabel the whole length of the bacterial filament, including the section internalized by the macrophage.

8. Wash cells three times with 1× PBS and mount the coverslip onto glass slides using fluorescent mounting media.

9. Acquire confocal z-stacks series to observe the entire filament.

10. Using imaging analysis software such as Image J (NIH) or an equivalent, measure the length of the filamentous targets labeled in the two rounds of immunolabeling. The difference in the total length stained after the second round of immuno-labeling and the length of the filament stained after the first round of immunolabeling can be used to indicate the length of the filament internalized by the macrophage.

11. To complement the approach described above in **steps 1–9**, the phagocytosis assay can be performed in cells expressing recombinant probes for plasma membrane markers such as PM-GFP (a chimeric GFP fused to a myristoylation/palmitoylation sequence from Lyn kinase) or GPI-GFP (glycosyl phosphati-dylinositol-GFP) to delineate the phagocytic cups [13, 14].

3.4 Remodeling and Maturation of Phagocytic Cups and Phagosomes

The transition from phagocytic cups to phagosomes requires the remodeling of the lipids and the associated proteins of the plasma membrane, rearrangement of the actin cytoskeleton, as well as the focal exocytosis of endomembranes at the phagocytic cup [15]. The study of these processes using spheroidal targets is challenging due to the fast pace at which these events take place. Given the long duration of the phagocytic cup stage for filamentous targets, greater spatial and temporal resolution can be achieved without the requirement of genetic and/or pharmacological approaches to stall the process in order to study these remodeling events.

3.4.1 Phagocytic Cup Remodelling

Using filamentous bacterial targets and the protocol described below, we studied the remodelling of actin as well as phosphoinositides (PIs) at the phagocytic cup using fixed and live-cell imaging as described elsewhere in [4].

1. Perform a phagocytosis assay as described in Subheading 3.2.

2. Fix cells using 4% PFA (v/v in 1× PBS) following the desired incubation periods and differentially immunolabel the filamentous targets as described in Subheading 3.3 to distinguish between the external and intracellular segments.

3. Stain cells using fluorophore-conjugated phalloidin to label F-actin and to visualize the phagocytic cups.

4. Mount the coverslips onto glass slides using fluorescent mounting media.

5. Acquire confocal z-stacks to analyze target internalization over time and actin distribution around the particle to assess actin remodelling.

6. To complement the fixed-cell imaging approach described in **steps 1–4** with live-cell imaging, perform a phagocytosis assay as outlined in Subheading 3.2 using fluorescent bacterial targets in cells stably or transiently expressing fluorescent actin probes (*see* **Note 15**). Similarly, to examine PI remodelling, cells can be transfected with recombinant probes to determine changes in PI levels (e.g. PLCδ-PH-GFP to visualize phosphatidylinositol 4,5-bisphosphate levels).

3.4.2 Phagocytic Cup Maturation

The canonical model of phagocytosis requires the nascent phagosome to separate from the plasma membrane for maturation to occur. However, similar to phagosome maturation, the phagocytic cups formed for filamentous bacteria fuse with endosomes and lysosomes. The large surface presented by the elongated phagocytic cups can be used for mechanistic studies of highly dynamic membrane fusion and fission events, protein complex formation and disassembly, and membrane lipid remodelling processes associated with phagosome maturation. Using antibodies, GFP chimeric constructs for early and late endosomal proteins or fluorogenic endosomal probes and following the protocol in Subheading 3.2, we assessed the maturation of phagocytic cups though fixed and live-cell imaging as described in [4].

Fluid phase uptake of DQ-BSA can be used to label endosomes and lysosomes and their delivery to the phagocytic cup can be followed by fixed or live-cell microscopy (*see* **Note 16**). This is shown in Fig. 1b where a cell with lysosomes labeled with DQ-BSA is internalizing a PFA fixed, RFP-expressing *L. pneumophila* filament. Cells were fixed 20 min after attachment was synchronized and the external segment of the filament was immunolabeled (blue). Images to the right show a higher magnification of the phagocytic cup labeled with DQ-BSA delivered to this compartment via endosomal fusion.

The steps described below are for labeling endosomal compartments with DQ-BSA but can be used for other fluid phase markers including fluorescent-labeled dextrans.

1. Dilute DQ-BSA in tissue culture media according to the manufacturer's instructions.

2. Replace the media used to culture cells with media containing DQ-BSA.

3. Incubate cells with the DQ-BSA for 1 h at 37 °C and 5 % CO_2.

4. Remove the media and wash cells twice with sterile 1× PBS warmed to 37 °C.

5. Incubate cells in fresh tissue culture media that is pre-warmed to 37 °C for 1 h at 37 °C and 5 % CO_2 for the desired period to chase the probes into early or late endosomes or lysosomes.

6. Perform a phagocytosis assay as described in Subheading 3.2.

7. Fix cells in 4 % PFA (v/v in 1× PBS) for 20 min at room temperature for fixed cell imaging or follow the method outlined in **step 5** of Subheading 3.3. For live-cell imaging, follow the method outlined in **step 2**.

3.4.3 Assessing the Luminal Environment of Phagocytic Cups and Phagosomes During Maturation

To assess the luminal environment of a phagocytic cup and its acidification and acquisition of hydrolytic properties during the transition to a phagosome, fluorescent probes can be cross-linked to filamentous bacterial targets for real-time microscopy-based assays. We successfully used this approach to analyze the acquisition of hydrolytic properties by phagocytic cups containing filamentous bacterial targets during their transition to phagosomes using live-cell microscopy [4].

Although the following method describes the cross-linking of DQ-BSA to filamentous targets, this technique can be modified to measure acidification, generation of ROS etc.

1. Resuspend 1×10^9 PFA-fixed filaments in 1× PBS with 25 mg/mL of carbodiimide cross-linker.

2. Mix by agitation at room temperature for 15 min.

3. Wash cells three times with 1 mL of 0.1 M sodium borate (pH 8.0) coupling buffer to remove excess cross-linker.

4. Use 500 μL of coupling buffer with 1.0 mg DQ-BSA (or another fluorogenic probe) and 1 mg/mL IgG to opsonize the targets for 12–16 h with gentle rotation at room temperature.

5. Spin the targets at $3000 \times g$ and wash three times with 250 mM glycine in 1× PBS (pH 7.2) to quench the probe. Vortex in-between wash steps for 5 min.

6. Wash the substrate-bound targets three times with 1× PBS.

7. Cool cells for 5 min at 15 °C by placing in a precooled centrifuge.

8. Add 50 µL of the opsonized targets to each well and synchronize binding by centrifugation at 15 °C and $300 \times g$ for 5 min.

9. Incubate the cells at 37 °C and 5 % CO_2 for an additional 5 min.

10. Remove media from the wells and wash three times with 1× PBS pre-warmed to 37 °C.

11. Replace with fresh tissue culture media pre-warmed to 37 °C and incubate cells at 37 °C and 5 % CO_2 for desired periods.

12. The cells can be visualized by live-cell microscopy or fixed for immunofluorescence imaging as outlined in Subheading 3.2.

4   Notes

1. Ciprofloxacin is an antibiotic that belongs to the fluoroquinolone family and prevents bacterial replication by inhibiting DNA gyrase. Usage at sublethal doses activates the bacterial SOS stress response, inducing filamentation without causing cell death [9, 10].

2. We did not observe any difference with respect to the stability of the filaments between storing the fixed bacteria at 4 °C resuspended in 4 % PFA (v/v in 1× PBS) or 1× PBS. If storing in PFA, wash the bacteria twice with 1× PBS as described in Subheading 3.1.2 before using for an experiment.

3. Warming to 37 °C helps to dissolve the ferric pyrophosphate for making the BYE and BCYE media.

4. Plates stored at 4 °C can be used for up to 4 months. In our experience, using plates older than 4 months reduces the proportion of longer filaments.

5. To ensure that the cultures are enriched in filamentous bacteria, use freshly streaked plates. Using older plates or plates stored at 4 °C to start cultures yields cultures enriched in bacillary forms of the bacteria.

6. BYE media stored at 4 °C can be used for up to 4 months. In our experience, using media older than 4 months reduces the proportion of longer filaments obtained in cultures.

7. Fixing with PFA preserves the filamentous bacteria, allowing for the study of effects of target morphology on phagocytosis, excluding any interference from potential bacterial effectors.

Live bacteria can be used to investigate the impact of filamentous morphology on pathogenicity.

8. Coverslips are stored in 70% ethanol for sterilization. Prior to use for plating cells, they are washed thoroughly in sterilized ultrapure water to remove ethanol. Alternatively, coverslips can be sterilized by autoclaving them prior to use.

9. The phagocytosis assays can be performed without opsonization of the targets. However, opsonization with IgG antibodies increases binding and internalization efficiencies.

10. Although an anti-Lp1 antibody is used to opsonize targets in this chapter, nonspecific IgG, such as donkey anti-IgG can be used as well.

11. Unlike the phagocytosis of spherical targets where target binding can be synchronized by cooling cells to 4 °C while inhibiting target uptake, the attachment of longer filamentous targets is inhibited at 4 °C. This could be due to actin-dependent reorientation of the targets needed to gain access to their short axes as described in [16] in order for phagocytosis to occur or the lack of the receptors needed for target attachment with high avidity. Therefore, the attachment is synchronized at 15 °C.

12. A late-exponential culture of *L. pneumophila* at OD 2.0 contains approximately 1.5×10^9 bacteria/mL. Using this approximation, cells are presented with fixed bacterial targets with a target-to-cell ratio of 150:1.

13. Cells are incubated with filamentous targets for 5 min following synchronization of binding to enhance attachment. If washed immediately following centrifugation, the efficiency with which longer filaments bind is greatly reduced.

14. If the fixed filamentous bacterial targets express a fluorescent protein such as GFP, cell membrane permeabilization and immunolabeling of internalized targets is not required.

15. If desired, external segments of the filamentous bacteria can be immunolabeled prior to performing live-cell imaging. Place cells on ice and replace media with cold media containing the fluorescent dye-conjugated secondary antibody of choice, against the primary antibody used to opsonize targets, for 5 min. Wash the unbound antibody and proceed with live-cell imaging. As labeling in the cold is only done for 5 min, we found that using a higher concentration (2–3 times greater than what is used for fixed-cell labeling) of secondary antibody provided better labeling.

16. DQ-BSA is a self-quenched protease substrate that is converted into fluorescent peptides once hydrolyzed by lysosomal proteases.

References

1. Champion JA, Mitragotri S (2006) Role of target geometry in phagocytosis. Proc Natl Acad Sci U S A 103(13):4930–4934. doi:10.1073/pnas.0600997103, PubMed PMID: 16549762, PubMed Central PMCID: PMC1458772

2. Wright SD, Silverstein SC (1984) Phagocytosing macrophages exclude proteins from the zones of contact with opsonized targets. Nature 309(5966):359–361

3. Swanson JA (2008) Shaping cups into phagosomes and macropinosomes. Nat Rev Mol Cell Biol 9(8):639–649. doi:10.1038/nrm2447, PubMed PMID: 18612320; PubMed Central PMCID: PMCPMC2851551

4. Prashar A, Bhatia S, Gigliozzi D, Martin T, Duncan C, Guyard C et al (2013) Filamentous morphology of bacteria delays the timing of phagosome morphogenesis in macrophages. J Cell Biol 203(6):1081–1097. doi:10.1083/jcb.201304095, PubMed PMID: 24368810, PubMed Central PMCID: PMC3871431

5. Champion JA, Mitragotri S (2009) Shape induced inhibition of phagocytosis of polymer particles. Pharm Res 26(1):244–249. doi:10.1007/s11095-008-9626-z, PubMed PMID: 18548338, PubMed Central PMCID: PMC2810499

6. Russell DG, Vanderven BC, Glennie S, Mwandumba H, Heyderman RS (2009) The macrophage marches on its phagosome: dynamic assays of phagosome function. Nat Rev Immunol 9(8):594–600. doi:10.1038/nri2591, PubMed PMID: 19590530, PubMed Central PMCID: PMC2776640

7. Yates RM, Hermetter A, Russell DG (2005) The kinetics of phagosome maturation as a function of phagosome/lysosome fusion and acquisition of hydrolytic activity. Traffic 6(5):413–420.doi:10.1111/j.1600-0854.2005.00284.x

8. Feucht A, Errington J (2005) ftsZ mutations affecting cell division frequency, placement and morphology in Bacillus subtilis. Microbiology 151(Pt 6):2053–2064. doi:10.1099/mic.0.27899-0

9. Harry E, Monahan L, Thompson L (2006) Bacterial cell division: the mechanism and its precison. Int Rev Cytol 253:27–94. doi:10.1016/S0074-7696(06)53002-5

10. Prashar A, Bhatia S, Tabatabaeiyazdi Z, Duncan C, Garduno RA, Tang P et al (2012) Mechanism of invasion of lung epithelial cells by filamentous Legionella pneumophila. Cell Microbiol 14(10):1632–1655. doi:10.1111/j.1462-5822.2012.01828.x

11. Bos J, Zhang Q, Vyawahare S, Rogers E, Rosenberg SM, Austin RH (2015) Emergence of antibiotic resistance from multinucleated bacterial filaments. Proc Natl Acad Sci U S A 112(1):178–183. doi:10.1073/pnas.1420702111, PubMed PMID: 25492931, PubMed Central PMCID: PMC4291622

12. Steinberg BE, Grinstein S (2007) Assessment of phagosome formation and maturation by fluorescence microscopy. Methods Mol Biol 412:289–300. doi:10.1007/978-1-59745-467-4_19

13. Teruel MN, Blanpied TA, Shen K, Augustine GJ, Meyer T (1999) A versatile microporation technique for the transfection of cultured CNS neurons. J Neurosci Methods 93(1):37–48

14. Nichols BJ, Kenworthy AK, Polishchuk RS, Lodge R, Roberts TH, Hirschberg K et al (2001) Rapid cycling of lipid raft markers between the cell surface and Golgi complex. J Cell Biol 153(3):529–541, PubMed PMID: 11331304, PubMed Central PMCID: PMC2190578

15. Lee WL, Mason D, Schreiber AD, Grinstein S (2007) Quantitative analysis of membrane remodeling at the phagocytic cup. Mol Biol Cell 18(8):2883–2892. doi:10.1091/mbc.E06-05-0450, PubMed PMID: 17507658, PubMed Central PMCID: PMC1949373

16. Moller J, Luehmann T, Hall H, Vogel V (2012) The race to the pole: how high-aspect ratio shape and heterogeneous environments limit phagocytosis of filamentous Escherichia coli bacteria by macrophages. Nano Lett 12(6):2901–2905. doi:10.1021/nl3004896

Chapter 22

Growing and Handling of *Mycobacterium tuberculosis* for Macrophage Infection Assays

Evgeniya V. Nazarova and David G. Russell

Abstract

Macrophage survival assays are a critical component of any *Mycobacterium tuberculosis* research program. Here we describe the methods that we use routinely for infection of macrophages of various origins. The protocols are efficient, relatively simple and are accepted widely. We provide users with methods for the infection of small numbers of macrophages—more suitable for microscopy; and for larger numbers of macrophages—for flow cytometry analysis or extraction for biochemical characterization.

Key words Phagocytosis, Macrophage, *Mycobacterium tuberculosis*, Infection

1 Introduction

Although we have learned a lot about the biology of *Mycobacterium* spp. that are cultured in broth rich with nutrients, there is a critical need to understand how the pathogen adapts to more natural environments—such as its host macrophage. Macrophages are among the very first cells that will encounter *Mycobacterium tuberculosis* (Mtb) following its inhalation into the lung and they play crucial role in the outcome of infection. Therefore in vitro models of macrophage infection represent a powerful tool for studying host–pathogen interaction. However it is important to remember that this bacterium is a human pathogen and must be handled under biosafety level 3 containment (*see* **Note 1**). Using this model the dynamics of transcriptional response of Mtb to the host environment has been successfully probed [1–3]. Through the use of fluorescently labeled reporter Mtb we have been able to look at the localization of bacteria, and the availability of host lipids within the macrophage [4]. Flow cytometry and functional physiological assays have enabled us to assess impact of infection on relative phagosomal function and metabolic states of both host and pathogen [4]. Finally we have performed an extensive empirical screen for small molecules capable of impairing the

Roberto Botelho (ed.), *Phagocytosis and Phagosomes: Methods and Protocols*, Methods in Molecular Biology, vol. 1519, DOI 10.1007/978-1-4939-6581-6_22, © Springer Science+Business Media New York 2017

survival of Mtb in its host macrophage and identified many inhibitors that blocked growth in intracellular Mtb but had no effect on bacteria in rich broth [5]. These experimental data all support the contention that macrophage infection and survival assays are critical to expanding our understanding of the biology of the infectious agent as well as its host.

Here we describe the methods that we use for efficient infection of macrophages of differing host origins. One of the biggest technical challenges for these infection assays is the extremely hydrophobic nature of the bacterial cell surface due to its high lipid content. To counteract this we use gelatin-containing infection medium to break up clumps of bacteria, in addition to the multiple passing of bacterial suspensions through syringe needles.

We include protocols for both small and large scale infections. The small scale protocols are intended for analysis by methods such as confocal microscopy. Whereas the large scale infection protocols are designed for isolation of macrophages for lipid or protein extraction, bacterial and host RNA and DNA isolation, electron microscopy, flow cytometry, or isolation of bacteria after macrophage infection. It is intended that these methods may be modified to accommodate the specific needs of your assay and appropriate scale of choice.

2   Materials

2.1 Cells, Reagents, and Buffers

1. Macrophages: Bone marrow-derived murine macrophages (BMMØ) are our cells of choice for these assays. BMMØ are derived from the bone marrow extracted from the femur, tibia, and ilium of euthanized mice and maintained in Dulbecco's modified Eagle's medium (DMEM) supplemented with 10% fetal bovine serum (FBS), 2 mM L-glutamine, 1 mM sodium pyruvate, and 10% L-cell conditioned media (BMMØ media). J774 cells (available from the American Type Culture Collection, Rockville, MD) are maintained in DMEM supplemented with 10% FBS and 2 mM L-glutamine + 1 mM sodium pyruvate.

2. Sterile glass slide with lid and with chambers of preferred format. Typically, 8-well multi-chamber slide with glass coverslip (Ibidi or other) bottom and with 1 cm^2 growth area per well is used if infected macrophages will be examined by confocal microscopy.

3. Sterile 75 cm^2 polystyrene tissue culture flask with filtered cap. Sterile 150 cm^2 polystyrene tissue culture flask with filtered cap also can be used (both from Corning or other) (see **Note 2**).

4. BMDM medium: DMEM supplemented with 10% fetal calf serum, 10% L-cell conditioned medium, 2 mM L-glutamine, 1 mM sodium pyruvate. Filter-sterilize. Store at 4 °C for no longer than 2 months.

5. HMDM medium: DMEM supplemented with 10% human serum, 2 mM L-glutamine, 1 mM sodium pyruvate. Filter-sterilize. Store at 4 °C for no longer than 2 months.

6. Antibiotics: 100 U/ml penicillin and 100 μg/ml streptomycin.

7. 25 cm^2 and/or 75 cm^2 polystyrene tissue culture flask with filtered cap or any sterile flask covered with filtered cap that supports growth of up to 10 ml and 40 ml of bacterial culture, respectively, with approximately 50 ml and 230 ml of free space inside the flask.

8. Middlebrook 7H9 OADC medium: 4.7 g of the Middlebrook 7H9 Broth dehydrated base powder, 2 ml glycerol, 0.05% tyloxapol, 100 ml of Middlebrook OADC Enrichment, 900 ml of distilled water. Filter-sterilize. Store at room temperature.

9. 15 ml conical tube(s).

10. Basal uptake buffer (BUB): 2.25 g glucose, 2.5 g bovine serum albumin, 0.5 ml gelatin, 50 mg $CaCl_2$, 50 mg $MgCl_2$, 500 ml phosphate-buffered saline (PBS). Filter-sterilize, store at 4 °C for no longer than 6 months (*see* **Note 3**).

11. Sterile 1 ml tuberculin syringe with 25 gauge needle.

2.2 Instruments

1. CO_2 incubator.

2. Spectrophotometer with ability to measure absorbance at 600 nm.

3. Centrifuge that allows to spin 15 ml-conical tubes at $3300 \times g$.

3 Methods

3.1 Macrophage Preparation

3.1.1 Small Scale, for Microscopy

1. Seed macrophages into glass slide with chambers of the preferred format at about 80% confluency in macrophage media without antibiotics. Typically, 8-well multi-chamber slide with glass coverslip bottom is used to seed 1.8×10^5 murine BMDM in 0.2 ml of macrophage media per well (*see* **Note 4**).

2. Incubate at 37 °C and 7% CO_2 overnight. Since murine BMDM can still slowly replicate, you need to assume that there is 2×10^5 cells per well on the day of infection.

3.1.2 Larger Scale: 75 or 150 cm^2 Tissue Culture Flasks

Here we give an example how to perform an infection of macrophages seeded in one 75 cm^2 tissue culture flask. Double all the amounts if you wish to use 150 cm^2 tissue culture flask.

1. Seed 1.5×10^7 macrophages in 20 ml of macrophage media without antibiotics per flask.

2. Incubate at 37 °C and 7% CO_2 overnight.

3.2 Preparation of Bacterial Cultures

1. Inoculate 8 ml Middlebrook 7H9 OADC medium in a 25 cm^2 tissue culture flask with approximately 0.1–0.2 ml of frozen *M. tuberculosis* glycerol stock.

2. Incubate at 37 °C in standing culture for 7–10 days, until culture reaches logarithmic phase (*see* **Note 5**).

3. Pass into required volume of Middlebrook 7H9 OADC medium depending on scale of infection at starting optical density at 600 nm (OD_{600}) of about 0.05. Use 25 cm^2 flasks for volumes not more than 10 ml, and 75 cm^2 flasks for volumes up to 40 ml (*see* **Note 6**).

4. Incubate at 37 °C in standing culture for 7 days, culture should reach OD_{600} of approximately 0.6 (*see* **Note 7**).

3.3 Infection Protocol

3.3.1 Small Scale, for Microscopy

1. Measure OD600 of bacterial culture. Spin down equivalent of 2 ml of culture with OD600 = 0.6 in 15 ml conical tube at 3300 × g for 10 min. For example, if your culture has an OD600 of 0.5, you need to spin down 2.4 ml (0.6 × 2 ml/0.5 = 2.4 ml).

2. Assuming that OD600 = 0.6 gives 10^8 bacteria/ml, you will have 2×10^8 bacteria in the pellet.

3. Resuspend bacterial pellet in 2 ml of BUB, so that you have 1×10^8 bacteria per ml. Place this suspension in a new 15 ml centrifuge tube (*see* **Note 8**).

4. Using a 1 ml tuberculin syringe with 25 gauge needle, insert the syringe into the Falcon tube with extreme care and pass the bacterial suspension in and out of the syringe 12–15 times (*see* **Note 9**). Proceed to the next step immediately, so that bacteria do not have chance to clump (*see* **Note 10**).

5. Add required volume of bacterial suspension to the macrophages to achieve desired multiplicity of infection (MOI) (Table 1) (*see* **Note 11**). In order to minimize dilution or avoid over-filling the well it is advisable not exceed 40 µl of BUB. If volume is too small to pipette, dilute bacterial suspension as necessary.

6. Mix bacteria with macrophage media in the well by gentle pipetting up and down 5 times (*see* **Note 12**).

Table 1
Calculations of bacterial amounts needed to infect macrophages at desired MOI

	MOI 1:1	MOI 2:1	MOI 4:1	MOI 5:1	MOI 10:1
Amount of bacteria needed to be added to 2×10^5 cells per well	2×10^5	4×10^5	8×10^5	1×10^6	2×10^6
Volume of bacterial suspension needed to be added to 2×10^5 cells per well, µl	2	4	8	10	20

7. Incubate at 37 °C and 7% CO_2 for 2–5 h.

8. Replace media used for infection with 0.22 ml of fresh macrophage media without antibiotics. If assay will exceed 2 days, change media every other day.

3.3.2 Larger Scale: 75 cm² Tissue Culture Flask

Our calculations are designed to generate an MOI of 4:1. Adjust amounts accordingly if you need to use different MOI.

1. Measure OD of bacterial culture. Spin down equivalent of 1 ml of culture with OD600 = 0.6 in 15 ml conical tube at $3300 \times g$ for 10 min. For example, if your culture has an OD600 of 0.5, you need to spin down 1.2 ml $(0.6 \times 1 \text{ ml}/0.5 = 1.2 \text{ ml})$.

2. Assuming that OD600 = 0.6 gives 10^8 bacteria/ml, you will have 1×10^8 bacteria in the pellet.

3. Resuspend bacterial pellet in 1.5 ml of BUB. Place this suspension in a new 15 ml Falcon tube in an appropriate holder (*see* **Note 8**).

4. Using a 1 ml tuberculin syringe with 25 gauge needle insert the syringe into the Falcon tube with extreme care and pass the bacterial suspension in and out of the syringe 12–15 times (*see* **Note 9**). Proceed to the next step immediately, so that bacteria do not have chance to clump (*see* **Note 10**).

5. Add 3.5 ml of BUB, mix well. Now you have 2×10^7 bacteria per ml.

6. Transfer 3 ml of bacterial suspension into 75 cm² flask containing macrophages in 20 ml of macrophage media. Mix well by either pipetting up and down multiple times or gentle shaking (*see* **Note 13**). Use Table 2 to perform infections at desired MOI.

Table 2

Calculations of bacterial amounts needed to infect macrophages at desired MOI

	MOI 1:1	MOI 2:1	MOI 4:1	MOI 5:1	MOI 10:1
Amount of bacteria needed to be added to 1.5×10^7 cells in 75 cm² flask	1.5×10^7	3×10^7	6×10^7	7.5×10^7	1.5×10^8
Original volume of bacteria at OD600 0.6 needed to be pelleted, ml	0.5	0.5	1	1	2
Total volume of BUB that needs to be added to bacterial pellet, ml	10	5	5	4	4
Concentration of bacteria in BUB suspension, bacteria/ml	5×10^6	1×10^7	2×10^7	2.5×10^7	5×10^7
Volume of bacterial suspension needed to be added to 1.5×10^7 cells in 75 cm² flask, ml	3	3	3	3	3

7. Incubate at 37 °C and 7% CO_2 for 2–5 h.

8. Replace media used for infection with 20 ml of fresh macrophage media without antibiotics. If assay will exceed 2 days, change media every other day.

---

4 Notes

1. *Mycobacterium tuberculosis* is a human pathogen and must be handled under biosafety level 3 containment by trained personnel. Such facilities have to be maintained under registered institutional safety and employee health monitoring programs. Some investigators use nonpathogenic *Mycobacterium* spp., such as *Mycobacterium smegmatis*, as a surrogate but the relevance of data generated with nonpathogenic organisms is open to obvious concerns regarding validity.

2. In case of 150 cm² tissue culture flasks double all the amounts described for 75 cm² tissue culture flask.

3. If gelatin becomes solidified, pre-warm it at 55–65 °C overnight. Due to its high viscosity, a syringe would be the easiest way to measure it out.

4. If you are using human monocyte-derived macrophages, seed 3×10^5 monocytes per well in media containing antibiotics 7–10 days before infection, change to media without antibiotics on day 7 of incubation.

5. Cultures can be grown shaking to avoid the generation of hypoxic conditions. In that case, the rate of growth will be twice as fast, so adjust timing accordingly. We prefer to use standing cultures so that the surface glycocalyx of Mtb is less disturbed during culture.

6. It is important to pass bacterial cultures after the inoculum has grown, since the initial culture frequently contains clumps of bacteria. Therefore avoid transferring these clumps directly into the culture when you perform this procedure.

7. We recommend using cultures at logarithmic stage for infection, since they are more homogeneous.

8. Try not to touch sides of the new tube with any part of pipette for safety reasons.

9. It is important to use a 15 ml conical tube because the tuberculin syringe fits perfectly inside of it without touching the bottom. When attaching needle to the syringe be careful not to touch the main part of syringe below the plunger to avoid contamination of bacterial suspension. Keep the syringe in the tube throughout this entire process to minimize the potential for self-inoculation. This process can be performed with blunt syringe needles.

10. If you wish to perform infection with different strains of Mtb, you should not syringe more than four bacterial suspensions before proceeding to the next step to avoid re-clumping of bacteria.

11. We have noticed that different strains of Mtb behave differently in macrophage infection and induce differing levels of cytotoxicity. When infecting murine BMDM, CDC1551 can be used at higher MOI such as 5 and 10 for 7–10 days of infection without killing host cells, whereas H37Rv and Erdman should not be used at MOI higher than 5, otherwise macrophages die within 3–5 days. In our experience HMDM are less tolerant of Mtb infection than murine BMDM, so we usually use MOI for CDC1551 not higher than 5 (0.5–3 is preferred), and for Erdman and H37Rv not higher than 1 to maintain infected macrophage monolayers for at least 5 days.

12. Make sure that you do not generate air bubbles in the wells while pipetting to avoid spillage and contamination.

13. If you are infecting multiple flasks, you can mix macrophage media with bacterial BUB-suspension at the same ratios in the separate large flask for all of the samples. Then replace the media in each macrophage-containing flask with this bacterial BUB-macrophage media suspension. Handling the suspension in this manner provides greater consistency between samples.

Acknowledgments

E.V.N. and D.G.R. are supported by the following awards from the National Institutes of Health, USA; AI067027, AI118582, and HL055936 to D.G.R.

References

1. Schnappinger D, Ehrt S, Voskuil MI, Liu Y, Mangan JA, Monahan IM, Dolganov G, Efron B, Butcher PD, Nathan C, Schoolnik GK (2003) Transcriptional adaptation of *Mycobacterium tuberculosis* within macrophages: insights into the phagosomal environment. J Exp Med 198:693–704

2. Rohde KH, Veiga DF, Caldwell S, Balázsi G, Russell DG (2012) Linking the transcriptional profiles and the physiological states of *Mycobacterium tuberculosis* during an extended intracellular infection. PLoS Pathog 8, e1002769

3. Tan S, Sukumar N, Abramovitch RB, Parish T, Russell DG (2013) *Mycobacterium tuberculosis* responds to chloride and pH as synergistic cues to the immune status of its host cell. PLoS Pathog 9, e1003282

4. Podinovskaia M, Lee W, Caldwell S, Russell DG (2013) Infection of macrophages with *Mycobacterium tuberculosis* induces global modifications to phagosomal function. Cell Microbiol 15:843–859

5. VanderVen BC, Fahey RJ, Lee W, Liu Y, Abramovitch RB, Memmott C, Crowe AM, Eltis LD, Perola E, Deininger DD, Wang T, Locher CP, Russell DG (2015) Novel inhibitors of cholesterol degradation in *Mycobacterium tuberculosis* reveal how the bacterium's metabolism is constrained by the intracellular environment. PLoS Pathog 11, e1004679

Chapter 23

Mycobacterium tuberculosis: Readouts of Bacterial Fitness and the Environment Within the Phagosome

Shumin Tan, Robin M. Yates, and David G. Russell

Abstract

Macrophages fulfill most of their microbicidal duties in their phagosomes following uptake of microbes. However, some microbes, such as *Mycobacterium tuberculosis*, have evolved mechanisms to subvert the normal maturation process of their phagocytic compartment to limit the hostility of this environment. The experimental analysis of this process and its subsequent impact on bacterial fitness is technically demanding and has required the development of a broad range of readouts to correlate function and outcome. In this chapter we detail two technically divergent platforms to measure the environment within the phagosomal compartment that contains Mtb in the short term, and more long-term readouts of bacterial fitness and Mtb's reaction to host-derived stresses. The readouts are all fluorescence-based and are adaptable to measurement by a range of platforms, including spectrofluorometry, confocal microscopy, and flow cytometry.

Key words Macrophage, *Mycobacterium tuberculosis*, Phagosome, Phagocyte

---

1 Introduction

It has been known for many years that the success of pathogenic *Mycobacterium* spp. is linked to their ability to block the fusion of the phagosomes in which they reside with the lysosomal network of their host cell [1–3]. Early studies used electron microscopy to document this behavior and only since the mid-90s have more physiological readouts been exploited to characterize the nature of the Mtb-containing phagosomes. In 1994, we used fluorescein to measure the pH of the phagosomes containing *Mycobacterium avium* and found them to equilibrate to pH 6.2–6.4 [4]. This analysis was repeated for *Mycobacterium tuberculosis* in 2004 [5]. Examination of the trafficking of a range of ligands such as transferrin and cholera toxin B, and the relative distribution of several host cell proteins such as the transferrin receptor, the early endosomal marker EEA1 and the pro-enzyme form of the lysosomal hydrolase cathepsin D led to the conclusion that the Mtb-containing phagosome behaved like an early endosome [6–10]. The vacuole

Roberto Botelho (ed.), *Phagocytosis and Phagosomes: Methods and Protocols*, Methods in Molecular Biology, vol. 1519,
DOI 10.1007/978-1-4939-6581-6_23, © Springer Science+Business Media New York 2017

was highly dynamic and fused readily with other early endosomal compartments. Measurement of the intravacuolar pH was a useful indicator of phagosome maturation and, by inference, the relative "fitness" of the bacterium. Activation of the macrophage with inflammatory cytokines such as interferon-γ (IFN-γ) overcomes this blockage and delivers the bacilli to compartments with a reduced pH of 5.2 [10, 11]. Finally, mutant bacteria defective in the modulation of their phagosome showed reduced survival inside murine macrophages [5].

Despite its usefulness there are distinct limitations to the pH assay. It is achieved through the labeling of reactive amino groups on the bacterial surface with the succinimidyl ester form of carboxyfluorescein. This label comes off the bacterium within hours of uptake by macrophages and is therefore only useful for analysis within the first few hours of infection in tissue culture models. Over the years we have been developing alternative readouts of the intracellular environment experienced by Mtb, and its impact on bacterial fitness. Transcriptional profiling of the response of Mtb following internalization by the host macrophage has allowed us to identify specific environmental cues detected by Mtb that induce defined regulons [12–15]. pH and chloride concentration are interdependent parameters that vary during phagosome maturation—as the pH drops the Cl- ion concentration increases [12, 15]. Mtb exhibits an overlapping transcriptional response to both stimuli. Promoter elements that are responsive to these environmental cues were placed in replicating plasmids upstream of the *gfp* gene and introduced into Mtb. These early reporter bacterial strains could be induced to express GFP under appropriate environmental conditions in the test tube and showed levels of expression that correlated with the inferred status of their phagosome [12, 15].

More recently, we have developed a new generation of reporter Mtb strains that constitutively express mCherry and express GFP under immune-regulated environmental stresses, such as nitric oxide (*hpsX'::GFP, smyc'::mCherry*), or that expresses GFP as a fusion partner with the single stranded binding protein (SSB) (SSB-GFP, *smyc'::mCherry*) [15, 16]. The latter construct had been developed previously in *Escherichia coli*, and used to score bacterial replication [17]. Formation of the replisome leads to the bacteria possessing green fluorescent foci during replication of the bacterial chromosome. In Mtb these foci persist for 76–80% of the division cycle of the bacterium [16]. Therefore by scoring the number of red bacilli with green foci, one can determine the relative replication status of a bacterial population. We have used these reporter strains to probe the relative fitness of bacterial populations in in vivo experimental mouse infection under differing immune conditions [16]. These reporter strains have the opposite limitation to the CF-SE-labeled Mtb in that some of these readouts are best observed in the context of longer-term infection, and maintaining the health of host cells in tissue culture for an appropriate duration is extremely challenging. These reporter strains are thus much more suited to in vivo infection models lasting several weeks.

These assays, developed over the past 20 years, have the capacity to probe the host environment and its impact on bacterial fitness in short-term infections in tissue culture, or in long-term in vivo infections. As mentioned, all these assays have caveats or qualifiers, and none of the assays are universally applicable. In this chapter we present the methods behind the assays and highlight their application under differing experimental conditions.

2 Materials

2.1 Cells, Reagents, and Buffers

1. Macrophages: Bone marrow-derived murine macrophages (BMMØ) are our cells of choice for these assays (*see* **Note 1**). BMMØ are derived from the bone marrow extracted from the femur, tibia, and ilium of euthanized mice and maintained in Dulbecco's modified Eagle's medium (DMEM) supplemented with 10% fetal bovine serum (FBS), 2 mM L-glutamine, 1 mM sodium pyruvate, and 10% L-cell conditioned media (BMMØ media).

2. If we have to use a cell line we use J774 cells (available from the American Type Culture Collection, Rockville, MD), which are maintained in DMEM supplemented with 10% FBS and 2 mM L-glutamine + 1 mM sodium pyruvate.

3. *Mycobacterium tuberculosis* is maintained in liquid culture in Middlebrook 7H9 medium (BD, Franklin Lakes, NJ).

4. Coverslips: Clean 0.13 mm × 12.5 mm × 25 mm cover glass (*see* **Note 2**). Sterilize by autoclaving.

5. Cuvette buffer: Tissue culture tested phosphate-buffered saline (PBS) pH 7.2 adjusted to contain 1 mM $CaCl_2$, 2.7 mM KCl, 0.5 mM $MgCl_2$, 5 mM dextrose, 10 mM HEPES, and 0.1% cold water teleost gelatin (*see* **Note 3**).

6. Binding dish: A microbiological petri dish containing a square piece of Parafilm adhered to the dish by a couple of drops of water.

7. The amine reactive fluorescent reagent: 5-(and-6)-carboxyfluorescein, succinimidyl ester (mixed isomers) (CF-SE) (Life Technologies Corporation). Dissolve in high quality anhydrous dimethyl sulfoxide (DMSO) at 5 mg/ml before use. Stock solutions can be aliquoted and stored at −20 °C. Protect reagents from light and moisture.

8. Coupling buffer: 0.1 M sodium borate in ddH_2O, adjusted to pH 8.0 with 10 M NaOH. Filter-sterilize through 0.22-μm filter.

9. Tween 80 detergent.

10. Reference pH buffers (pH 4.5–5.5): 0.15 M potassium acetate. Adjust pH to 4.5, 5.0, and 5.5 with 10 M NaOH.

11. Reference pH buffers (pH 6–7.5): 0.1 M piperazine-N,N'-bis(2-ethanesulfonate (PIPES), 0.1 M KCl. Adjust pH to 6.0, 6.5, 7.0, and 7.5 with 5 M HCl.

12. Nigericin (InvivoGen, San Diego, CA).

13. Reporter strains of Mtb were generated by transformation of the parental strain with pCherry3 replicating plasmid that expresses mCherry under regulation of the *smyc* promoter [18], together with the reporter construct [15, 16]. In this chapter we use Mtb transformed with plasmids containing *hspX'::GFP, smyc'::mCherry* to report on immune-mediated stress due to iNOS expression at sites of infection, and SSB-GFP, *smyc'::mCherry* encoding a fusion protein of the single stranded binding protein SSB with GFP to indicate the relative replication rates of Mtb under differing immune pressures [15, 16]. Additional details regarding these bacterial strains are provided in **Note 4**.

14. Alexa Fluor 647 conjugated phalloidin, Alexa Fluor 514 conjugated goat anti-rabbit IgG and DAPI (Life Technologies Corporation). Stored in DMSO at –20 °C in aliquots of 10 mg/ml.

15. Rabbit anti-murine iNOS antibody (BD Transduction Laboratories, San Jose, CA).

16. Mounting medium containing anti-fade.

17. High vacuum grease (Dow Corning, Auburn, MI).

2.2 Instruments

1. For pH measurements we utilized a temperature controlled-spectrofluorometer with variable excitation and emission monochromators. The authors use the QMSE4 model spectrofluorometer from Photon Technologies International (Lawrenceville, NJ) equipped with a thermostat-controlled four-chambered turret for simultaneous measurement of four experimental variables. The QMSE4 is interfaced with a PC-compatible computer and is managed by Felix32 software (Photon Technologies International).

2. For confocal microscopy we used a Leica SP5 spectral confocal laser-scanning system with an inverted microscope (Leica Microsystems GmbH, Germany). Leica Application Suite Advanced Fluorescence (LAS-AF 2.6) was utilized for image acquisition. Volocity image analysis software (PerkinElmer Life Sciences, USA) was used for image analysis and quantification of fluorescence signals.

3. Data are finally analyzed and displayed using standard mathematical software such as Microsoft Excel® or MATLAB®.

3   Methods

3.1   Measurement of the pH of Mtb-Containing Phagosomes (All Manipulations Conducted Under BSL3 Conditions)

Mycobacterium tuberculosis is a human pathogen and must be handled under Biosafety Level 3 containment by trained personnel. Such facilities have to be maintained under registered institutional safety and employee health monitoring programs. Some investigators use nonpathogenic *Mycobacterium* spp., such as *Mycobacterium smegmatis*, as a surrogate but the relevance of data generated with nonpathogenic organisms is open to obvious concerns regarding validity.

3.1.1   Fluorescent Labeling of Mtb

1. A mid-log phase culture of Mtb Erdman is harvested by centrifugation at $2000 \times g$ for 12 min at 4 °C. The medium discarded into a bleach container and the bacteria washed 2× with PBS supplemented with 0.1 % Tween 80. The bacteria are resuspended in Coupling buffer with 0.1 % Tween 80 to a concentration of 5×10^8/ml. 5 μl of CF-SE stock solution is added to the bacterial suspension and the tube is covered with aluminum foil and incubated on a nutator mixer for 20 min at room temperature.

2. Following incubation the bacteria are harvested by centrifugation in a tabletop microcentrifuge with sealed rotor. The bacteria are washed 3× with Cuvette buffer, adjusted to 2.5×10^8/ml in Cuvette buffer and stored in the dark on ice while preparing the spectrofluorometer.

3.1.2   Macrophage Monolayer Preparation and Handling

1. Fully differentiated BMMØ monolayers are grown to confluency in untreated petri dishes. Growth media is removed and replaced with cold PBS pH 7.2 (without Ca^{2+} and Mg^{2+}) and incubated at 4 °C for 10 min to facilitate BMMØ detachment from the plastic. BMMØ are then gently dislodged with a rubber policeman and harvested by centrifugation at $230 \times g$ at 4 °C for 10 min.

2. Sterile, clean 12.5×25 mm coverslips are placed in a sterile quadrant petri dish (2 per quadrant) using fine-point forceps that have been dipped in 70 % ethanol and flamed (*see* **Note 5**).

3. BMMØ are gently resuspended in 1 ml BMMØ media, counted using a hemocytometer, and diluted in BMMØ media to a density of ~1.25×10^6 macrophages/ml.

4. 10 ml of BMMØ suspension is added to the petri dish and incubated at 37 °C for 24 h to allow a monolayer to establish on the coverslips. Care should be exercised to prevent excessive movement of the coverslips in the petri dish. The BMMØ monolayer-covered coverslips (subsequently referred to as monolayers) are then ready for infection.

A.

B.

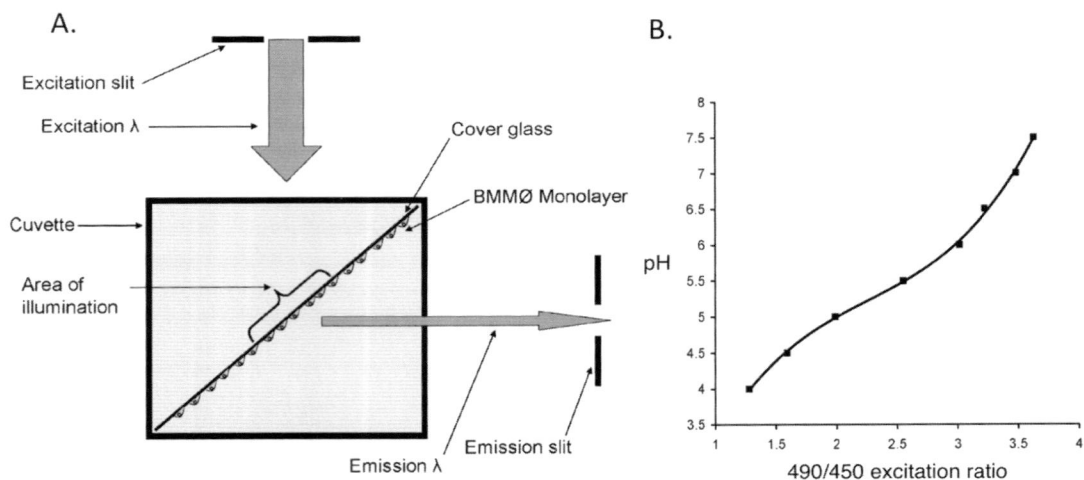

Fig. 1 Measuring phagosomal pH using a spectrofluorometer. (**a**) Diagram illustrating the orientation of the BMMØ monolayer in the spectrofluorometer. Cover glass is oriented to achieve a 45° incidence with the excitation beam (excitation λ), and with the cells on the side facing the emission slit. (**b**) Excitation ratio versus pH standard curve generated by CF-SE-labeled beads in buffers of known pH [19, 20]. The curve of best-fit can be described by the equation $y = 0.4233x^3 - 2.8693x^2 + 7.3391x - 1.5901$ (where $x = 490/450$ excitation ratio and $y = $ pH)

3.1.3 Spectrofluorometer Setup and Operation

1. The spectrofluorometer should be set up according to manufacturer's directions such that optimal measurements can be taken using the desired wavelengths (*see* **Note 6**).

2. Clean quartz cuvettes containing Cuvette buffer are inserted into a thermostat-controlled sample holder and warmed to 37 °C prior to the loading of the monolayers (*see* **Note 7**).

3. Using fine-point forceps, a monolayer-covered coverslip is grasped at one end (*see* **Note 8**) and dipped four times into a 50-ml tube containing Cuvette buffer to remove BMMØ growth media and loosely adhered cells. The coverslip is then placed in the cuvette with a vertical orientation (length of the coverslip parallel to the long axis of the cuvette), on the diagonal (width of the coverslip at 45° to the short axes of the cuvette) and with the cellular side facing the emission slit (Fig. 1a).

4. At this point, the background measurements of each monolayer are acquired for the wavelengths required.

5. At the conclusion of the background determination, the coverslips are carefully removed from the cuvettes using forceps and placed with the cellular side up on the Parafilm in the binding dish inside a biosafety cabinet.

6. 90 μl of CF-SE-labeled Mtb suspension is carefully laid over each monolayer. The bacterial suspension is incubated with the cell monolayer at room temperature for 5 min (*see* **Note 9**).

7. The coverslips are then dipped ten times in Cuvette buffer at room temperature to remove unbound bacteria and are placed in the cuvettes with the same orientation. The cuvettes are sealed with caps, removed from the biosafety cabinet and placed in the heated cuvette holder in the spectrofluorometer chamber.

8. Emission at 520 nm is recorded with excitation alternating between 450 and 490 nm (*see* **Note 10**). Typically an integration time of 1 s per data point is optimal. Data are routinely collected for a 3 h time period to follow the acidification of the Mtb-containing phagosomes.

3.1.4 Conversion of Excitation Ratio to pH

1. At the conclusion of the experiment the cuvettes are removed from the spectrofluorometer and transferred to the biosafety cabinet. The Cuvette buffer is removed and replaced with the pH reference buffers 5.5, 6.0, 6.5, and 7.0 with 10 μM nigericin (*see* **Note 11**). The cuvettes are returned to the spectrofluorometer and calibration readings at these defined pHs are acquired for 5 min at 450/520 nm and 490/520 nm.

2. Extensive calibration of CF-SE-labeled beads has been conducted and documented in an earlier chapter in this series [19]. The CF-SE-bead 490/450 excitation ratio for each pH standard (minus background) is calculated and plotted against pH. Confirm that the pKa of the bound carboxyfluorescein is close to the pH that is of most interest and that the curve generated is reasonably flat over the pH range required (*see* **Notes 11** and **12**).

5. The polynomial equation that best describes the curve is calculated (Fig. 1b). Usually a third or fourth order polynomial is sufficient. This can be done with standard mathematical software such as Microsoft Excel® or MATLAB®.

6. This equation is used to convert the real time phagosomal 490/450 excitation ratios into pH units and these values are plotted against time (Fig. 2) (*see* **Note 10**).

7. Figure 2 shows the pH plots from the phagosomes containing wild-type Mtb in comparison to those containing a mutant defective in *rv3378c* that was isolated in a genetic screen to identify genetic loci required for successful modulation of phagosome maturation [5]. The wild-type bacteria maintain their vacuoles at pH 6.4, whereas the Rv3378c mutant is delivered into lower pH compartments, approximately pH 5.8. This behavior is reflected in the acquisition of lysosomal tracers, Fe^{2+}-dextran (Fig. 2c), and a reduction in the survival and/or growth of the mutants inside their host macrophages (Fig. 2d).

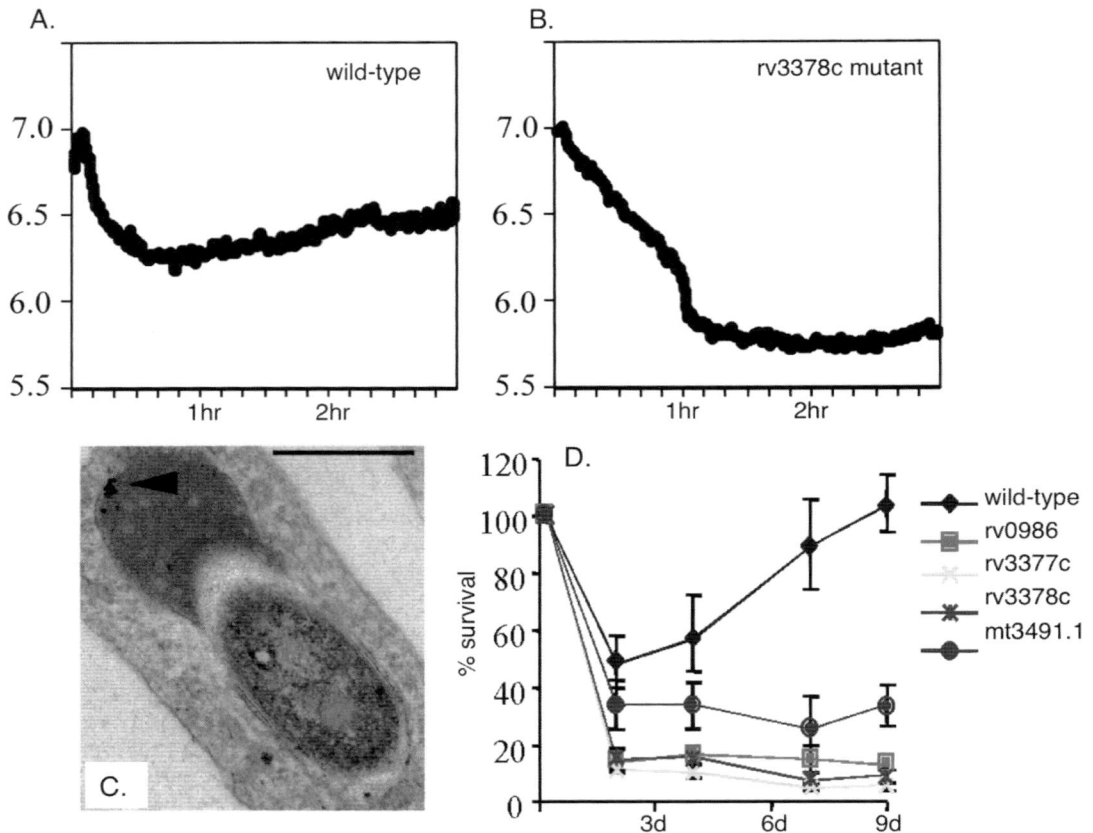

Fig. 2 Measuring the pH of Mtb-containing phagosomes. (a) Graphs showing the acidification of the phagosomes of BMMØ infected with CF-SE-labeled wild-type Mtb [5]. The phagosomes containing wild-type viable bacteria acidify to pH 6.4 within the time frame of the assay (3 h). (b) In contrast, the phagosomes containing the *rv3378c* mutant are able to acidify further, down to pH 5.8. (c) Electron microscopy examination of macrophages infected with the mutant bacteria demonstrated that these phagosomes exhibited an increased association with lysosomes, defined by their dense cargo, and the presence of an iron dextran tracer (*arrowhead*). (d) The survival kinetics of wild-type Mtb compared to the mutants isolated as defective in the regulation of their phagosome indicates that these mutants are unable to enter into growth phase following uptake by the BMMØ

3.2 Use of Reporter Mtb Strains to Probe Bacterial Fitness In Vivo

3.2.1 Preparation of Reporter Mtb Strains

1. Reporter strains of Mtb, generated as detailed previously (*see* Note 4), are grown to mid-log ($OD_{600} \sim 0.6$; this corresponds to $\sim 10^8$ bacteria/ml) in Middlebrook 7H9 medium (BD, Franklin Lakes, NJ) with 50 µg/ml hygromycin B to ensure maintenance of the replicating plasmids An equal volume of 20 % glycerol is added to the bacterial culture and mixed (final glycerol concentration = 10 %). This suspension is aliquoted in 1 ml volumes into 1.5 ml screw cap cryovials and stored at –80 °C until required. Prior to initial use, an aliquot of each strain is thawed and serial dilutions plated on Middlebrook 7H10 agar (BD) with 50 µg/ml hygromycin B, and colony-forming units (CFUs) counted after 3–4 weeks to provide a definitive quantification of viable bacteria in each frozen stock.

2. At the time of challenge, an aliquot of reporter bacteria for each strain required is thawed, passed 5–8× through a tuberculin syringe, and diluted in PBS containing 0.05 % Tween 80 to obtain a final concentration of 1000 CFUs in 25 µl. Serial dilutions of the suspension are also plated on Middlebrook 7H10 agar with 50 µg/ml hygromycin B to verify the inoculum numbers.

3.2.2 Infection of Mice with the Bacterial Reporter Strains

1. We use intranasal challenge as our route of infection. This method has significant advantages as it is inexpensive and easy to use, and generates the infected tissue we require for analysis. However it is important to bear in mind the limitations of this method for modeling infections, for the reason discussed in **Note 13**.

2. 6–8 week old C57BL/6J female mice (Jackson Laboratories, Bar Harbor, ME) are used routinely for infection studies, *see* also **Note 14**. Mice are placed individually in an induction chamber and exposed to 2 % isoflurane delivered by a Harvard Systems anesthesia system (Harvard Biosciences, Inc., Holliston, MA).

3. During this step the mouse is monitored visually, and when lightly anesthetized the mouse is removed, held with its head inclined and 25 µl total of bacterial suspension is carefully delivered down both nares. The mouse is placed in a cage and monitored for its recovery from anesthesia.

4. At various time intervals post-challenge mice are sacrificed for analysis. Mice are euthanized by CO_2 inhalation.

5. To obtain colony-forming unit counts, the left lung lobe and the accessory lobe of the right lung are manually homogenized in a Whirl-Pak tissue homogenizer bag (Nasco, Fort Atkinson, WI) with 0.5 ml PBS containing 0.05 % Tween 80. The homogenate is removed into a clean screw-cap tube, and the residual contents in the homogenizer bag rinsed with an additional 0.5 ml PBS containing 0.05 % Tween 80. This is then added to the tube containing the initial 0.5 ml homogenate (total = 1 ml homogenate). The homogenate is serial diluted and plated on Middlebrook 7H10 Agar ±50 µg/ml hygromycin B.

6. For image analysis of the infected tissue, the rest of the right lung lobes are fixed in 4–5 ml of 4 % paraformaldehyde in PBS in a 15 ml conical tube, for immunofluorescence microscopy and histological examination. The tubes are surface decontaminated and removed from the BSL-3 facility, and the samples left in fixative overnight at room temperature, protected from light, in a sealed container, prior to immunofluorescence staining.

7. Thick sections of the fixed lung lobe tissue (~0.5 mm) are cut by hand with a razor (*see* **Note 15**), then placed in a 1.5 ml

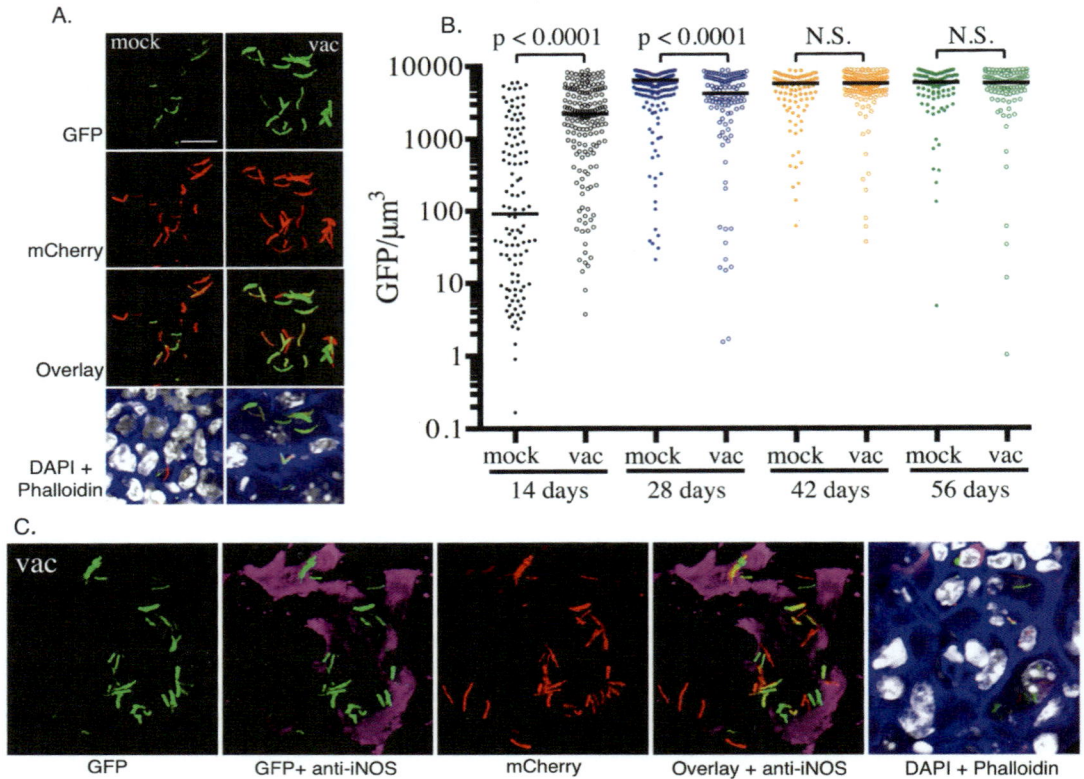

Fig. 3 Measuring the induction of *hspX'*-driven expression of GFP. (a) Illustrates the levels of expression of *hspX'* promoter dependent GFP at 14 days post-infection in naïve mice (mock) and mice vaccinated with heat-killed Mtb (vac), as reported previously [16]. At 14 days the vaccinated mice have a robust immune response, while this response is not fully developed in the naïve mice. (b) Images are acquired using the Leica LAS software and are quantified by Volocity. The bacterial volume is defined by the mCherry signal, and the total GFP signal intensity is measured for each bacterial volume. *Each dot* on the scatter plot represents a single bacterium or a cluster of bacteria that cannot be separated. The average level of expression of GFP at 14 days is markedly lower in naïve mice than in vaccinated mice. By 28 days, high levels of induction of GFP expression are observed in both naïve and vaccinated groups. The *horizontal bars* represent the median value for each group and *p*-values were generated using a Mann–Whitney statistical test. (c) The *hspX* promoter is regulated by the *dosR* regulon, which is activated by hypoxia and nitric oxide. Nitric oxide is generated by the inducible nitric oxide synthase (iNOS), which is expressed in mouse macrophages in the presence of IFN-γ and TNF-α. Probing the murine lung tissue with an anti-iNOS antibody reveals the presence of the host enzyme in regions that contain GFP-expressing Mtb

microcentrifuge tube and blocked and permeabilized by incubating for 1 h at room temperature in 500 µl PBS + 3% BSA + 0.1% Triton X-100 (blocking buffer) (*see* **Note 16**).

8. For immunohistochemistry (Fig. 3), samples are incubated with primary antibodies overnight at 4 °C, washed 3 × 5 min with blocking buffer at room temperature, then incubated with secondary antibodies for 2 h at room temperature. After washing 3 × 5 min with blocking buffer, samples are mounted

with mounting medium containing antifade (*see* **Note 17**). All staining and washing steps are carried out on a rocking platform, with the samples protected from light. We routinely stain 2 sections/sample in the same 1.5 ml microcentrifuge tube, in 200 μl volumes of antibody solutions. Wash steps are carried out in 500 μl volumes. All antibodies are diluted in blocking buffer. Rabbit anti-iNOS was used at 1:100, and Alexa Fluor 514 goat anti-rabbit IgG used at 1:200 for secondary detection. Alexa Fluor 647 conjugated phalloidin (1:50) was used for visualization of the actin cytoskeleton and nuclei were visualized with DAPI (1:500).

9. Samples are imaged with a Leica SP5 spectral confocal microscope (*see* **Note 18**), and z-stacks reconstructed into 3D using Volocity software. We routinely image a depth of 10 μm, with 0.5 μm steps.

10. For quantification of *hspX'*::GFP signal (Fig. 3), the volume of each bacterium was measured via the mCherry channel, with the corresponding sum of the GFP signal for that bacterium simultaneously acquired, to obtain GFP/μm^3 values. Settings for the GFP channel are maintained throughout the imaging of samples within each experimental set to allow comparison of values (*see* **Note 19**).

11. For quantification of SSB-GFP signal (Fig. 4), bacteria with or without SSB-GFP foci are scored manually by inspection of 3D reconstructed and extended focus images in Volocity. At least 100 bacteria, taken across multiple images, should be quantified for each variable under experimentation. Statistical differences between data sets are determined by a nonparametric Mann–Whitney test.

4 Notes

1. Primary BMMØ are generally preferred for their enhanced phagocytic proficiency and adhesion.

2. 0.13 × 12.5 × 25 mm cover glass is not commercially available. Cover glass can be custom ordered from ProSciTech (Thuringowa, QLD, Australia). Alternatively, 25 × 25 mm cover glass is available from Fisher Scientific (Pittsburgh, PA, USA) and can be cut in half by diamond pencil in house.

3. Gelatin is used as an alternative to FBS for spectrofluorometric assays as it has low autofluorescence with excitation wavelengths above 450 nm. If assays are expected to take greater than 6 h, FBS is preferred for sustained macrophage viability.

4. Reporter Mtb strains are generated using the replicating plasmid pCherry3, which encodes *mCherry* under regulation of the

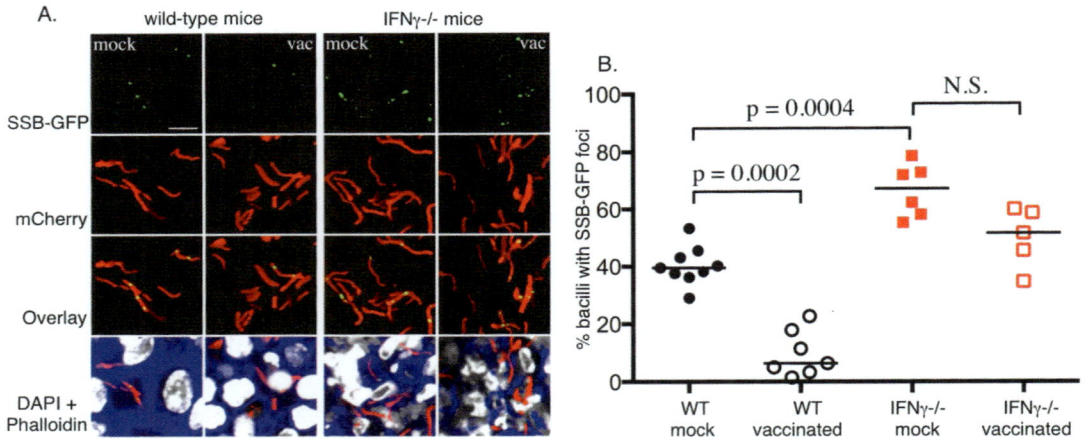

Fig. 4 Using a SSB-GFP fusion protein to assess bacterial replication in tissue. Naïve (mock) and vaccinated (vac) wild-type and IFN-γ-deficient mice were challenged with Mtb expressing mCherry and SSB-GFP, as reported previously [16]. The mice were sacrificed at 28 days and the lung tissue analyzed by confocal microscopy. (a) Shows confocal images from the infected lung tissue demonstrating the presence of green foci among some of the bacteria in the section. The presence or absence of GFP foci within the bacteria was scored manually. (b) A scatter plot of the % of GFP foci-positive bacteria in each mouse in each group. The *dots each* represent a single mouse. The *horizontal bars* represent the median value for each group and *p*-values were obtained with a Mann–Whitney statistical test. The *graph* indicates that the number of SSB-GFP foci-positive bacteria was inversely proportional to the robustness of the immune response. This suggests that replication was best controlled in vaccinated mice, and least controlled in mice deficient in the macrophage-activating cytokine IFN-γ

constitutive promoter *smyc*. We use replicating plasmids because they ensure high levels of expression of fluorescent protein. While integrating plasmids might represent a more stable solution the loss in signal is often too great. Expression of these fluorescent protein constructs can come at some loss to fitness so it is important to assess the maintenance of the plasmid by replica plating bacteria from infected mice on agar plates with and without hygromycin B selection (50 μg/ml) and confirming that they still express GFP under appropriate conditions.

5. Arrange coverslips so as not to overlap, taking care not to overcrowd them in the petri dish as coverslips can move after monolayers have been established and damage to BMMØ can occur. Alternatively, coverslips can be separated from each other using partitioned petri dishes or 6 well plates.

6. Some general considerations are: the focusing of illumination on the sample, the addition of long-pass and short-pass filters, and the adjustment of excitation and emission slit widths to maximize signal to noise ratio and to minimize photobleaching.

7. Cuvette buffer should be of a similar temperature to the cuvettes at addition. This prevents bubble formation that can create unwanted scatter of light.

8. Coverslips should only be grasped by forceps at the uppermost edge to avoid damaging the area of the monolayer that is illuminated in the spectrofluorometer.

9. The Mtb suspension has an approximate multiplicity of infection (MOI) of 25:1. This is at vast excess and is required to guarantee binding of adequate numbers of CF-SE-labeled bacteria to generate sufficient signal return to ensure the accuracy of the readings. In actuality, under these conditions, we observe around 5 bacilli bound per macrophage following washing of the coverslips.

10. Carboxyfluorescein has two wavelengths of maximal excitation that lead to fluorescence emission at 520 nm. Excitation at 450 nm induces a lower signal but the signal is pH insensitive and acts as an internal standard. In contrast, excitation at 490 nm results in a stronger fluorescent signal at 520 nm, but this signal is pH sensitive and is quenched by the protonation of the dye at lower pH.

11. Standard curves are generated using phagocytosed Mtb on BMMØ monolayers using the ionophore nigericin (10 μM) in K^+-containing buffers of known pH. This eliminates the possibility of modification of the fluorochrome's pKa by the intracellular environment. For convenience we routinely generate standard curves using a suspension of CF-SE-labeled Mtb. In each experimental setup however, it should be determined that these curves are equivalent to those generated with Mtb-containing monolayers treated with nigericin. Standard pH curves should be generated at the conclusion of every phagosomal pH experiment. Subtle changes to components of the experiment such as degree of CF-SE labeling, slit width and PMT voltage, can have profound effects on the relationship between excitation ratio and actual pH.

12. The pKa of free carboxyfluorescein-SE is 6.4. We have found that when bound to proteins or to the surface of Mtb, the pKa is shifted to ~5.5 making it particularly useful in the generation of phagosomal pH profiles. However, should the pKa of the CF-SE conjugates be inappropriately high thus rendering measurement of lower pH values inaccurate, then Oregon green-SE should be used instead of carboxyfluorescein-SE.

13. Intranasal infection has significant advantages for in vivo Mtb infection experiments in its convenience and low cost. However, it is important to recognize that this delivery method is relatively nonphysiological compared to an aerosol delivery, because Mtb infection in nature is initiated by the inhalation of extremely small numbers of bacilli into the deep lung.

14. In addition to wild-type mice we have also examined immune-compromised C57BL/6J mice that were defective in expression

of IFN-γ, and immune-enhanced mice that were vaccinated with heat-killed Mtb 4 weeks prior to challenge [16].

15. We do not process the samples with cryostat or paraffin-based sectioning, thus minimizing manipulation of the tissue, and enabling thick section imaging as desired. It is important to obtain tissue sections that are even, to allow proper mounting of the sample after staining. To do this, we make an initial cut through the middle of the lung lobe, then subsequently slice from this cut face to obtain the tissue sections.

16. If no primary antibody stain is being done, this first 1 h incubation step can be skipped, and the sections placed directly into the solution of secondary antibodies (e.g., phalloidin and DAPI) in blocking buffer for the 2 h incubation at room temperature.

17. To ensure that the sample is not compressed on mounting, we use inert vacuum grease to make a "chamber" in which the sample sits. Placing the vacuum grease in a 10 ml syringe allows for easy dispensing. The coverslip is then gently and evenly pressed down such that the sample is held in place on the slide, without being unduly compressed. Use of a mounting medium that does not harden allows for re-mounting of the sample as needed. An alternative to the use of vacuum grease is to use SecureSeal™ imaging spacers (Electron Microscopy Sciences, Hatfield, PA).

18. Use of a spectral confocal system is particularly helpful for 5-color imaging, to enable proper separation of the various fluorophores used—in particular here, for the separation of GFP versus Alexa Fluor 514 versus mCherry signal. We also use sequential scans, to further minimize signal bleed-through.

19. For quantification of *hspX*::GFP signal, it is also important that saturation in the GFP channel be minimized, to allow the greatest dynamic range and accurate measurements. It is thus best to first test settings on both samples with the lowest and highest signals, to obtain parameters that will be acceptable across all samples to be compared, so that the same settings can be used throughout the experiment and valid comparisons made.

Acknowledgments

S.T. and D.G.R. are supported by the following awards from the National Institutes of Health, USA; AI114952 (S.T.) and AI067027, AI118582, and HL055936 (D.G.R.). R.M.Y. is supported by awards from the Canadian Institutes of Health Research and Natural Sciences and the Engineering Research Council of Canada.

References

1. Draper P, Hart PD, Young MR (1979) Effects of anionic inhibitors of phagosome-lysosome fusion in cultured macrophages when the ingested organism is Mycobacterium lepraemurium. Infect Immun 24:558–561

2. Hart PD, Armstrong JA (1974) Strain virulence and the lysosomal response in macrophages infected with Mycobacterium tuberculosis. Infect Immun 10:742–746

3. Hart PD, Young MR (1991) Ammonium chloride, an inhibitor of phagosome-lysosome fusion in macrophages, concurrently induces phagosome-endosome fusion, and opens a novel pathway: studies of a pathogenic mycobacterium and a nonpathogenic yeast. J Exp Med 174:881–889

4. Sturgill-Koszycki S, Schlesinger PH, Chakraborty P, Haddix PL, Collins HL, Fok AK, Allen RD, Gluck SL, Heuser J, Russell DG (1994) Lack of acidification in Mycobacterium phagosomes produced by exclusion of the vesicular proton-ATPase. Science 263:678–681

5. Pethe K, Swenson DL, Alonso S, Anderson J, Wang C, Russell DG (2004) Isolation of Mycobacterium tuberculosis mutants defective in the arrest of phagosome maturation. Proc Natl Acad Sci U S A 101:13642–13647

6. Clemens DL, Horwitz MA (1995) Characterization of the Mycobacterium tuberculosis phagosome and evidence that phagosomal maturation is inhibited. J Exp Med 181:257–270

7. Clemens DL, Horwitz MA (1996) The Mycobacterium tuberculosis phagosome interacts with early endosomes and is accessible to exogenously administered transferrin. J Exp Med 184:1349–1355

8. Russell DG, Dant J, Sturgill-Koszycki S (1996) Mycobacterium avium- and Mycobacterium tuberculosis-containing vacuoles are dynamic, fusion-competent vesicles that are accessible to glycosphingolipids from the host cell plasmalemma. J Immunol 156:4764–4773

9. Sturgill-Koszycki S, Schaible UE, Russell DG (1996) Mycobacterium-containing phagosomes are accessible to early endosomes and reflect a transitional state in normal phagosome biogenesis. EMBO J 15:6960–6968

10. Via LE, Fratti RA, McFalone M, Pagan-Ramos E, Deretic D, Deretic V (1998) Effects of cytokines on mycobacterial phagosome maturation. J Cell Sci 111(Pt 7):897–905

11. Schaible UE, Sturgill-Koszycki S, Schlesinger PH, Russell DG (1998) Cytokine activation leads to acidification and increases maturation of Mycobacterium avium-containing phagosomes in murine macrophages. J Immunol 160:1290–1296

12. Abramovitch RB, Rohde KH, Hsu FF, Russell DG (2011) aprABC: a Mycobacterium tuberculosis complex-specific locus that modulates pH-driven adaptation to the macrophage phagosome. Mol Microbiol 80: 678–694

13. Rohde KH, Abramovitch RB, Russell DG (2007) Mycobacterium tuberculosis invasion of macrophages: linking bacterial gene expression to environmental cues. Cell Host Microbe 2:352–364

14. Rohde KH, Veiga DF, Caldwell S, Balazsi G, Russell DG (2012) Linking the transcriptional profiles and the physiological states of Mycobacterium tuberculosis during an extended intracellular infection. PLoS Pathog 8, e1002769

15. Tan S, Sukumar N, Abramovitch RB, Parish T, Russell DG (2013) Mycobacterium tuberculosis responds to chloride and pH as synergistic cues to the immune status of its host cell. PLoS Pathog 9, e1003302

16. Sukumar N, Tan S, Aldridge BB, Russell DG (2014) Exploitation of Mycobacterium tuberculosis reporter strains to probe the impact of vaccination at sites of infection. PLoS Pathog 10, e1004394

17. Reyes-Lamothe R, Possoz C, Danilova O, Sherratt DJ (2008) Independent positioning and action of Escherichia coli replisomes in live cells. Cell 133:90–102

18. Carroll P, Schreuder LJ, Muwanguzi-Karugaba J, Wiles S, Robertson BD, Ripoll J, Ward TH, Bancroft GJ, Schaible UE, Parish T (2010) Sensitive detection of gene expression in mycobacteria under replicating and non-replicating conditions using optimized far-red reporters. PLoS One 5, e9823

19. Yates RM, Russell DG (2008) Real-time spectrofluorometric assays for the lumenal environment of the maturing phagosome. Methods Mol Biol 445:311–325

20. Yates RM, Russell DG (2005) Phagosome maturation proceeds independently of stimulation of toll-like receptors 2 and 4. Immunity 23:409–417

Chapter 24

Using Flow Cytometry to Analyze *Cryptococcus* Infection of Macrophages

Robert J. Evans, Kerstin Voelz, Simon A. Johnston, and Robin C. May

Abstract

Flow cytometry is a powerful analytical technique, which is increasingly being used to study the interaction between host cells and intracellular pathogens. Flow cytometry is capable of measuring a greater number of infected cells within a sample compared to alternative techniques such as fluorescence microscopy. This means that robust quantification of rare events during infection is possible. Our lab and others have developed flow cytometry methods to study interactions between host cells and intracellular pathogens, such as *Cryptococcus neoformans*, to quantify phagocytosis, intracellular replication, and non-lytic expulsion or "vomocytosis" from the phagosome. Herein we describe these methods and how they can be applied to the study of *C. neoformans* as well as other similar intracellular pathogens.

Key words *Cryptococcus neoformans*, *Cryptococcus gattii*, Flow cytometry, Macrophage, Infection, Mycology

1 Introduction

1.1 Cryptococcus neoformans and Cryptococcus gattii

The *Cryptococcus* genus is part of the basidiomycete phylum of the fungal kingdom of life. The genus contains over 50 described species; however, almost all human and veterinary cases of cryptococcal infection are caused by just two species—*Cryptococcus neoformans* and *Cryptococcus gattii*. Although these two species are closely related, they present with different pathologies during infection. *C. neoformans* is an opportunistic pathogen of individuals with existing immune deficiencies whereas *C. gattii* can infect immune competent hosts. During infection both fungal species interact with host phagocytes in the lungs and phagocytosis by host alveolar macrophages provides an intracellular niche for *Cryptococcus* to replicate [1]. However, occasionally cryptococci can escape from the macrophage via a process known as non-lytic expulsion or "vomocytosis" [2–4]. The central role of macrophages during cryptococcal infection makes this host pathogen interaction a key area of *Cryptococcus* research. In vitro cell culture is often used to study this interaction

Roberto Botelho (ed.), *Phagocytosis and Phagosomes: Methods and Protocols*, Methods in Molecular Biology, vol. 1519, DOI 10.1007/978-1-4939-6581-6_24, © Springer Science+Business Media New York 2017

as in many cases, the interaction between *Cryptococcus* species and macrophages cultured in vitro can be used as a reliable indicator of virulence in vivo [5, 6].

Methods to quantify cryptococcal parasitism of macrophages with flow cytometry have been previously published by our lab and others [3, 5, 7]. Typically, such methods exploit *Cryptococcus* strains with a genomic fluorescent tag [7], or a combination of antibody and cell dye stains [3, 5]. Herein we describe our previously published method for flow cytometry analysis of cryptococcal phagocytosis by macrophages and subsequent *Cryptococcus* replication within the phagosome [7].

1.2 Creating Fluorescently Tagged Cryptococcus strains

Our protocol relies on the use of *Cryptococcus* strains that have been genetically modified to express a fluorescent marker protein such as green fluorescent protein (GFP). During the development of this assay we created two fluorescent strains in the H99 (*C. neoformans var. grubii* serotype A, genotype VNI) and R265 (*C. gattii* serotype B, genotype VGII) genetic backgrounds. These two strains are both common "wild type" reference strains used by many *Cryptococcus* researchers.

The step-by-step generation the H99-GFP and R265-GFP used in this study can be found in our previous publication [7], but, in brief, cryptococci were transformed via biolistic delivery with an insertion cassette containing the GFP gene flanked upstream by the *Cryptococcus* JEC21 actin promoter *act1* and downstream by the JEC21 tryptophan terminator *trp1* (Fig. 1).

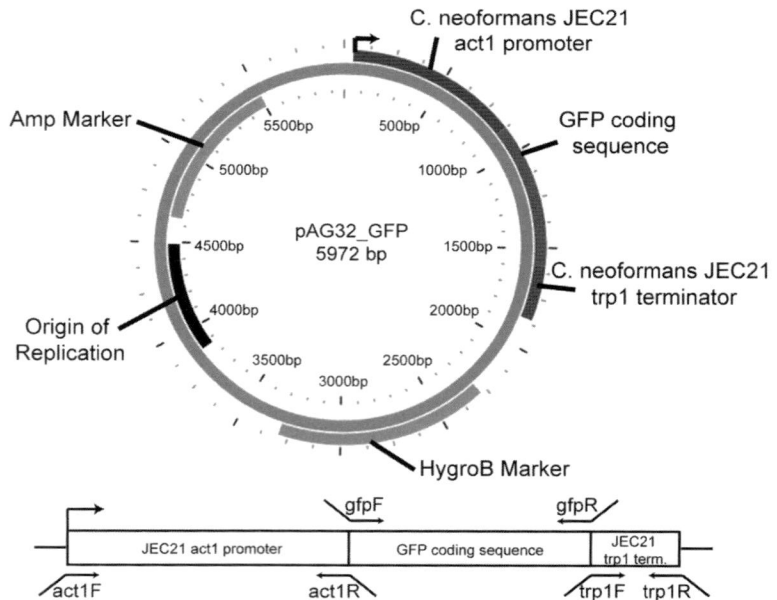

Fig. 1 Schematic layout of pAG32_GFP plasmid used to create H99-GFP and R265-GFP [7]

For an investigator seeking to make their own fluorescently tagged strains a number of factors must be considered once stable transformants have been produced to ensure valid results in the flow cytometry assay. It must be ensured that the insertion of the transgene and/or its expression has not significantly altered the physiology of the fluorescent mutant in comparison to its wild type parent. This process is important as it helps to ensure the validity of all future findings using the transformed strain (*see* **Note 1**).

2 Materials

2.1 Cells and Strains

1. J774.A1 cells were acquired from the European Collection of Authenticated Cell Cultures (ECACC).

2. The genetic background for the fluorescent *Cryptococcus neoformans* mutants used was H99 *C. neoformans var grubii.* serotype A.

2.2 Specialist Equipment

1. Class II laminar flow hood: *Cryptococcus neoformans* is a class II organism and as such should be worked on inside a class II safety hood to protect the user. Additionally, cells in culture are easily susceptible to outside contamination; therefore, any work with uninfected cells should be performed within a laminar flow hood.

2. Tissue culture incubator: mammalian cells such a J774 cells should be grown in a specialist incubator that can control heat, humidity and CO_2 levels. J774 cells are grown at 37 °C with 5% CO_2 and 95% relative humidity.

3. Hemocytometer: to determine the correct concentration of *Cryptococcus neoformans* cells for infection a hemocytometer counting chamber should be used. A BS 748 standard hemocytometer with a chamber depth of 0.1 mm is recommended.

4. Flow cytometer—this protocol requires a flow cytometer that is capable of exciting and measuring GFP (excitation wavelength 395 nm, and detection wavelength 509 nm).

2.3 Media (Macrophage)

1. Serum supplemented DMEM: Dulbecco's Modified Eagle's medium, low glucose, 10% v/v fetal bovine serum (FBS), 1% v/v 10,000 units penicillin and 10 mg/ml streptomycin, 1% v/v 200 mMl-glutamine. Keep sterile store at 4 °C.

2. Serum free DMEM: Dulbecco's Modified Eagle's medium, low glucose, 1% v/v 10,000 units penicillin and 10 mg/ml streptomycin, 1% v/v 200 mMl-glutamine. Keep sterile and store at 4 °C.

2.4 Media (Cryptococcus)	1. YPD (Yeast, peptone, dextrose) liquid growth media: 1 % w/v peptone, 1 % w/v yeast extract, and 2 % w/v d-(+)-glucose. Autoclave to sterilize.
	2. YPD agar: Liquid YPD growth media + 2 % w/v agar. Autoclave to sterilize, pour into 9 cm petri dishes.

2.5 Miscellaneous Reagents

1. Phorbol 12-myristate 13-acetate (PMA): 1 mg/ml PMA in dimethyl sulfoxide (DMSO), store at –20 °C aliquoted.

2. Phosphate buffered saline (1× PBS): 8 g/L sodium chloride, 0.2 g/L potassium chloride, 1.15 g/L disodium hydrogen phosphate, 0.2 g/L potassium dihydrogen phosphate, in deionized H_2O pH 7.3, autoclave to sterilize and store at room temperature.

3. 18B7 antibody: 10 mg/ml in 100 % glycerol (a kind gift from Arturo Casadevall, Johns Hopkins Bloomberg School of Public Health, Maryland, USA), mouse IgG against *Cryptococcus* capsule polysaccharide glucuronoxylomannan, store at –20 °C aliquoted.

4. Accutase: use at concentration suggested by manufacturer, store at –20 °C aliquoted.

5. Fixing media: 2 % w/v formaldehyde, 2 % v/v FBS in 1× PBS, store at –20 °C aliquoted, once thawed discard within 2 weeks.

3 Protocol

3.1 Experimental Design

1. *C. neoformans* samples to be analyzed by this protocol need to be fluorescently tagged. Additionally, a sample that contains macrophages infected with a nonfluorescent *C. neoformans* strain is also need as a control for the flow cytometer (*see* Subheading 3.4). Ideally this nonfluorescent strain should be the same genetic background as the fluorescent *Cryptococcus* strain.

2. For each condition, four duplicate infections need to be prepared (e.g., four separate wells) for measurement at each time point. The following time points are recommended for the calculation of intracellular proliferation—0 h (2 h post infection, immediately after washing away extracellular yeast), 18 h (20 h post infection), 24 h (26 h post infection), and 48 h (50 h post infection) (*see* **Note 2**).

3.2 Cryptococcus Preparation

Long term, *Cryptococcus* strains can be stored at –80 °C in glycerol stocks or using Biobank storage beads. For experimentation *Cryptococcus* strains are grown on YPD agar at 25 °C and stored at 4 °C.

1. Prepare overnight cultures of the fluorescently tagged *Cryptococcus* strain from YPD agar stock plates 24 h prior to the start of the assay. Grow overnight cultures in 2 ml YPD broth with constant rotational movement to prevent sedimentation.

2. On the day of experiment, transfer 1 ml of the overnight culture into a sterile 1.5 ml microcentrifuge tube and wash 3 times with sterile PBS.

3. Once washed, count the overnight culture with a hemocytometer (a 1:20 dilution of the overnight culture is usually sufficient for accurate counting) and dilute to a concentration of 1×10^7 *Cryptococcus* cells per 1 ml in PBS in 1.5 ml microcentrifuge tubes.

4. Opsonize *Cryptococcus* for 1 h with 10 µg/ml anti-capsular 18B7 antibody (a kind gift from Arturo Casadevall, Albert Einstein College of Medicine, New York, USA); for best results put microcentrifuge tubes on a rotator. Alternatively, *C. neoformans* can be opsonized with pooled human serum (separated from donor blood).

3.3 Macrophage Infection

The J774 murine macrophage cell line is used here as an example; however, this protocol could easily be adapted for other in vitro cell lines or in vitro cultured primary monocytes/macrophages. While the basic principles remain the same, when using this protocol for other cell types it will be important to adjust the media and growing conditions to best suit the cells used.

1. Before experimentation maintain J774 macrophages in serum supplemented (10% FBS) DMEM media for at least four passages (from liquid nitrogen storage) before use, only use J774 cells between passages 4 and 15. Maintain J774 cells in T75 tissue culture flasks and incubate at 37 °C with 5% CO_2 (*see* **Note 2**).

2. 24 h prior to the start of the experiment, take a confluent T75 flask of J774 cells and seed into a 24-well plate at a concentration of 1×10^5 cells per well in 1 ml serum supplemented DMEM. Incubate the plate for 24 h at 37 °C, 5% CO_2.

3. 1 h before the start of the experiment, activate the seeded macrophages by removing the media and replacing with 1 ml serum free DMEM media supplemented with 150 ng/ml phorbol 12-myristate 13-acetate (PMA) per well. Incubate for 1 h at 37 °C, 5% CO_2.

4. After 1 h incubation, remove the media from each well and replace with 1 ml serum-free DMEM.

5. Infect the macrophage monolayer by adding 100 µl washed and opsonized fluorescently tagged *C. neoformans* (1×10^7 *Cryptococcus* per ml, *see* **step 3** in Subheading 3.1). Incubate for 2 h at 37 °C, 5% CO_2.

6. After 2 h incubation, aspirate the media from each well and wash gently with 37 °C PBS to remove extracellular, non-phagocytosed cryptococci. Repeat this wash step, checking periodically under a tissue culture microscope to check for remaining extracellular *Cryptococcus* cells (*see* **Notes 3** and **4**).

7. Once extracellular cryptococci cells have been washed away, add 250 μl accutase to each well and incubated for 15 min at 37 °C. After this incubation, gently pipette the accutase to disassociate infected macrophages from the growing surface and to create a single cell suspension.

3.4 Flow Cytometer Setup

The setup for each flow cytometer is different. This assay was developed using a FACSCaliber (BD Biosciences), it should be possible to perform this assay using any flow cytometer that can detect GFP.

1. Take the cell suspensions from the previous step. Fix the samples by adding an equal volume of fixing media (2 % formaldehyde, 2 % fetal bovine calf serum in PBS) to the accutase samples (*see* **Note 5**).

2. To setup the flow cytometer, first calibrate the instrument using a nonfluorescent control—(macrophages infected with a nonfluorescent *C. neoformans* strain). Using the dot plot output adjust the forward scatter (FSC-H) and side scatter (SSC-H) parameters to ensure that all populations of interest are visible. Using the same nonfluorescent control also adjust the GFP detection channel (in this case FL1-H) using a histogram output to set the negative fluorescent signal to a baseline value (on most instruments this is three $\log_{10}$ from the detection maximum) (*see* **Note 6**).

3. Analyze each sample, making sure to keep the same calibration settings throughout the experiment. To enable reliable comparison between samples, collect events for each sample over a fixed period. Ideally 10,000 events should be collected for each sample to provide reliable data (*see* **Note 7**).

3.5 Data Analysis

1. *Phagocytosis*—Display the collected data on a dot plot with FSC-H on the "X" axis and FL1-H on the "Y" axis (as in Fig. 2). The rate of phagocytosis can be calculated by comparing the number of macrophages with intracellular yeast (Fig. 2 "Region2") to the number of macrophages without intracellular yeast (Fig. 2 "Region3").

2. *Intracellular proliferation*—For each strain, collect data for a series of time points post infection—0 h (2 h post infection, immediately after washing away extracellular yeast), 18 h (20 h post infection), 24 h (26 h post infection), and 48 h (50 h post infection). Display the collected data for each time point on a dot plot with FSC-H on the

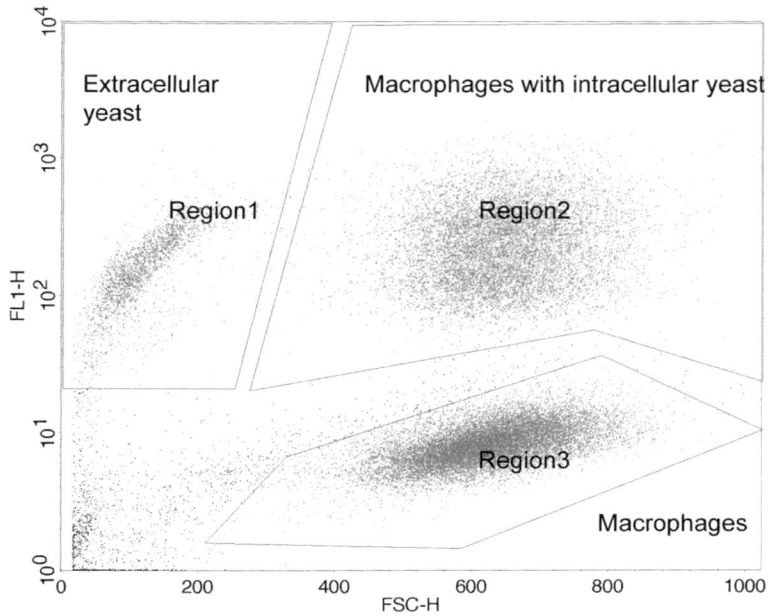

Fig. 2 Dot plot (FL1-H vs. FSC-H) generated from flow cytometry data collected from a sample of macrophages infected with a GFP tagged *Cryptococcus* strain. Three cell populations have been "gated"—Region1 Extracellular yeast (GFP high, FSC low), Region2 Macrophages with intracellular yeast (GFP high, FSC high) and Region3 macrophages that do not contain intracellular yeast (GFP low, FSC high)

"*X*" axis and FL1-H on the "*Y*" axis (as in Fig. 2). Draw a gate around the region of the plot containing macrophages with intracellular yeast (Fig. 2 "Region2"). For each time point calculate the geometric mean fluorescent intensity of events within this gate. To calculate the intracellular proliferation rate (IPR) take the time point with the highest mean fluorescent intensity (usually 18 h or 24 h), divide this mean fluorescent intensity by the mean fluorescent intensity of the events at the 0 h time point.

4   Notes

1. To calculate the intracellular proliferation rate, the *Cryptococcus* count at the maximal time point (e.g., the time point when *Cryptococcus* count is highest) is divided by the count at the baseline (0 h) time point. Generally, the maximal time point is either 18 h or 24 h. At later time points the fungal burden within macrophages decrease due to macrophage lysis.

2. To confirm a newly transformed strain behaves like the original during infection, a number of tests should be performed. Firstly, the location of the insertion should be determined—e.g., Southern blotting, genome sequencing of flanking regions. Ensure that the transformed fragment has not inserted into a (known) gene. Additionally check how many copies of the fragment have inserted into the strain (multiple insertions are possible), multiple insertions help to increase a strain's fluorescent signal; however, excessive expression of transgenes can lead to suboptimal growth due to increased cellular stress. Secondly the transformed strain should be tested in a number of culture conditions and in cell culture (preferably in the same cell line which will be used for the flow cytometry assay) to check for differences to the wild type, which again could have been caused due to insertion and/or expression of the transgene. To see the full range of tests we performed on our H99-GFP and R265-GFP strains please refer to our previous publication [7]. Finally when using the strain for flow cytometry analysis it must be confirmed that the fluorescent signal of the strain is resolvable from the background autofluorescence which all cells display. In the case of our strains the fluorescence signal was strong and easily discernible from non-transformed strains using spectral confocal microscopy [7].

3. For best results never split the cells less than 1/8, while passaging as sparse cell numbers affects cellular viability and potentially also immunological responses.

4. Three to four washes are usually sufficient to remove extracellular cryptococci, gentle tapping of the plate can be performed between washes to encourage *Cryptococcus* cells to unstick from the plate surface.

5. While washing, care must be taken to avoid washing away macrophages—avoid pipetting liquid directly onto the growing surface, always pipette liquid down the edge of the well.

6. The protocol we describe herein is a one-color assay—needing only GFP excitation (395 nm) and emission (509 nm) filters. The filter set used for GFP is the same used for fluorescein isothiocyanate (FITC)—one of the most ubiquitous fluorophores in flow cytometry. Thus, almost all flow cytometers should be capable of analyzing samples produced using the below protocol. Additional colors can be added to the protocol to allow the measurement of additional parameters such as phagocytosed/non-phagocytosed cells or to measure macrophage or *Cryptococcus* phenotypes such as phagosomal acidification, cell viability etc.

7. When preparing macrophage samples for flow cytometry, different volumes of accutase and fixing solution can be used to

optimize sample concentration. For flow cytometry the sample concentration should be adjusted such that the number of events per second detected by the instrument is in its optimal range, this range differs between instruments. Always ensure that equal volumes of accutase and fixing media are used.

References

1. Feldmesser M, Kress Y, Novikoff P, Casadevall A (2000) Cryptococcus neoformans is a facultative intracellular pathogen in murine pulmonary infection. Infect Immun 68:4225–4237, PMC101732
2. Ma H, Croudace JE, Lammas DA, May RC (2006) Expulsion of live pathogenic yeast by macrophages. Curr Biol 16:2156–2160
3. Nicola AM, Robertson EJ, Albuquerque P, Derengowski Lda S, Casadevall A (2011) Nonlytic exocytosis of Cryptococcus neoformans from macrophages occurs in vivo and is influenced by phagosomal pH. MBio 2(4). doi:10.1128/mBio.00167,11. Print 2011
4. Alvarez M, Casadevall A (2006) Phagosome extrusion and host-cell survival after Cryptococcus neoformans phagocytosis by macrophages. Curr Biol 16:2161–2165
5. Alanio A, Desnos-Ollivier M, Dromer F (2011) Dynamics of Cryptococcus neoformans-macrophage interactions reveal that fungal background influences outcome during cryptococcal meningoencephalitis in humans. MBio 2(4). doi:10.1128/mBio.00158,11. Print 2011
6. Ma H, Hagen F, Stekel DJ, Johnston SA, Sionov E, Falk R et al (2009) The fatal fungal outbreak on Vancouver Island is characterized by enhanced intracellular parasitism driven by mitochondrial regulation. Proc Natl Acad Sci U S A 106:12980–12985, Pmc2722359
7. Voelz K, Johnston SA, Rutherford JC, May RC (2010) Automated analysis of cryptococcal macrophage parasitism using GFP-tagged cryptococci. PLoS One 5(12), e15968

INDEX

A

Actin 1, 2, 4–7, 11, 43, 55, 57, 59, 62,
    66–67, 72, 79–81, 94, 113, 125, 132–134, 139, 201, 202,
    210, 211, 249, 258, 259, 266, 275, 286, 298, 318, 319,
    322, 343, 350
Antibodies 20, 22, 39, 44, 48–51, 55,
    57, 72, 109, 114, 115, 117–120, 125, 148, 153, 157,
    165–167, 196, 211, 305, 312, 319, 322, 342, 346
Antigen presentation 10, 79, 145–149,
    151–167, 202, 241, 250
Antigen processing 146, 215
Antigens 3, 10, 50, 79, 80, 145–149,
    151–167, 186, 202, 215, 241, 249
Apoptosis 3, 25, 31–33, 38, 57, 76, 185,
    265, 267, 268, 280
Automated image analysis 60, 228
Autophagy 146, 148, 155, 156, 165, 250, 286

B

Bacteria 1, 38, 44, 56, 79, 113, 125,
    165, 174, 201, 279, 285, 297, 311, 325, 334
Bacterial killing assay 300, 306–308
Biohazard 88
Blood 2, 18–23, 26, 27, 29, 38, 44,
    45, 83, 85, 88, 120, 121, 127, 131–132, 134, 147, 153,
    160, 195, 212, 299, 300, 303, 306, 353
Bone marrow-derived macrophages (BMM) 171–178,
    216, 229, 243, 301

C

Caenorhabditis elegans 2, 4, 7, 9, 10, 265–270, 272–281
Cell fractionation v
Cell line 26, 27, 29–31, 34, 38, 44, 49,
    58, 60, 65, 71, 73, 82, 83, 96, 114, 126, 128–131, 136,
    137, 151, 167, 172, 192, 195, 216, 236, 237, 243, 250,
    286, 292, 305, 315, 335, 353, 356
Cell lysis 18, 19, 243, 245
Chemical functionalization 95
Confocal microscopy 29, 46, 49, 50, 60, 63,
    65–67, 74, 81, 84, 87, 114, 116, 119, 121, 149, 153,
    157–160, 165, 171, 175, 177, 182, 269, 276, 277,
    313, 326, 336, 343, 344, 356
Cryptococcus 349–356

D

Cytokines 27, 30, 38, 44, 48, 56, 87,
    89, 147, 151–152, 165, 238, 334, 344
Cytokinesis 6

D

Dendritic cells (DCs) 2, 43, 120, 146, 149,
    151, 153–155, 164, 228

E

Efferocytosis 25–40, 185
Endosomes ....v, 7–10, 31, 56, 57, 80,
    113, 150, 159, 202, 215, 216, 241, 249, 266, 268, 286,
    317, 319, 320, 333

F

Fc-receptors 3, 20, 44, 120, 195, 196
Filamentous bacteria 311–322
Flow cytometry .... v, 17–24, 114, 117–119,
    122, 165, 170, 189, 297, 325, 326, 349–356
Fluorescence 17, 31, 49, 52, 57, 69, 82, 97, 114,
    149, 179, 187, 203, 216, 234, 273, 297, 336, 345, 356
    microscopy 50, 60, 61, 65, 79–90,
    97, 99, 114, 130, 185–197, 297, 298
    quantification 238, 336
Fluorescence resonance energy transfer
    (FRET) 125–133, 135–142
Fluorescent probes 58, 59, 94, 119, 121,
    122, 150, 167, 176, 190, 191, 208, 312, 313, 320
Fluorescent proteins 57, 66, 80, 81, 84,
    89, 136, 158, 298, 313, 322, 344, 350
Force 4, 7, 34, 87, 95, 97, 101, 102, 104, 109, 110, 210

G

GTPases 4, 5, 43, 125, 126, 133, 202, 249, 267

H

High-throughput imaging .... v, 216

I

Image analysis ....v, 40, 46, 84, 87–88, 99,
    107–108, 170, 178, 193, 201–212, 228, 230, 336, 341
ImageJ 46, 52, 84, 87, 98, 99, 107,
    172, 178, 188, 211

Roberto Botelho (ed.), *Phagocytosis and Phagosomes: Methods and Protocols*, Methods in Molecular Biology, vol. 1519,
DOI 10.1007/978-1-4939-6581-6, © Springer Science+Business Media New York 2017

Immunity..298
Immunofluorescence 43–52, 57, 60–62,
    113–122, 148, 150, 152–156, 160, 170, 212, 316,
    321, 341
Infection assays.........................203, 207, 209–210, 325–331
Innate immunity... 17, 79, 185,
    241, 249, 297
Intracellular pathogens 169, 174, 286

L

LC3 ....145, 147–167
Legionella 10, 82, 312, 313
Leukocytes .... v, 23, 24, 147, 153, 165
Live-cell imaging.................................... v, 38, 62, 65–66, 82,
    126, 142, 189, 232, 236, 319, 320, 322
Lysosomes .... v, 7–11, 31, 43, 56, 80, 93, 94,
    113, 145, 146, 150, 159, 160, 186, 187, 190, 202, 215,
    216, 241, 249, 266–268, 281, 285, 317, 319, 320, 340

M

Macrophage...................................... 2, 24, 26, 43, 55, 79, 93,
    114, 126, 146, 169, 186, 201, 216, 228, 241, 250, 286,
    297, 311, 325, 334, 349
Magnetic tweezers...93–111
Mass spectrometry..249–262
Membrane fusion 93, 146, 227, 319
Mice 19, 23, 170, 172, 181, 219, 221,
    224, 231, 326, 335, 341–345
Microscopy .... v, 17, 26, 29, 39,
    43–52, 56, 60, 79–89, 96, 100, 110, 114–117, 119,
    121, 131, 138, 139, 153, 157, 158, 166, 185, 186,
    188–197, 207, 228, 245, 265, 267, 269–270, 277,
    297, 298, 308, 312, 313, 316, 319–321, 326, 327,
    333, 336, 340, 341, 344, 346, 356
Multiplex imaging 58, 232
Mycobacteria...171, 174–181

N

Neutrophils..2, 17–24, 28, 33, 37, 38,
    43, 79, 93, 120, 191, 297, 312

O

Organelle manipulation...95

P

Phagocytic cups...............................5–7, 37, 55, 56, 103, 110,
    133, 134, 139, 189, 202–204, 311, 316–321
Phagocytic index 49, 50, 63, 73, 316
Phagocytic particles..212
Phagocytosis......................................v, 17, 25, 43, 55, 79, 81,
    93, 113, 125, 145, 169, 185, 194, 201, 223, 241, 249,
    265, 297, 311, 349
Phagosomes...4, 43, 55, 79, 93, 113,
    126, 146, 169, 186, 202, 215, 227, 241, 249, 266, 286,
    298, 311, 333, 350
    damage 286
    isolation 115, 117–118,
        228, 241–247
    proteomics ....v, 170, 249–262
pH sensitive fluorescence probes 191
Primary leukocytes 147

R

Ratiometric imaging..................................... 99, 131, 133, 139
Reactive oxygen species (ROS)............................ 56, 57, 59,
    68–70, 75, 215, 216, 219, 221–223, 227, 228, 320
Redox reactions 216, 218–223

S

Salmonella 10, 203, 206, 207,
    209–210, 212, 285–293, 313–314
Silica particles...................................55–63, 65–75, 217, 229,
    230, 234–236, 238
Spectrophotometry................................... 299, 300, 308, 327

T

Transfection.....................................27, 29–31, 38, 44, 46, 51,
    59, 65, 71, 81, 83–84, 86–89, 110, 127–129, 133,
    135–137, 159, 165, 171, 172, 176–177, 182

W

Western blotting... v, 51, 131, 148,
    170, 241–247

X

Xenophagy...286

Printed in Great Britain
by Amazon